OXFORD READINGS IN PHILOSOPHY

THE PHILOSOPHY OF SCIENCE

Published in this series

The Problem of Evil, edited by Marilyn McCord Adams and
Robert Merrihew Adams
The Philosophy of Artificial Intelligence, edited by Margaret A. Boden
The Philosophy of Artificial Life, edited by Margaret A. Boden
Self-Knowledge, edited by Quassim Cassam
Perceptual Knowledge, edited by Jonathan Dancy
The Philosophy of Law, edited by R. M. Dworkin
Environmental Ethics, edited by Robert Elliot
Theories of Ethics, edited by Philippa Foot
The Philosophy of History, edited by Patrick Gardiner
The Philosophy of Mind, edited by Jonathan Glover
Scientific Revolutions, edited by Ian Hacking
The Philosophy of Mathematics, edited by W. D. Hart
Conditionals, edited by Frank Jackson
The Philosophy of Time, edited by Robin Le Poidevin and
Murray MacBeath
The Philosophy of Religion, edited by Basil Mitchell
Meaning and Reference, edited by A. W. Moore
A Priori Knowledge, edited by Paul K. Moser
The Philosophy of Science, edited by David Papineau
Political Philosophy, edited by Anthony Quinton
Explanation, edited by David-Hillel Ruben
The Philosophy of Social Explanation, edited by Alan Ryan
Propositions and Attitudes, edited by Nathan Salmon and Scott Soames
Consequentialism and its Critics, edited by Samuel Scheffler
Applied Ethics, edited by Peter Singer
Causation, edited by Ernest Sosa and Michael Tooley
Theories of Rights, edited by Jeremy Waldron
Free Will, edited by Gary Watson
Demonstratives, edited by Palle Yourgrau

Other volumes are in preparation

THE PHILOSOPHY
OF SCIENCE

edited by

DAVID PAPINEAU

OXFORD UNIVERSITY PRESS

1996

Oxford University Press, Walton Street, Oxford OX2 6DP

Oxford New York
Athens Auckland Bangkok Bombay
Calcutta Cape Town Dar es Salaam Delhi
Florence Hong Kong Istanbul Karachi
Kuala Lumpur Madras Madrid Melbourne
Mexico City Nairobi Paris Singapore
Taipei Tokyo Toronto
and associated companies in
Berlin Ibadan

Oxford is a trade mark of Oxford University Press

Published in the United States
by Oxford University Press Inc., New York

Introduction and selection © Oxford University Press 1996

British Library Cataloguing in Publication Data
Data available

Library of Congress Cataloging in Publication Data
Papineau, David, 1947–
The philosophy of science / David Papineau.
(Oxford readings in philosophy)
Includes bibliographical references and index.
1. Science—Philosophy. 2. Realism. I. Title. II. Series.
Q175.P3375 1996 501—dc20 95-49209
ISBN 0-19-875164-8
ISBN 0-19-875165-6 (pbk.)

1 3 5 7 9 10 8 6 4 2

Typeset by Best-set Typesetter Ltd., Hong Kong
Printed in Great Britain
on acid-free paper by
Bookcraft (Bath) Ltd
Midsomer Norton, Avon

CONTENTS

INTRODUCTION

THE EPISTEMOLOGY OF SCIENCE

The philosophy of science can usefully be divided into two broad areas. The *epistemology* of science deals with the justification of claims to scientific knowledge. The *metaphysics* of science investigates philosophically puzzling features of the world described by science. In effect, the epistemology of science asks whether scientific theories are true, whereas the metaphysics of science considers what it would tell us about the world if they were.

The essays in this collection will be concerned with the epistemology of science. They will ask whether we are justified in believing scientific theories, and what attitude we should take to them if we can't. Perhaps the metaphysics of science has as much claim to the title of philosophy of science as the epistemology of science. However, it is a heterogeneous area that is resistant to anthologizing. Problems in the metaphysics of science tend either to merge into general metaphysics (like the analysis of causation, or probability, or laws of nature) or to fall within the subject area of specific sciences (like questions about quantum indeterminacy or about the units of natural selection). The epistemology of science, by contrast, deals with problems which arise for science in general, rather than specific sciences, yet at the same time are distinguishable from problems arising in other areas of philosophy.

What is more, in the last decade or so, a number of philosophers have done much to consolidate and unify work within the epistemology of science. In the first half of this century, the dominant tradition in the epistemology of science was the logical empiricism of Rudolf Carnap and Carl Hempel, which used the techniques of formal logic and mathematics to analyse the structure of scientific theories and to formulate theories of scientific explanation and confirmation (cf. P. H. Nidditch, ed., *The Philosophy of Science*, Oxford Readings in Philosophy, 1968). However, in

I would like to thank Stathis Psillos for discussing this introduction with me, and also for his help in selecting the material for this volume.

the 1960s this logic-based approach was challenged by the historically
orientated work of N. R. Hanson, T. S. Kuhn, and Paul Feyerabend,
who appealed to detailed case-studies from the history of science to argue
that the presuppositions of logical empiricism were fatally flawed (cf. Ian
Hacking (ed.) *Scientific Revolutions*, Oxford Readings in Philosophy,
1981).

This led to a schism in studies of scientific theorizing. Some philosophers
continued to work within the tradition of Carnap and Hempel, formalizing
ideal patterns of scientific reasoning. But many became convinced that this
formal approach bore little relation to the reality of scientific practice, and
turned instead to historical analysis for insight into the structure of science.
One by-product of this historical turn was that many philosophers of
science, and even more historians and sociologists of science, became
sceptical about any objective standards of scientific rationality, and came
to view theory choices as nothing but expressions of social and institutional
pressures.

The essays in this collection, along with much other recent work in the
epistemology of science, help to bridge this gap between the logical empiricists and the historicists. Contemporary epistemologists of science have
learned from Kuhn and others that real science is less rule-bound than
logical empiricism supposed, but they do not automatically conclude that
it lacks rationality altogether. Or, to put the point the other way round,
while contemporary epistemologists of science retain the logical empiricists' concern with issues of scientific objectivity, they are no longer committed to the logical empiricists' overly formal account of what objectivity
requires.

REALISM AND ITS ANTITHESES

Much recent debate in the epistemology of science centres on the issue of
scientific realism. However, discussions of scientific realism, and in particular of the alternatives to realism, are beset by terminological confusion.
Some initial clarification will be helpful.

Suppose we take realism, for any body of putative knowledge, scientific
or not, to involve the conjunction of two theses: (1) *an independence thesis*:
our judgements answer for their truth to a world which exists independently of our awareness of it; (2) *a knowledge thesis*: by and large, we can
know which of these judgements are true.

Realism, as so defined, is threatened by an obvious internal tension:
if the world is independent of our awareness of it, then how can we gain

secure knowledge of it? Various resolutions of this tension are possible. Realists will seek some way of simultaneously upholding both the knowledge thesis and the independence thesis. But there are two traditional alternatives to realism, defined by their rejection of one of these theses. The *idealist or verificationist* tradition abandons the independence thesis, arguing that the very notion of some further world, beyond the world as we perceive it, is incoherent. *Sceptics*, by contrast, abandon the knowledge thesis, and accept that we cannot know the truth about the world.

In contemporary epistemology of science, scepticism is the main alternative to realism. This marks a contrast with the epistemology of everyday knowledge. When mainstream philosophers consider our everyday knowledge of objects like trees and tables, they take the most serious alternative to realism to be some version of idealism or verificationism. Thus, for example, 'phenomenalists' argue that it is impossible for a concept to stand for anything except a pattern of sensations, and that therefore it is a confusion to think that human judgements can answer to anything beyond the world as it appears in sense perception. Given this, phenomenalists can then argue that knowledge of trees and tables is unproblematic, on the grounds that there is no difficulty knowing about our own sensations.

It is true, of course, that many introductory textbooks in mainstream epistemology discuss scepticism alongside phenomenalism as alternatives to realism. But nobody outside the philosophy classroom seriously questions our knowledge of medium-sized physical objects like trees and tables. In mainstream epistemology, sceptical arguments about trees and tables reflect back on our assumptions: since we obviously do know about tables, an argument that such knowledge is impossible challenges us to find the flaw in our reasoning.

By contrast, scepticism in the epistemology of science is by no means just a philosophical exercise. For it is not at all obvious that we know about the entities postulated by modern scientific theories, such as gravitational waves or neutrinos. After all, we never have any direct sensory evidence for these entities. And the track record of past theories which postulated similar unobservable entities is not good. Considerations like these have persuaded a significant number of contemporary philosophers of science that knowledge of such entities really is unattainable.

Scepticism has not always been the main alternative to a realist epistemology of science. In the past, many philosophers of science, from J. S. Mill to Rudolf Carnap in the *Aufbau* (1928), opted for a phenomenalist account of the content of scientific claims, arguing that terms like 'mass', 'charge', and 'force' were properly understood as standing for complexes of observable circumstances. Indeed, this was arguably the dominant view until well into this century. But the rise of modern micro-physics, with its talk of atoms and fields, has made this view problematic. Moreover, Carnap's attempts to reduce such theoretical talk to observational claims ran into technical obstacles. In any case, the view is highly counter-intuitive. It is one thing to maintain that claims about trees are really claims about sensations. But it is hard to take this thesis seriously for claims about unobservable objects like electrons.

In consequence, nearly all contemporary philosophers of science accept that science aims literally to describe an unobservable world of microscopic particles and intangible waves. And a significant number draw the sceptical conclusion that science cannot succeed in this aim: since the world which science aims to decribe is beyond the reach of human perception, we have no reason to think that its theories are true.

These issues can be obscured by terminology. Contemporary sceptical opponents of scientific realism call themselves 'instrumentalists' or 'fictionalists' or 'constructive empiricists', and would probably object to the simple epithet 'sceptic'. Even so, and unlike earlier phenomenalist philosophers of science like Mill or Carnap, these contemporary philosophers of science all accept that scientific theories aim literally to portray an unobservable world, and conclude that for this reason we would be wrong to *believe* any scientific theories. If these philosophers differ from simple sceptics, it is only in adding to this rejection of belief the further thought that scientific theories can nevertheless be useful 'instruments' or 'fictions' for predictive purposes, and to this extent can be 'accepted' as working tools.

Another source of terminological confusion is the term 'anti-realism'. This term was first coined by Michael Dummett to describe a position in the idealist-verificationist tradition. Dummett's 'anti-realism' does not aim to construct the world out of sensations, in the style of simple phenomenalism; but even so, it insists, with the idealist-verificationist tradition, that our judgements cannot possibly answer to conditions that are beyond the power of humans to verify. On the other hand, philosophers of science, especially in the United States, have come to use the term 'anti-realism' to stand for sceptical views, and in particular for the sceptical attitude towards scientific theories that is the most common contemporary alternative to scientific realism.

People can of course define their terms as they wish, but there is much scope for misunderstanding here. Note in particular that 'anti-realism' in Dummett's sense directly contradicts 'anti-realism' in the sense of

American philosophers of science. Dummett's 'anti-realism', like more traditional idealism and verificationism, seeks to *uphold* our claims to knowledge, by arguing that such claims should not be read as answering to a world beyond our ken. By contrast, American 'anti-realism' wants to *reject* any scientific claims to knowledge about the unobservable world, precisely on the grounds that such claims *do* answer to a world beyond our ken. The only feature common to these two views is their rejection of the conjunction of (1) and (2) that I used earlier to define realism. However, where Dummett's 'anti-realism' upholds the knowledge thesis by rejecting the independence thesis, American 'anti-realism'does exactly the opposite.

TABLE 1.

	SCEPTICISM (constructive empiricism; fictionalism; instrumentalism; 'anti-realism', US style)	REALISM	IDEALISM (verificationism; phenomenalism; 'anti-realism', à la Dummett)
WORLD	Independent	Independent	Dependent
KNOWLEDGE	No	Yes	Yes

A number of these points about scientific realism are at issue in the first two papers in this collection. Arthur Fine (Ch. I) argues in favour of a position he calls 'the natural ontological attitude' ('NOA'). This position consists in what he claims are a set of truisms about science common to both 'realism' and 'anti-realism'. Central among these truisms, argues Fine, is the doctrine that well-confirmed scientific theories should be accepted as true. Fine argues that the mistake made by both 'realists' and 'anti-realists' is to add overblown metaphysical theses about the nature of truth and reality to their homely core of agreed assumptions.

Alan Musgrave (Ch. II) agrees with Fine that NOA is the right attitude to scientific theorizing. But he objects that NOA is itself a species of 'realism', not a core that can be agreed between 'realists' and 'anti-realists'. After all, as he understands the term, 'anti-realists' do *not* accept well-confirmed scientific theories as true. Musgrave cites Bas van Fraassen and Larry Laudan as two prominent contemporary 'anti-realists' who are quite explicit in their insistence that it is wrong to believe in the truth of any scientific theory about the underlying structure of the unobservable world.

What is more, argues Musgrave, there is no reason why a 'realist', as he takes himself to be, should add any dubious metaphysics to NOA's acceptance of theories as true. It is enough for scientific 'realism', in Musgrave's sense, that we should accept that science tells us the truth about the unobservable world. Further metaphysical views about the nature of truth and reality are unnecessary additions to this basic 'realist' stance.

In large part, this apparent dispute between Fine and Musgrave is simply a result of their addressing different debates. Fine is not interested in the sceptical option, but is instead thinking of the dispute between realism and anti-realism in Dummett's sense—that is, of how far the world itself can be distinguished from the way it appears to human beings. This is the issue that NOA is neutral on. Fine thinks that both the realist, who insists that our judgements answer for their truth to a world of verification-independent facts, and the Dummett-style anti-realist, who denies this, are making heady metaphysical claims that sober philosophers of science would do well to avoid.

Musgrave is happy to agree, but he thinks that Fine's NOA is itself a form of realism. This is because he is primarily concerned with scepticism as the alternative to 'realism'. From his point of view, Fine's NOA and the two views it mediates between may as well all be deemed versions of 'realism', since they all maintain, against scepticism, that we ought to accept the best scientific theories about unobservables as true.

In terms of Table 1, we can say that Fine is concerned only with the two right-hand columns, and wishes to maintain that we should embrace only those truisms which are common to both. Musgrave, on the other hand, is concerned with the dispute between the left-hand column and the two right-hand ones (even though he agrees with Fine in favouring the common core of the right-hand columns against the left).[1]

One obvious moral of this pair of papers is that readers should be wary of terminology. 'Realism' and 'anti-realism' are just two examples. A number of related terms, including 'empiricist', 'positivist', 'pragmatist', and 'instrumentalist', vary in meaning as used by different philosophers. Careful readers of the papers in this collection will note how authors introduce such terms and will attend to any other indications of the meanings they attach to them.

---

[1] This diagnosis of the debate is not entirely clear-cut. Fine begins his paper by rejecting a number of arguments which are normally used to defend belief in scientific theories against scepticism, and to this extent at least he seems to see scepticism as the alternative to scientific realism. But later in the paper he appears to switch tack, when he says explicitly that his 'anti-realists' accept theories as true (see p. 36).

THE UNDERDETERMINATION OF THEORY BY EVIDENCE

One central challenge to scientific realism comes from the underdetermination of theory by observational evidence. Suppose that two theories T_1 and T_2 are *empirically equivalent*, in the sense that they make the same observational predictions. Then no body of observational evidence will be able to decide conclusively between T_1 and T_2.

Note that the problem with such theories is not just that the choice between T_1 and T_2 is underdetermined by *current* observational evidence. If current evidence fails to decide between two theories, then the obvious response is to suspend belief for the time being, and to seek out experiments that will decide between them. But with genuinely empirically equivalent theories this option is not available. If *all* the observational predictions of T_1 and T_2 are identical, then there is no experiment that can conclusively eliminate one in favour of the other.

The thesis of *the underdetermination of theory by evidence* asserts that we will always be left with empirically equivalent theories, however much evidence we gather. There are two persuasive arguments for this thesis. One stems from the so-called Duhem–Quine thesis, which asserts that any theory can retain its central assumptions in the face of any anomalous evidence by making adjustments to less central assumptions. Suppose we start with two competing theories T_1 and T_2, and look to future evidence to decide between them, as suggested in the last paragraph. Given the Duhem–Quine thesis, it follows that, even after any amount of future evidence, we will still have two theories T_1' and T_2', derived from the original pair by the successive revisions occasioned by this evidence, and which this evidence fails to decide between.

A more direct argument for the inevitable existence of empirically equivalent theories starts with given T_1, and then points out that we can always gerrymander a different T_2 which makes exactly the same predictions. The simplest version of this strategy will simply take T_2 to assert all the observational claims made by T_1 but to deny the existence of any of the unobservable mechanisms postulated by T_1. More interesting versions of the argument do not just eliminate the unobservable mechanisms postulated by T_1; they replace them with extra 'self-correcting' structures designed to yield exactly the same observational appearances. (For example, if T_1 is some dynamic theory, make T_2 the theory that the universe is accelerating at 1 ft/sec² in a given direction, and add a universal force acting on all bodies to produce this acceleration. The result will be that T_1 and T_2 predict exactly the same observable relative motions.)

The paper by Lawrence Sklar (Ch. III) addresses the issues raised by the underdetermination of theory by evidence. Sklar is initially attracted to the thought that any pair of empirically equivalent theories are really just notational variants of each other, the same theory framed in different words, like Newton's *Principia* written in Latin and in English. But he recognizes that this position lacks plausibility. It is in effect a version of the old phenomenalist view of scientific theories (the 'positivist' view for Sklar) according to which claims about apparently unobservable entities like electrons are really claims about observable phenomena: for note that empirically equivalent theories will automatically be notational variants only if it is impossible to make meaningful (and hence contradictable) claims about any reality behind the observable appearances.

Sklar's paper is concerned to explore the options left open once we accept the underdetermination of theory by evidence (though in the end he is doubtful that any of these options is acceptable). One option is to embrace scepticism, on the grounds that we ought never to believe any theory, if no empirical evidence can conclusively eliminate its empirically equivalent alternatives. Another is to seek ways of discriminating between empirically equivalent theories, arguing that even when a number of theories make the same observational predictions, one can be more belief-worthy than the others.

The paper by Bas van Fraassen (Ch. IV) embraces the former option. According to van Fraassen's 'constructive empiricism' (which is elaborated in greater detail in van Fraassen 1980), we ought never to believe in the truth of any theory which goes beyond the observable phenomena. At most, we should believe that such a theory is 'empirically adequate'—that is, that it is correct in what it says about the observable portion of the world. Much of van Fraassen's article is concerned to show that this notion of empirical adequacy, and the associated distinction between the observable and the unobservable, is better explained within the 'semantic' account of scientific theories, which identifies theories with sets of models, than within the more traditional 'syntactic' account of theories as sets of sentences. (In connection with the observable–unobservable distinction, it is worth observing that a sceptic like van Fraassen asks rather less of the observable–unobservable distinction than a phenomenalist like Mill or Carnap: for, where phenomenalists maintain that we cannot *meaningfully talk* about unobservables, the scientific sceptic holds only that beings with our limited perceptual abilities ought not to *believe* any unobservable claims.)

The alternative, non-sceptical response to underdetermination is to argue that we can have grounds for believing one among a set of empirically equivalent theories. At first sight this might seem unpromising. If nothing we observe can rule out alternative theories, then how can we possibly be justified in selecting one for belief? But this last argument implicitly imposes very high standards on justified belief. It assumes that we are entitled to believe only the *logical* consequences of our observations (since it moves immediately from the existence of alternative theories consistent with our observations to the inadmissibility of belief in any particular theory). However, this an unreasonably strong requirement on justified belief. Nor is it one that most contemporary scientific sceptics would wish to impose. For it would imply (as van Fraassen notes at the end of his article) that we are not even entitled to believe in a theory's empirical adequacy, since this in itself requires us to make an inductive leap beyond the logical consequences of our finite stock of observational data.

So perhaps there is room for a non-sceptical response to under-determination. The idea, roughly, would be that, among the empirically equivalent theories consistent with our observational data, some are better explanations than others in virtue of their greater simplicity or elegance or unifying power, and that these virtues are indications that those theories are true. (Compare the way in which we restrict ourselves to 'projectible' generalizations when we perform simple enumerative inductions from finite data at the observable level.)

A preference for the 'best explanation' certainly seems to be part of scientific practice. As we saw above, there are always empirical equivalents for any given theory, based either on 'self-correcting' mechanisms or on the rejection of unobservable mechanisms altogether. But few practising scientists would regard the availability of such clumsy alternatives as a good reason for disbelieving normal theories. Given that the explanations of the observable data given by normal theories are far more elegant, scientists are generally happy to embrace the normal theories as true.

Still, even if 'inference to the best explanation' is part of intuitive scientific practice, it is not necessarily a good form of inference. Sceptical philosophers of science will argue that scientists go astray whenever they commit themselves to the truth of their best explanations. One obvious objection to inference to the best explanation is considered in Peter Lipton's paper 'Is the Best Good Enough?' (Ch. V). Suppose we allow that scientists standardly make accurate *comparative* judgements when they judge that T_1 is more likely to be true than $T_2, \ldots, T_n$. This alone won't ensure that they will get at the truth, because there may yet be even more likely theories among those they haven't yet thought of. It would be silly to infer that the 'best' explanation is true, if it is only the 'best' among those that have been considered so far. In the seventeenth century, Newton's theory was far and away the best account of gravitational motion on offer. But that was only because nobody had yet been able to formulate general relativity.

In response to this argument, Lipton suggests that it is sometimes possible to use comparative scientific judgements to reach the absolute assessments that scientific realism requires. For we can always ensure that our survey of theoretical options is exhaustive by including as a 'catch-all' T_{n+1}, the negation of all the theories $T_1, \ldots, T_n$ thought of so far. True, scientists will sometimes judge, when they have done this, that T_{n+1} is the most likely to be true—that is, that the truth most likely lies among the theories they haven't yet thought of. But in other cases they will rank T_{n+1} below the theories they have thought of, and hence conclude that their 'best' theory is absolutely likely to be true.

THE PESSIMISTIC META-INDUCTION FROM PAST FALSITY

Lipton's paper thus defends inference to the best (of all possible) explanations against one possible objection. However, the underlying questions about the philosophical status of this kind of inference remain. Note that Lipton starts with the assumption that scientists can at least make sound *comparative* judgements of truth-likelihood. But sceptics need not accept this. More generally, sceptics can invoke the underdetermination of theory by evidence to question any link between explanatory goodness (simplicity, elegance, unificatory power) and truth. If there are always alternative theories consistent with the evidence, then what guarantee can there possibly be that the most explanatory theory will generally be true?

This sceptical challenge raises a number of delicate philosophical issues about truth, rationality, and the onus of argument, to which I shall return in the next section. But first I shall consider a rather simpler argument against inferring the truth of the best explanation. This appeals not to the subtleties of underdetermination, but to direct empirical evidence of past theoretical failures. If we survey cases where scientists have embraced their best explanations as true, these explanations have normally turned out false: consider, for example, Ptolemaic astronomy, the caloric theory of heat, the ether theory of electromagnetism, and so on. Given this poor past record of best explanations, ought we not to conclude that inference to the best explanation will in general lead us to falsehoods, rather than truths?

This 'pessimistic meta-induction from past falsity' lies behind Larry Laudan's 'A Confutation of Convergent Realism' (Ch. VI). Laudan attacks the argument that if some scientific theory is 'successful', in the sense of generating confirmed predictions across a variety of contexts, then the best of all possible explanations of that success is the truth of the theory ('It would be a miracle if everything worked as the theory predicted, yet the theory were false'). He surveys a wide range of theories which were arguably successful in this sense, and points out that such success is not normally explained by the truth of the theory involved. For nearly all such theories—he gives a long list from history—have been recognized with hindsight to be false.

A possible realist response (anticipated by Laudan) is to argue only for the *approximate* truth of successful theories, rather than their unqualified truth. No doubt predictively successful theories are not always true in every precise detail. But perhaps they can still be assumed to be *close* to the truth, and in general closer to the truth than their predecessors.

One objection to this ploy is that the notion of 'approximate truth' is extremely difficult to articulate clearly. A sustained tradition of research on the notion of 'verisimilitude' (see the bibliography at the end of this volume) has made it clear that there can be no interest-independent measure of a theory's distance from the truth. However, even apart from this technical difficulty, there is a more obvious objection to the realist appeal to approximate truth: namely, that most past predictively successful theories are not even close to the truth, on any intuitively plausible reading of 'approximate truth'. Laudan's objection to realism is not just that past theories turn out to be wrong on points of detail. It is rather that they tend to be radically at variance with the truth, commiting themselves to a range of explanatory entities (like crystal spheres or caloric fluid or the ether) that have no counterparts whatsoever in reality.

John Worrall (Ch. VII) explores a different realist response to Laudan's argument. He concedes that predictively successful past theories standardly contain fundamental errors. But he argues that this does not require a blanket rejection of all scientific claims about the unobservable mechanisms behind the observable predictions. In Worrall's view, the lessons of history bear differentially on different components in scientific theories. More specifically, he argues that history shows that past theories are characteristically wrong about the *nature* of the unobservable realm, but not about the *structure* of its behaviour. In his central example, Worrall argues that nineteenth-century scientists were wrong to think of electromagnetic radiation as embodied in an 'ether', but quite right about the mathematical equations governing electromagnetism. Worrall draws the general moral that we should believe in the structure of unobservable reality postulated by successful theories, but avoid committing ourselves to any claims about the nature of that reality.

The general strategy exemplified by Worrall's paper seems to offer the best hope for realism. In the face of past theoretical failures, realists need to show that some parts of the failed theories fared better than others. If they can then identify some principled difference between the good parts and the bad parts, they can recommend belief in the good parts of current theories.

It is a further question, however, whether Worrall's specific way of drawing the distinction does the trick. Some philosophers of science would argue that, since our intellectual access to unobservable entities is always mediated by a structure of theoretical assumptions rather than by direct insight into their nature, Worrall's restriction of belief to structural claims is in fact no restriction at all (cf. Psillos 1995). If this is right, then realists need to find some better way of distinguishing the parts of theories which are likely to be discredited from the parts which are worthy of belief.

TRUTH AND RATIONALITY

Let us suppose that realism can succeed in blocking the pessimistic meta-induction in some such way. Realists will not then embrace our current theories *in toto* as the best explanation of predictive success, but only those parts which the evidence of history indicates to be genuinely responsible for such success.

However, realism still needs to deal with the other sceptical challenge mentioned earlier. By inferring the best explanation of predictive success, realists suppose that the best (simplest, most elegant, most unifying) explanation is likely to be true. Sceptics challenge this supposition, pointing out that more than one unobservable explanation will always be consistent with any body of observational evidence. Since we have no independent access to the unobservable realm, how can we possibly know that the 'better' explanation is generally the true one?

There are two possible responses to this challenge, corresponding to the middle and right-hand columns in Table 1. Both are 'realist' in Musgrave's sense, in that they both reject scepticism and uphold belief in scientific theories. But in Dummett's (and Fine's) sense, one is 'realist' and the other 'anti-realist', since they disagree on whether truth involves conformity to an independent reality.

The Dummettian 'anti-realist' option is illustrated in the paper by Brian Ellis (Ch. VIII). If we think of truth as correspondence to an independent reality, argues Ellis, then there is no alternative to van Fraassen's scepticism, for there is no way of establishing that theories with explanatory virtues are generally true. However, this gap between explanatory goodness and truth can be bridged, urges Ellis, if we take the truth to be by definition the view we ought rationally to hold, given all possible observable evidence. For, if it is a norm of rationality that we should believe the most explanatory (simple, elegant, unifying) among underdetermined theories, then the best such theory (given total evidence) will by definition be a true theory.

Ellis's position, which he characterizes as 'internal realism', stands or falls with his 'pragmatic' account of truth as whatever is justifiable by rational norms. Ellis in effect takes rational norms as given a priori, and then defines truth in terms of these a priori norms. The resulting position is anti-realist in Dummett's sense. It doesn't make judgements answer for their truth to a realm of sensation, in the style of traditional phenomenalist anti-realism, but it does deny that judgements answer to anything independent of the rational norms of human thought.

Ellis's 'pragmatic' account of truth is by no means uncontroversial. Note that it is not an *argument* for such an account of truth that it would deliver an answer to the sceptical challenge. (Analogously, that we *could* know about trees, if they *were* just patterns of perceptions, is not a good reason for thinking that trees *are* patterns of perceptions.) Many philosophers prefer the alternative view, that truth should be defined as conformity to an independent reality rather than in terms of the norms of human rationality.

It would take us too far afield to discuss the proper analysis of the concept of truth here. Let us instead consider whether the non-pragmatic view of truth has any answer to the sceptical challenge. Ellis says that if truth is conformity to an independent reality, then there is no way to link explanatory virtue with truth. But this is too quick. A non-pragmatist account of truth will not supply any *conceptual* link between truth and explanatory virtue, since it does not define truth in terms of explanatory virtue. But it still leaves open the possibility of an *empirical* connection between explanatory virtue and truth. If such an empirical link could be defended, then this would offer a different way of defending inferences to the best explanation against the sceptical challenge.

The second contribution to this volume by Larry Laudan (Ch. IX) bears on this possibility. Laudan is not concerned to defend realism against scepticism. But he shows how principles of scientific theory choice,

such as inference to the best explanation, might be vindicated by empirical generalizations which connect theoretical virtues with the goals of
science.

Laudan's starting-point is an older debate about the relevance of the
history of science to the philosophy of science. The 'historical turn' instigated by Hanson, Kuhn, and Feyerabend in the 1960s led philosophers of
science to test methodological proposals against empirical evidence from
the history of science. If Newton could be shown to have violated Carnap's
methodology, say, then this would be taken to count against Carnap's
methodology. However, the logic of this kind of argument is obscure. Why
should a methodologist's precepts have to fit the empirical facts of
Newton's practice? As Laudan observes, even if we accept that Newton is
a member of the scientific élite, we need not therewith accept that every
aspect of his actual practice is methodologically exemplary.

Laudan offers a different explanation of the relevance of historical
evidence to methodology. Suppose we distinguish the *aims* from the *means*
of scientific practice. Aims may include finding true theories or predictively reliable theories or theories that offer proofs of God's existence.
Given some such end, x, the practice of choosing theories that display a
certain putative virtue, y, can be viewed as a means to that end. This
suggests that methodological principles have the form 'If you want x, you
ought to choose theories with y'. That is, methodological principles have
the form of *hypothetical* imperatives, specifying a means to an end. As
such, they can be assessed on empirical grounds, like all similar means–
ends recommendations: does the empirical evidence show that y is in fact
an effective means to x? More specifically, in the present context, does the
historical record show that theories with feature y are a good route to aim
x? From Laudan's point of view, Newton's practice is methodologically
relevant not because of Newton's special status, but simply because it
provides cases which might help to show whether choosing y generally
leads to x.

Though Laudan leaves the question open in this second article, he does
not in fact think that truth (as opposed to predictive reliability, say) is a
sensible aim for science. Since he defends the pessimistic meta-induction,
he thinks that the historical evidence demonstrates the ineffectiveness of
any methodological means by which scientists might attempt to reach the
truth (cf. Laudan 1984: 137).

However, if we are able to block the pessimistic induction along the lines
outlined in the last section, then perhaps we can press Laudan's means–
ends view of scientific rationality into the service of realism. Perhaps
relevant historical evidence will enable us to identify certain methodologi-

cal strategies as effective routes to *true* theories. A methodology for science of this kind would have affinities with the reliabilist tradition in mainstream epistemology, which holds that a belief qualifies as knowledge if it has been produced by a reliable method—that is, by a method which, as a matter of empirical fact, generally produces true beliefs. This tradition in effect inverts the approach exemplified by Ellis: instead of defining truth in terms of a priori norms, it identifies the rationally correct methodology as that which, as a matter of empirical fact, provides an effective route to the truth.

Richard Boyd has developed this kind of approach to the philosophy of science in a series of articles over the last two decades. Boyd offers an empirical argument for the reliability of modern scientific methods: namely, that the only good explanation for the predictive success of science is that modern science in general provides an effective route to the (approximate) truth. He points out that the procedures by which scientists devise and test new theories are standardly informed by background assumptions provided by already established theories. Boyd maintains that these procedures for developing new theories could scarcely be expected to yield predictive success unless these established theories are by and large true (cf. Lipton, this volume, p. 100).

In the article reprinted here (Ch. X), Boyd explains how this argument can withstand the apparently sceptical implications of the pessimistic meta-induction. He concludes the article by addressing the charge that his position is circular. Note that Boyd moves from the empirical evidence that science is predictively successful, via an inference to the best explanation, to the conclusion that modern scientific method is an effective route to the truth. But the legitimacy of this kind of inference is precisely what his sceptical opponents deny. The sceptic's original challenge was to the legitimacy of first-order inferences to the best explanation, such as the inference that the atomic theory of matter is needed in order to explain the data of chemical experiments. Boyd is now in effect responding to this sceptical challenge with a meta-inference of the same form: the best explanation of science's general predictive success is that modern science generally gets at the truth. Boyd's opponents have not been slow to object that this begs the issue. After all, if they don't accept that the atomic theory of matter is needed to explain the chemical data, why should they accept that science's truth is needed to explain its predictive success? (Cf. Fine, this volume, pp. 24–5; Laudan, this volume, pp. 133–5.)

Boyd's response is that this last inference is not to be considered in isolation, but as part of an overall 'realist package'. This package should be compared as a whole with the overall alternatives. If the realist package is philosophically more defensible overall than the alternatives, then the charge of circularity falls away. Realists of Boyd's stripe take themselves to have an answer to the sceptical challenge to the link between explanatory virtue and truth. They say that the predictive success of science is good empirical evidence for this link. True, it is good evidence only by realist lights. But if realism can be shown to be right on overall philosophical grounds, then these are the right lights.

Philosophers like Laudan and Boyd advocate a *naturalized* philosophy of science. Instead of seeking to identify principles of scientific rationality on some a priori basis, they look to empirical information about the effectiveness of different scientific practices to decide the right methodology for science. It is worth observing that one important by-product of this 'naturalist turn' is the prospect of a reconciliation between philosophers and sociologists of science.

The last two decades have been marked by an explosion of exciting work in the sociology of science. Traditionally, sociologists were content to study the external aspects of science, like the growth of scientific institutions or the structure of scientific education. But recent sociological work has turned inwards to scientific theorizing itself, and aims to show how social influences and interactions play a decisive role in the resolution of specific theoretical debates.

This kind of sociology has been widely regarded as undermining any epistemological analysis of science, on the grounds that epistemology deals in a priori standards of rationality, while sociological studies seem to show that scientific theory choices aren't governed in this way at all, but rather by scientific power struggles and opportunistic manœuvres.

This rejection of epistemology presupposes, however, that epistemology traffics solely in a priori principles of theory evaluation. By contrast, if epistemology of science is conducted in the naturalist mode, then the conflict disappears. Naturalized philosophers of science have no axe to grind for a priori methodological principles over social processes. They can happily accept that theory choices are often determined by social processes. The only normative question they will then want to ask is whether those processes are an effective means to scientific aims. Nor is this a question they can answer by themselves. For it is an empirical question, not an a priori one, and the philosophers will therefore need the help of the sociologists and historians of science to answer it. (For further reading on this issue, see the sections on naturalized philosophy of science and on sociology of science in the selected bibliography at the end of this volume.)

CONFIRMATION THEORY AND BAYESIANISM

It is becoming increasingly common for issues in the epistemology of science to be discussed within the framework of Bayesian confirmation theory. Confirmation theory seeks to quantify questions of theoretical belief. Instead of asking simply *whether* we ought to believe theory T given evidence E, confirmation theorists ask *how much* we ought to believe T given E.

Bayesian confirmation theorists argue that such questions are best formulated as questions about subjective probabilities. Suppose that you attach an initial subjective probability Prob(H) to hypothesis H, and a conditional probability of Prob(H/E) to H on the supposition that evidence E is true. Then E will confirm H for you if your Prob(H/E) is greater than your initial Prob(H). Accordingly, when you observe E, you should increase your probability for H to your old Prob(H/E).

A simple theorem of the probability calculus ('Bayes's theorem') states that Prob(H/E) = Prob(E/H)/Prob(E) × Prob(H). This means that E confirms H for Bayesians—that is, Prob(H/E) is greater than Prob(H)—if Prob(E/H) is greater than Prob(E).

This makes intuitive sense. If datum E is surprising in itself (Prob(E) is low) but is what you would expect if hypothesis H were true (Prob(E/H) is high), then E is surely good support for H. The apparent bending of light near the sun is a striking fact in itself, but is just what is predicted by the theory of general relativity. So its observation in 1919 gave strong support to Einstein's theory.

Not all philosophers who work on confirmation theory are Bayesians. Clark Glymour, for one, has developed an alternative 'bootstrapping' account of confirmation in his *Theory and Evidence* (1980). In the chapter reprinted in this volume (Ch. XII), however, he concentrates on criticisms of the Bayesian alternative. Glymour first explains the basic structure of the Bayesian approach (and in particular the 'Dutch-book argument' for supposing that degrees of belief are a species of probability), and then points to various weaknesses.

For Glymour, the central flaw in Bayesianism is the subjectivity of its probabilities. Bayesianism places no substantial constraints on the probabilities that enter into confirmation relations, other than that they reflect actual degrees of belief. How likely is H, or E, or E given H? For the Bayesian these are questions about the subjective attitudes of individuals.

Glymour argues that this gives Bayesianism a spurious plausibility. Since

Bayesianism *per se* does not restrict the probabilities that enter into confirmation relations, it can always postulate as 'reasonable' any probabilities it may need to replicate standard forms of scientific reasoning. Glymour objects that this does not *explain* the worth of those forms of reasoning (after all, just the same trick could replicate any other forms of reasoning). He also criticizes specific attempts made by Bayesians to identify additional a priori principles governing reasonable degrees of belief.

Wesley Salmon's 'Rationality and Objectivity in Science *or* Tom Kuhn Meets Tom Bayes' (Ch. XI) offers a naturalistic response to the objection that Bayesian probabilities are arbitrary. Instead of seeking a priori principles governing reasonable subjective probabilities, Salmon suggests that we should look to the history of science to tell us about the empirical frequencies with which hypotheses of various kinds have turned out to be successful.

Salmon argues that this move can effect a reconciliation between the formalist tradition of Carnap and Hempel and the historicist approach instigated by Kuhn's *The Structure of Scientific Revolutions* (1962). Kuhn argues that scientists' judgements of plausibility play a crucial role in deciding their attitudes to hypotheses. Many of Kuhn's followers (but not Kuhn himself) have inferred that scientific judgements are consequently arbitrary. Salmon suggests that this charge of arbitrariness is best rebutted by construing such judgements of plausibility as Bayesian probabilities, and then grounding these probabilities in the empirical frequencies with which the hypotheses of the kind in question have proved successful.

It is worth observing that Salmon is open to a charge of circularity analogous to that brought against Boyd. Just as Boyd appealed to empirical evidence concerning scientific practice to vindicate scientists' preference for certain kinds of explanation, so Salmon appeals to similar evidence to vindicate the use of Bayesian reasoning. And, just as Boyd's vindication was a special case of the kind of inference he was trying to defend, so Salmon's inference, from the empirical evidence to a conclusion about probabilities, can be viewed as a special case of Bayesian reasoning. It would be an interesting exercise to consider whether Boyd's reply to this charge of circularity will work for an objectivist Bayesian like Salmon.

ARE THEORIES IMPORTANT?

Most of the articles in this collection are concerned with the truth or otherwise of scientific theories. However, some contemporary philosophers of science have argued that this is not necessarily the crux of debates about scientific realism. Nancy Cartwright, in *How the Laws of Physics Lie* (1983), and Ian Hacking, in *Representing and Intervening* (1983), have elaborated a position that they call 'entity realism', as opposed to 'theory realism'.

This view recognizes that the modern physical sciences are responsible for any number of striking physical effects, from lasers and optic fibres to electron microscopes and superconductors. It denies, however, that these physical effects provide convincing support for fundamental physical theory. Cartwright argues that the standard 'derivations' of these effects from fundamental theory are mediated by *ad hoc* assumptions, mathematical short cuts, and fudge factors. Since these devices are in general unmotivated by the fundamental theory, the derivation fails to provide any inductive support for the fundamental theory. The basic theory effectively does no work in the derivations, by comparison with the simplifications, argues Cartwright, and so deserves no credit.

However, this does not imply the non-existence of the subatomic particles and other unobservable entities implicated in the relevant physical effects. For Cartwright and Hacking, the success of our attempts to manipulate these entities is ample testimony to their existence. If these entities were not real, we could not use them to produce physical effects. Hacking encapsulates the view in his well-known slogan 'If you can spray them, they exist'.

Entity realism presents an interesting alternative to orthodox, theory-based realism. But it faces some obvious challenges. For a start, it is possible to doubt Cartwright's claim that fundamental physical theory is effectively redundant in the mathematical analysis of physical effects. Certainly simplifications play an important role in such analyses. But many of these simplifications are themselves guided by theoretical considerations, and so arguably add to the credit of the theory when they succeed. Another challenge is to the notion that we can separate commitment to entities from commitment to theories. To a large extent we think about unobservable entities in science as entities that play certain theoretical roles. This makes it difficult to see how we can accept the entities without accepting at least some of the surrounding theory.

In the paper reprinted in this volume (Ch. XIII), Cartwright takes a somewhat less radical tack. Here she is prepared to concede that the abstract laws of fundamental physics receive inductive support from physical effects. However, she denies that such laws need be universally applicable. Even if they accurately characterize the behaviour of physical phenomena in certain precisely controlled laboratory contexts, it does not follow that they govern all phenomena.

Everyday systems may dance to their own tunes, independently of the forces and equations of basic physics. Cartwright observes that the possibility of such 'emergent' patterns is a familiar theme in biological thinking. But she wants to go beyond this, and deny that even physical systems need be governed by fundamental physical laws. Perhaps the behaviour of a falling dollar bill escapes the laws of physics as much as the behaviour of a biological organism. Cartwright urges that we replace the reductionist picture of a unified system founded on a few basic laws with a patchwork of many laws each of limited range.

One issue raised by Cartwright's paper is whether different philosophical morals may apply in different areas of science. Perhaps we should be fundamentalists in physics but not in biology. Or perhaps we should be theory realists in chemistry, entity realists in geology, and outright sceptics in palaeobiology. At the beginning of this introduction I said that the epistemology of science deals with problems which arise for science in general. Certainly, most of the pieces in this collection have sought morals that will apply across the scientific board. But perhaps a more fine-grained approach would be worth the extra effort. Now that we are clear about the epistemological options on offer, there is no obvious reason why we should expect the same alternative to apply to every scientific discipline.

REFERENCES

Carnap, R. (1928). *Der logische Aufbau der Welt*. English translation: *The Logical Structure of the World*. Berkeley: University of California Press, 1967.

Cartwright, N. (1983). *How the Laws of Physics Lie*. Oxford: Clarendon Press.

Glymour, C. (1980). *Theory and Evidence*. Princeton: Princeton University Press.

Hacking, I. (1983). *Representing and Intervening*. Cambridge: Cambridge University Press.

Hacking, I. (ed.) (1981). *Scientific Revolutions*. Oxford: Oxford University Press.

Kuhn, T. (1962). *The Structure of Scientific Revolutions*. Chicago: University of Chicago Press.

Laudan, L. (1984). *Science and Values*. Berkeley: University of California Press.

Nidditch, P. (ed.) (1968). *Philosophy of Science*. Oxford: Oxford University Press.

Psillos, S. (1995). 'Is Structural Realism the Best of Both Worlds?' *Dialectica*, 49: 15–46.

van Fraassen, B. (1980). *The Scientific Image*. Oxford: Clarendon Press.

I

THE NATURAL ONTOLOGICAL ATTITUDE

ARTHUR FINE

> Let us fix our attention out of ourselves as much as possible; let us
> chase our imagination to the heavens, or to the utmost limits of the
> universe; we never really advance a step beyond ourselves, nor can
> conceive any kind of existence, but those perceptions, which have
> appear'd in that narrow compass. This is the universe of the imagin-
> ation, nor have we any idea but what is there produced.
>
> Hume, *Treatise*, book I, part II, section VI

Realism is dead. Its death was announced by the neo-positivists, who
realized that they could accept all the results of science, including all the
members of the scientific zoo, and still declare that the questions raised by
the existence claims of realism were mere pseudo-questions. Its death was
hastened by the debates over the interpretation of quantum theory, where
Bohr's non-realist philosophy was seen to win out over Einstein's passionate realism. Its death was certified, finally, as the last two generations of
physical scientists turned their backs on realism and have managed, nevertheless, to do science successfully without it. To be sure, some recent
philosophical literature has appeared to pump up the ghostly shell and to
give it new life. I think these efforts will eventually be seen and understood
as the first stage in the process of mourning, the stage of denial. But I think
we shall pass through this first stage and into that of acceptance, for realism
is well and truly dead, and we have work to get on with, in identifying a
suitable successor. To aid that work, I want to do three things in this essay.
First, I want to show that the arguments in favour of realism are not sound,
and that they provide no rational support for belief in realism. Then I want
to recount the essential role of non-realist attitudes for the development of
science in this century, and thereby (I hope) to loosen the grip of the idea
that only realism provides a progressive philosophy of science. Finally, I
want to sketch out what seems to me a viable non-realist position, one that

Reprinted from J. Leplin (ed.), *Scientific Realism* (Berkeley: University of California Press, 1984), 83–107, by permission of the author and the publisher.

is slowly gathering support and that seems a decent philosophy for post-realist times.[1]

1. ARGUMENTS FOR REALISM

Recent philosophical argument in support of realism tries to move from the success of the scientific enterprise to the necessity for a realist account of its practice. As I see it, the arguments here fall on two distinct levels. On the ground level, as it were, one attends to particular successes, such as novel, confirmed predictions, striking unifications of disparate-seeming phenomena (or fields), successful piggybacking from one theoretical model to another, and the like. Then we are challenged to account for such success, and told that the best and, it is slyly suggested, perhaps, the *only* way of doing so is on a realist basis. I do not find the details of these ground-level arguments at all convincing. Neither does Larry Laudan (1984), and, fortunately, he has provided a forceful and detailed analysis which shows that not even with a lot of handwaving (to shield the gaps in the argument) and charity (to excuse them) can realism itself be used to explain the very successes to which it invites our attention. But there is a second level of realist argument, the methodological level, that derives from Popper's (1972) attack on instrumentalism, which he attacks as being inadequate to account for the details of his own, falsificationist methodology. Arguments on this methodological level have been skilfully developed by Richard Boyd (1981, 1984), and by one of the earlier Hilary Putnams (1975). These arguments focus on the methods embedded in scientific practice, methods teased out in ways that seem to me accurate and perceptive about ongoing science. We are then challenged to account for why these methods lead to scientific success, and told that the best and, (again) perhaps, the only truly adequate way of explaining the matter is on the basis of realism.

I want to examine some of these methodological arguments in detail to display the flaws that seem to be inherent in them. But first I want to point out a deep and, I think, insurmountable problem with this entire strategy of defending realism, as I have laid it out above. To set up the problem, let

---

[1] In the final section, I call this post-realism 'NOA'. Among recent views that relate to NOA, I would include Hilary Putnam's 'internal realism', Richard Rorty's 'epistemological behaviorism', the 'semantic realism' espoused by Paul Horwich, parts of the 'Mother Nature' story told by William Lycan, and the defence of common sense worked out by Joseph Pitt (as a way of reconciling W. Sellars's manifest and scientific images). For references, see Putnam 1981, Rorty 1979, Horwich 1982, Lycan 1985, 1988, and Pitt 1981.

me review the debates in the early part of this century over the foundations of mathematics, the debates that followed Cantor's introduction of set theory. There were two central worries here, one over the meaningfulness of Cantor's hierarchy of sets in so far as it outstripped the number-theoretic content required by Kronecker (and others); the second worry, certainly deriving in good part from the first, was for the consistency (or not) of the whole business. In this context, Hilbert devised a quite brilliant programme to try to show the consistency of a mathematical theory by using only the most stringent and secure means. In particular, if one were concerned over the consistency of set theory, then clearly a set-theoretic proof of consistency would be of no avail. For if set theory were inconsistent, then such a consistency proof would be both possible and of no significance. Thus, Hilbert suggested that finite constructivist means, satisfactory even to Kronecker (or Brouwer), ought to be employed in metamathematics. Of course, Hilbert's programme was brought to an end in 1931, when Gödel showed the impossibility of such a stringent consistency proof. But Hilbert's idea was, I think, correct, even though it proved to be unworkable. Metatheoretic arguments must satisfy more stringent requirements than those placed on the arguments used by the theory in question, for otherwise the significance of reasoning about the theory is simply moot. I think this maxim applies with particular force to the discussion of realism.

Those suspicious of realism, from Osiander to Poincaré and Duhem to the 'constructive empiricism' of van Fraassen,[2] have been worried about the significance of the explanatory apparatus in scientific investigations. While they appreciate the systematization and coherence brought about by scientific explanation, they question whether acceptable explanations need to be true and, hence, whether the entities mentioned in explanatory principles need to exist.[3] Suppose they are right. Suppose, that is, that the usual explanation-inferring devices in scientific practice do not lead to principles that are reliably true (or nearly so), nor to entities whose existence (or near existence) is reliable. In that case, the usual abductive methods that lead us to good explanations (even to 'the best explanation') cannot be counted on to yield results even approximately true. But the strategy that leads to realism, as I have indicated, is just such an ordinary sort of abductive inference. Hence, if the non-realist were correct in his doubts, then such an inference to realism as the best explanation (or the

---

[2] Van Fraassen 1980. See esp. pp. 97–101 for a discussion of the truth of explanatory theories. To see that the recent discussion of realism is joined right here, one should contrast van Fraassen with Newton-Smith 1981, esp. ch. 8.

[3] Cartwright 1983 includes some marvellous essays on these issues.

like), while possible, would be of no significance exactly as in the case of a consistency proof using the methods of an inconsistent system. It seems, then, that Hilbert's maxim applies to the debate over realism: to argue for realism, one must employ methods more stringent than those in ordinary scientific practice. In particular, one must not beg the question as to the significance of explanatory hypotheses by assuming that they carry truth as well as explanatory efficacy.

There is a second way of seeing the same result. Notice that the issue over realism is precisely the issue of whether we should believe in the reality of those individuals, properties, relations, processes, and so forth used in well-supported explanatory hypotheses. Now what *is* the hypothesis of realism, as it arises as an explanation of scientific practice? It is just the hypothesis that our accepted scientific theories are approximately true, where 'being approximately true' is taken to denote an extra-theoretical relation between theories and the world. Thus, to address doubts over the reality of relations posited by explanatory hypotheses, the realist proceeds to introduce a further explanatory hypothesis (realism), itself positing such a relation (approximate truth). Surely anyone serious about the issue of realism, and with an open mind about it, would have to behave inconsistently if he were to accept the realist move as satisfactory.

Thus, both at the ground level and at the level of methodology, no support accrues to realism by showing that realism is a good hypothesis for explaining scientific practice. If we are open-minded about realism to begin with, then such a demonstration (even if successful) merely begs the question that we have left open ('Need we take good explanatory hypotheses as true?'). Thus, Hilbert's maxim applies, and we must employ patterns of argument more stringent than the usual abductive ones. What might they be? Well, the obvious candidates are patterns of induction leading to empirical generalizations. But, to frame empirical generalizations, we must first have some observable connections between observables. For realism, this must connect theories with the world by way of approximate truth. But no such connections are observable and, hence, suitable as the basis for an inductive inference. I do not want to labour the points at issue here. They amount to the well-known idea that realism commits one to an unverifiable correspondence with the world. So far as I am aware, no recent defender of realism has tried to make a case based on a Hilbert strategy of using suitably stringent grounds, and, given the problems over correspondence, it is probably just as well.

The strategy of arguments for realism as a good explanatory hypothesis, then, *cannot* (logically speaking) be effective for an open-minded non-believer. But what of the believer? Might he not, at least, show a kind of internal coherence about realism as an overriding philosophy of science, and should that not be of some solace, at least for the realist?[4] Recall, however, the analogue with consistency proofs for inconsistent systems. That sort of harmony should be of no solace to anyone. But for realism, I fear, the verdict is even harsher. For, so far as I can see, the arguments in question just do not work, and the reason for that has to do with the same question-begging procedures that I have already identified. Let me look closely at some methodological arguments in order to display the problems.

A typical realist argument on the methodological level deals with what I shall call the problem of the 'small handful'. It goes like this. At any time, in a given scientific area, only a small handful of alternative theories (or hypotheses) are in the field. Only such a small handful are seriously considered as competitors, or as possible successors to some theory requiring revision. Moreover, in general, this handful displays a sort of family resemblance in that none of these live options will be too far from the previously

---

[4] Some realists may look for genuine support, and not just solace, in such a coherentist line. They may see in their realism a basis for general epistemology, philosophy of language, and so forth (as does Boyd 1981, 1984). If they find in all this a coherent and comprehensive worldview, then they might want to argue for their philosophy as Wilhelm Wien argued (in 1909) for special relativity: 'What speaks for it most of all is the inner consistency which makes it possible to lay a foundation having no self-contradictions, one that applies to the totality of physical appearances' (quoted by Gerald Holton, 'Einstein's Scientific Program: Formative Years', in Woolf (ed.) 1980: 58). In so far as the realist moves away from the abductive defence of realism to seek support, instead, from the merits of a comprehensive philosophical system with a realist core, he marks as a failure the bulk of recent defences of realism. Even so, he will not avoid the critique pursued in the text. For although my argument above has been directed, in particular, against the abductive strategy, it is itself based on a more general maxim: namely, that the form of argument used to support realism must be more stringent than the form of argument embedded in the very scientific practice that realism itself is supposed to ground—on pain of begging the question. Just as the abductive strategy fails because it violates this maxim, so too would the coherentist strategy, should the realist turn from one to the other. For, as we see from the words of Wien, the same coherentist line that the realist would appropriate for his own support is part of ordinary scientific practice in framing judgements about competing theories. It is, therefore, not a line of defence available to the realist. Moreover, just as the truth-bearing status of abduction is an issue dividing realists from various non-realists, so too is the status of coherence-based inference. Turning from abduction to coherence, therefore, still leaves the realist begging the question. Thus, when we bring out into the open the character of arguments *for* realism, we see quite plainly that they do not work. See Fine 1986b for a more detailed discussion.

In support of realism there seem to be only those 'reasons of the heart' which, as Pascal says, reason does not know. Indeed, I have long felt that belief in realism involves a profound leap of faith, not at all dissimilar from the faith that animates deep religious convictions. I would welcome engagement with realists on this understanding, just as I enjoy conversation on a similar basis with my religious friends. The dialogue will proceed more fruitfully, I think, when the realists finally stop pretending to a rational support for their faith, which they do not have. Then we can all enjoy their intricate and sometimes beautiful philosophical constructions (of, e.g., knowledge, or reference, etc.), even though to us, as non-believers, they may seem only wonder-full castles in the air.

accepted theories in the field, each preserving the well-confirmed features of the earlier theories and deviating only in those aspects less confirmed. Why? Why does this narrowing down of our choices to such a small handful of cousins of our previously accepted theories work to produce good successor theories?

The realist answers this as follows. Suppose that the already existing theories are themselves approximately true descriptions of the domain under consideration. Then surely it is reasonable to restrict one's search for successor theories to those whose ontologies and laws resemble what we already have, especially where what we already have is well confirmed. And if these earlier theories were approximately true, then so will be such conservative successors. Hence, such successors will be good predictive instruments; that is, they will be successful in their own right.

The small-handful problem raises three distinct questions: (1) why only a small handful out of the (theoretically) infinite number of possibilities? (2) why the conservative family resemblance between members of the handful? and (3) why does the strategy of narrowing the choices in this way work so well? The realist response does not seem to address the first issue at all, for even if we restrict ourselves just to successor theories resembling their progenitors, as suggested, there would still, theoretically, always be more than a small handful of these. To answer the second question, as to why conserve the well-confirmed features of ontology and laws, the realist must suppose that such confirmation is a mark of an approximately correct ontology and approximately true laws. But how could the realist possibly justify such an assumption? Surely, there is no valid inference of the form 'T is well-confirmed; therefore, there exist objects pretty much of the sort required by T and satisfying laws approximating those of T'. Any of the dramatic shifts of ontology in science show the invalidity of this schema. For example, the loss of the ether from the turn-of-the-century electrodynamic theories demonstrates this at the level of ontology, and the dynamics of the Rutherford–Bohr atom *vis-à-vis* the classical energy principles for rotating systems demonstrates it at the level of laws. Of course, the realist might respond that there is no question of a strict inference between being well confirmed and being approximately true (in the relevant respects), but there is a probable inference of some sort. But of what sort? Certainly there is no probability relation that rests on inductive evidence here. For there is no independent evidence for the relation of approximate truth itself: at least, the realist has yet to produce any evidence that is independent of the argument under examination. But if the probabilities are not grounded inductively, then how else? Here, I think the realist may well try to fall back on his original strategy and suggest that being approximately true provides the best explanation for being well confirmed. This move throws us back to the ground-level realist argument, the argument from specific success to an approximately true description of reality, which Laudan (1984) has criticized. I should point out, before looking at the third question, that if this last move is the one the realist wants to make, then his success at the methodological level can be no better than his success at the ground level. If he fails there, he fails across the board.

The third question, and the one I think the realist puts most weight on, is why does the small-handful strategy work so well. The instrumentalist, for example, is thought to have no answer here. He must just note that it does work well and be content with that. The realist, however, can explain why it works by citing the transfer of approximate truth from predecessor theories to the successor theories. But what does this explain? At best, it explains why the successor theories cover the same ground as well as their predecessors, for the conservative strategy under consideration assures that. But note that here the instrumentalist can offer the same account: if we insist on preserving the well-confirmed components of earlier theories in later theories, then, of course, the later ones will do well over the well-confirmed ground. The difficulty, however, is not here at all but rather in how to account for the successes of the later theories in new ground, or with respect to novel predictions, or in overcoming the anomalies of the earlier theories. And what can the realist possibly say in this area except that the theorist, in proposing a new theory, has happened to make a good guess? For nothing in the approximate truth of the old theory can guarantee (or even make it likely) that modifying the theory in its less-confirmed parts will produce a progressive shift. The history of science shows well enough how such tinkering succeeds only now and again, and fails for the most part. This history of failures can scarcely be adduced to explain the occasional success. The idea that by extending what is approximately true, one is likely to bring new approximate truth is chimera. It finds support neither in the logic of approximate truth nor in the history of science. The problem for the realist is how to explain the *occasional success* of a strategy that *usually fails*.[5] I think he has no special resources with which to do this.

---

[5] I hope all readers of this essay will take this idea to heart. For in formulating the question as to how to explain why the methods of science lead to instrumental success, the realist has seriously misstated the explanandum. Overwhelmingly, the results of the conscientious pursuit of scientific inquiry are failures: failed theories, failed hypotheses, failed conjectures, inaccurate measurements, incorrect estimations of parameters, fallacious causal inferences, and so forth. If explanations are appropriate here, then what requires explaining is why the very same methods produce an overwhelming background of failures and, occasionally, also a pattern of successes. The realist literature has not yet begun to address this question, much less to offer even a hint of how to answer it.

In particular, his usual fallback on to approximate truth provides nothing more than a gentle pillow. He may rest on it comfortably, but it does not really help to move his cause forward.

The problem of the small handful raises three challenges: why small? why narrowly related? and why does it work? The realist has no answer for the first of these, begs the question as to the truth of explanatory hypotheses on the second, and has no resources for addressing the third. For comparison, it may be useful to see how well his arch-enemy, the instrumentalist, fares on the same turf. The instrumentalist, I think, has a substantial basis for addressing the questions of smallness and narrowness, for he can point out that it is extremely difficult to come up with alternative theories that satisfy the many empirical constraints posed by the instrumental success of theories already in the field. Often it is hard enough to come up with even one such alternative. Moreover, the common apprenticeship of scientists working in the same area certainly has the effect of narrowing down the range of options by channelling thought into the commonly accepted categories. If we add to this the instrumentally justified rule 'If it has worked well in the past, try it again', then we get a rather good account, I think, of why there is usually only a small and narrow handful. As to why this strategy works to produce instrumentally successful science, we have already noted that for the most part it does not. Most of what this strategy produces are failures. It is a quirk of scientific memory that this fact gets obscured, much as do the memories of bad times during a holiday vacation when we recount all our 'wonderful' vacation adventures to a friend. Those instrumentalists who incline to a general account of knowledge as a social construction can go further at this juncture and lean on the sociology of science to explain how the scientific community 'creates' its knowledge. I am content just to back off here and note that over the problem of the small handful, the instrumentalist scores at least two out of three, whereas the realist, left to his own devices, has struck out.[6]

I think the source of the realist's failure here is endemic to the methodological level, infecting all of his arguments in this domain. It resides, in the first instance, in his repeating the question-begging move from explanatory efficacy to the truth of the explanatory hypothesis. And in the second instance, it resides in his twofold mishandling of the concept of approximate truth: first, in his trying to project from some body of assumed approximate truths *to* some further and novel such truths, and second, in

[6] Of course, the realist can appropriate the devices and answers of the instrumentalist, but that would be cheating, and anyway, it would not provide the desired support of realism *per se*.

his needing genuine access to the relation of correspondence. There are no general connections of this first sort, however, sanctioned by the logic of approximate truth, nor secondly, any such warranted access. However, the realist must pretend that there are in order to claim explanatory power for his realism. We have seen those two agents infecting the realist way with the problem of the small handful. Let me show them at work in another methodological favourite of the realist, the 'problem of conjunctions'.

The problem of conjunctions is this. If T and T' are independently well-confirmed, explanatory theories, and if no shared term is ambiguous between the two, then we expect the conjunction of T and T' to be a reliable predictive instrument (provided, of course, that the theories are not mutually inconsistent). Why? challenges the realist, and he answers as follows. If we make the realist assumption that T and T', being well confirmed, are approximately true of the entities (etc.) to which they refer, and if the unambiguity requirement is taken realistically as requiring a domain of common reference, then the conjunction of the two theories will also be approximately true, and, hence, it will produce reliable observational predictions. QED.

But notice our agents at work. First, the realist makes the question-begging move from explanations to their approximate truth, and then he mistreats approximate truth. For nothing in the logic of approximate truth sanctions the inference from 'T is approximately true' and 'T′ is approximately true' to the conclusion that the conjunction 'T·T′' is approximately true. Rather, in general, the tightness of an approximation dissipates as we pile on further approximations. If T is within ε, in its estimation of some parameter, and T' is also within ε, then the only general thing we can say is that the conjunction will be within 2ε of the parameter. Thus, the logic of approximate truth should lead us to the opposite conclusion here; that is, that the conjunction of two theories is, in general, less reliable than either (over their common domain). But this is neither what we expect nor what we find. Thus, it seems quite implausible that our actual expectations about the reliability of conjunctions rest on the realist's stock of approximate truths.

Of course, the realist could try to retrench here and pose an additional requirement of some sort of uniformity on the character of the approximations, as between T and T'.[7] It is difficult to see how the realist could do this successfully without making reference to the distance between the approximations and 'the truth'. For what kind of internalist requirement

[7] Paul Teller has made this suggestion to me in conversation.

could possibly ensure the narrowing of this distance? But the realist is in no position to impose such requirements, since neither he nor anyone else has the requisite access to 'the truth'. Thus, whatever uniformity-of-approximation condition the realist might impose, we could still demand to be shown that this leads closer to the truth, not farther away. The realist will have no demonstration, except to point out to us that it all works (sometimes!). But that was the original puzzle.[8] Actually, I think the puzzle is not very difficult. For surely, if we do not entangle ourselves with issues over approximation, there is no deep mystery as to why two compatible and successful theories lead us to expect their conjunction to be successful. For in forming the conjunction, we just add the reliable predictions of one to the reliable predictions of the other, having antecedently ruled out the possibility of conflict.

There is more to be said about this topic. In particular, we need to address the question of why we expect the logical gears of the two theories to mesh. However, I think that a discussion of the realist position here would only bring up the same methodological and logical problems that we have already uncovered at the centre of the realist argument.

Indeed, this schema of knots in the realist argument applies across the board, and vitiates every single argument at the methodological level. Thus my conclusion here is harsh, indeed. The methodological arguments for realism fail, even though, were they successful, they would still not support the case. For the general strategy they are supposed to implement is just not stringent enough to provide rational support for realism. In the next two sections, I will try to show that this situation is just as well, for realism has not always been a progressive factor in the development of science, and, anyway, there is a position other than realism that is more attractive.

2. REALISM AND PROGRESS

If we examine the two twentieth-century giants among physical theories, relativity and the quantum theory, we find a living refutation of the realist's claim that only his view of science explains its progress, and we find some curious twists and contrasts over realism as well. The theories of relativity

[8] Niiniluoto 1982 contains interesting formal constructions for 'degree of truthlikeness' and related verisimilia. As conjectured above, they rely on an unspecified correspondence relation to the truth and on measures of the 'distance' from the truth. Moreover, they fail to sanction that projection, from some approximate truths to other, novel truths, which lies at the core of realist rationalizations.

are almost single-handedly the work of Albert Einstein. Einstein's early positivism and his methodological debt to Mach (and Hume) leap right out of the pages of the 1905 paper on special relativity.[9] The same positivist strain is evident in the 1916 general relativity paper as well, where Einstein (in section 3 of that paper) tries to justify his requirement of general covariance by means of a suspicious-looking verificationist argument which, he says, 'takes away from space and time the last remnants of physical objectivity' (Einstein et al. 1952: 117). A study of his tortured path to general relativity[10] shows the repeated use of this Machist line, always used to deny that some concept has a real referent. Whatever other, competing strains there were in Einstein's philosophical orientation (and there certainly were others), it would be hard to deny the importance of this instrumentalist/positivist attitude in liberating Einstein from various realist commitments. Indeed, on another occasion, I would argue in detail that without the 'freedom from reality' provided by his early reverence for Mach, a central tumbler necessary to unlock the secret of special relativity would never have fallen into place.[11] A few years after his work on general relativity, however, roughly around 1920, Einstein underwent a philosophical conversion, turning away from his positivist youth (he was 41 in 1920) and becoming deeply committed to realism (see Leplin (ed.) 1984, ch. 6). In particular, following his conversion, Einstein wanted to claim genuine reality for the central theoretical entities of the general theory, the four-dimensional space-time manifold and associated tensor fields. This is a serious business, for if we grant his claim, then not only do space and time cease to be real, but so do virtually all of the usual dynamical quantities.[12] Thus motion, as we understand it, itself ceases to be real. The current generation of philosophers of space and time (led by Howard Stein and John Earman) have followed Einstein's lead here. But, interestingly, not only do these ideas boggle the mind of the average man in the street (like you and me), they boggle most contemporary scientific minds as well.[13] That is, I believe the majority opinion among working, knowledgeable

---

[9] See Gerald Holton, 'Mach, Einstein, and the Sarah for Reality', in Holton 1973: 219–59. I have tried to work out the precise role of this positivist methodology in Fine 1986a: ch. 2. See also Fine 1981.

[10] Earman and Glymour 1978. The tortuous path detailed by Earman is sketched by B. Hoffmann 1972: 116–28. A non-technical and illuminating account is given by John Stachel 1979.

[11] I have in mind the role played by the analysis of simultaneity in Einstein's path to special relativity. Despite the important study by Arthur Miller (1981) and an imaginative pioneering work by John Earman et al. (1983), I think the role of positivist analysis in the 1905 paper has yet to be properly understood.

[12] Jones 1991 explains very nicely some of the difficulties here.

[13] I think the ordinary, deflationist attitude of working scientists is much like that of Steven Weinberg (1972).

scientists is that general relativity provides a magnificent organizing tool for treating certain gravitational problems in astrophysics and cosmology. But few, I believe, give credence to the kind of realist existence and non-existence claims that I have been mentioning. For relativistic physics, then, it appears that a non-realist attitude was important in its development, that the founder nevertheless espoused a realist attitude to the finished product, but that most who actually use it think of the theory as a powerful instrument, rather than as expressing a 'big truth'.

With quantum theory, this sequence gets a twist. Heisenberg's seminal paper of 1925 is prefaced by the following abstract, announcing, in effect, his philosophical stance: 'In this paper an attempt will be made to obtain bases for a quantum-theoretical mechanics based exclusively on relations between quantities observable in principle' (Heisenberg 1925: 879). In the body of the paper, Heisenberg not only rejects any reference to unobservables; he also moves away from the very idea that one should try to form any picture of a reality underlying his mechanics. To be sure, Schrödinger, the second father of quantum theory, seems originally to have had a vague picture of an underlying wavelike reality for his own equation. But he was quick to see the difficulties here and, just as quickly, although reluctantly, abandoned the attempt to interpolate any reference to reality.[14] These instrumentalist moves away from a realist construal of the emerging quantum theory were given particular force by Bohr's so-called philosophy of complementarity. This non-realist position was consolidated at the time of the famous Solvay conference, in October 1927, and is firmly in place today. Such quantum non-realism is part of what every graduate physicist learns and practises. It is the conceptual backdrop to all the brilliant successes in atomic, nuclear, and particle physics over the past fifty years. Physicists have learned to think about their theory in a highly non-realist way, and doing just that has brought about the most marvellous predictive success in the history of science.

The war between Einstein, the realist, and Bohr, the non-realist, over the interpretation of quantum theory was not, I believe, just a side-show in physics, nor an idle intellectual exercise. It was an important endeavour undertaken by Bohr on behalf of the enterprise of physics as a progressive science. For Bohr believed (and this fear was shared by Heisenberg, Sommerfeld, Pauli, and Born—and all the big guys) that Einstein's realism, if taken seriously, would block the consolidation and articulation of the new physics and, thereby, stop the progress of science. They were afraid, in particular, that Einstein's realism would lead the next generation

---

[14] See Wessels 1979 and Fine 1986a: ch. 5.

of the brightest and best students into scientific dead ends. Alfred Landé, for example, as a graduate student, was interested in spending some time in Berlin to sound out Einstein's ideas. His supervisor was Sommerfeld, and recalling this period, Landé (1974: 460) writes: 'The more pragmatic Sommerfeld . . . warned his students, one of them this writer, not to spend too much time on the hopeless task of "explaining" the quantum but rather to accept it as fundamental and help work out its consequences.'

The task of 'explaining' the quantum, of course, is the realist programme for identifying a reality underlying the formulae of the theory and thereby explaining the predictive success of the formulae as approximately true descriptions of this reality. It is this programme that I have criticized in the first part of this chapter, and this same programme that the builders of quantum theory saw as a scientific dead end. Einstein knew perfectly well that the issue was joined right here. In the summer of 1935, he wrote to Schrödinger: 'The real problem is that physics is a kind of metaphysics; physics describes "reality." But we do not know what "reality" is. We know it only through physical description. . . . But the Talmudic philosopher sniffs at "reality," as at a frightening creature of the native mind.'[15]

By avoiding the bogey of an underlying reality, the 'Talmudic' originators of quantum theory seem to have set subsequent generations on precisely the right path. Those inspired by realist ambitions have produced no predictively successful physics. Neither Einstein's conception of a unified field, nor the ideas of the de Broglie group about pilot waves, nor the Bohm-inspired interest in hidden variables has made for scientific progress. To be sure, several philosophers of physics, including another Hilary Putnam and myself, have fought a battle over the last decade to show that the quantum theory is at least consistent with some kind of underlying reality. I believe that Hilary has abandoned the cause, perhaps in part on account of the recent Bell inequality problem over correlation experiments, a problem that van Fraassen (1982) calls 'the Charybdis of realism'. My own recent work in the area suggests that we may still be able to keep realism afloat in this whirlpool.[16] But the possibility (as I still see it) for a realist account of the quantum domain should not lead us away from appreciating the historical facts of the matter.

One can hardly doubt the importance of a non-realist attitude for the development and practically infinite success of the quantum theory. Historical counterfactuals are always tricky, but the sterility of actual realist programmes in this area at least suggests that Bohr and company were

[15] Einstein to Schrödinger, 19 June 1935.
[16] See Fine 1982 for part of the discussion and also Fine 1986a: ch. 9.

right in believing that the road to scientific progress here would have been blocked by realism. The founders of quantum theory never turned on the non-realist attitude that served them so well. Perhaps that is because the central underlying theoretical device of quantum theory, the densities of a complex-valued and infinite-dimensional wave function, are even harder to take seriously than is the four-dimensional manifold of relativity. But now there comes a most curious twist. For just as the practitioners of relativity, I have suggested, ignore the *realist* interpretation in favour of a more pragmatic attitude toward the space-time structure, so the quantum physicists would appear to make a similar reversal and to forget their non-realist history and allegiance when it comes time to talk about new discoveries.

Thus, anyone in the business will tell you about the exciting period, in the fall of 1974, when the particle group at Brookhaven, led by Samuel Ting, discovered the J particle, just as a Stanford team at the Stanford Linear Accelerator Center, under Burton Richter, independently found a new particle they called ψ. These turned out to be one and the same, the so-called ψ/J particle (mass 3,098 MeV, spin 1, resonance 67 keV, strangeness 0). To explain this new entity, the theoreticians were led to introduce a new kind of quark, the so-called charmed quark. The ψ/J particle is then thought to be made up of a charmed quark and an anti-charmed quark, with their respective spins aligned. But if this is correct, then there ought to be other such pairs anti-aligned or with variable spin alignments, and these ought to make up quite new observable particles. Such predictions from the charmed-quark model have turned out to be confirmed in various experiments.

I have gone on a bit in this story in order to convey the realist feel to the way scientists speak in this area. For I want to ask whether this is a return to realism or whether, instead, it can somehow be reconciled with a fundamentally non-realist attitude.[17] I believe that the non-realist option is correct.

3. NON-REALISM

Even if the realist happens to be a talented philosopher, I do not believe that, in his heart, he relies for his realism on the rather sophisticated form

[17] The non-realism that I attribute to students and practitioners of the quantum theory requires more discussion and distinguishing of cases and kinds than I have room for here. It is certainly not the all-or-nothing affair I make it appear in the text. I carry out some of the required discussion in Fine 1986a: ch. 9. My thanks to Paul Teller and James Cushing, each of whom saw the need for more discussion.

of abductive argument that I have examined and rejected in the first section of this chapter, and which the history of twentieth-century physics shows to be fallacious. Rather, if his heart is like mine, then I suggest that a more simple and homely sort of argument is what grips him. It is this, and I will put it in the first person. I certainly trust the evidence of my senses, on the whole, with regard to the existence and features of everyday objects. And I have similar confidence in the system of 'check, double-check, check, triple-check' of scientific investigation, as well as the other safeguards built into the institutions of science. So, if the scientists tell me that there really are molecules, and atoms, and ψ/J particles, and, who knows, maybe even quarks, then so be it. I trust them, and thus must accept that there really are such things with their attendant properties and relations. Moreover, if the instrumentalist (or some other member of the species 'non-realistica') comes along to say that these entities and their attendants are just fictions (or the like), then I see no more reason to believe him than to believe that *he* is a fiction, made up (somehow) to do a job on me; which I do not believe. It seems, then, that I had better be a realist. One can summarize this homely and compelling line as follows: it is possible to accept the evidence of one's senses and to accept, *in the same way*, the confirmed results of science only for a realist; hence, I should be one (and so should you!).

What is it to accept the evidence of one's senses and, *in the same way*, to accept confirmed scientific theories? It is to take them into one's life as true, with all that implies concerning adjusting one's behaviour, practical and theoretical, to accommodate these truths. Now, of course, there are truths and truths. Some are more central to us and our lives, some less so. I might be mistaken about anything, but were I mistaken about where I am right now, that might affect me more than would my perhaps mistaken belief in charmed quarks. Thus, it is compatible with the homely line of argument that some of the scientific beliefs that I hold are less central than some, for example, perceptual beliefs. Of course, were I deeply in the charmed-quark business, giving up that belief might be more difficult than giving up some at the perceptual level. (Thus we get the phenomenon of 'seeing what you believe', well known to all thoughtful people.) When the homely line asks us, then, to accept the scientific results 'in the same way' in which we accept the evidence of our senses, I take it that we are to accept them both as true. I take it that we are being asked not to distinguish between kinds of truth or modes of existence or the like, but only among truths themselves in terms of centrality, degrees of belief, or such.

Let us suppose this understood. Now, do you think that Bohr, the archenemy of realism, could toe the homely line? Could Bohr, fighting for the

sake of science (against Einstein's realism), have felt compelled either
to give up the results of science or else to assign its 'truths' to some
category different from the truths of everyday life? It seems unlikely.
And thus, unless we uncharitably think Bohr inconsistent on this basic
issue, we might well come to question whether there is any necessary
connection moving us from accepting the results of science as true to being
a realist.[18]

Let me use the term 'anti-realist' to refer to any of the many different
specific enemies of realism: the idealist, the instrumentalist, the phenomenalist, the empiricist (constructive or not), the conventionalist, the
constructivist, the pragmatist, and so forth. Then, it seems to me that both
the realist and the anti-realist must toe what I have been calling 'the
homely line'. That is, they must both accept the certified results of science
as on a par with more homely and familiarly supported claims. That is not
to say that one party (or the other) cannot distinguish more from less well-
confirmed claims at home or in science; nor that one cannot single out
some particular mode of inference (such as inference to the best explanation) and worry over its reliability, both at home and away. It is just that
one must maintain parity. Let us say, then, that both realist and anti-realist
accept the results of scientific investigations as 'true', on a par with more
homely truths. (I realize that some anti-realists would rather use a different word, but no matter.) And call this acceptance of scientific truths the
'core position'.[19] What distinguishes realists from anti-realists, then, is what
they add on to this core position.

[18] I should be a little more careful about the historical Bohr than I am in the text. For
Bohr himself would seem to have wanted to truncate the homely line somewhere between the
domain of chairs and tables and atoms, whose existence he plainly accepted, and that of
electrons, where he seems to have thought the question of existence (and of realism, more
generally) was no longer well defined. An illuminating and provocative discussion of Bohr's
attitude toward realism is given by Paul Teller (1981). Thanks, again, to Paul for helping to keep
me honest.

[19] In this context, e.g., van Fraassen's 'constructive empiricism' would prefer the concept of
empirical adequacy, reserving 'truth' for an (unspecified) literal interpretation and believing in
that truth only among observables. I might mention here that in this classification Putnam's
internal realism comes out as anti-realist. For Putnam accepts the core position, but he would
add to it a Peircean construal of truth as ideal rational acceptance. This is a mistake, which I
expect that Putnam will realize and correct in future writings. He is criticized for it by Horwich
(1982), whose own 'semantic realism' turns out, in my classification, to be neither realist nor
anti-realist. Indeed, Horwich's views are quite similar to what is called 'NOA' below, and could
easily be read as sketching a philosophy of language compatible with NOA. Finally, the 'epistemological behaviorism' espoused by Rorty (1979) is a form of anti-realism that seems to me very
similar to Putnam's position, but achieving the core parity between science and common sense
by means of an acceptance that is neither ideal nor especially rational, at least in the normative
sense. (I beg the reader's indulgence over this summary treatment of complex and important
positions. I have been responding to Nancy Cartwright's request to differentiate these recent
views from NOA.) See Fine 1986a: ch. 8 for a discussion of these anti-realisms.

The anti-realist may add on to the core position a particular analysis of the concept of truth, as in the pragmatic and instrumentalist and conventionalist conceptions of truth. Or the anti-realist may add on a special analysis of concepts, as in idealism, constructivism, phenomenalism, and in some varieties of empiricism. These addenda will then issue in a special meaning, say, for existence statements. Or the anti-realist may add on certain methodological strictures, pointing a wary finger at some particular inferential tool, or constructing his own account for some particular aspects of science (e.g., explanations or laws). Typically, the anti-realist will make several such additions to the core.

What, then, of the realist; what does he add to his core acceptance of the results of science as really true? My colleague, Charles Chastain, suggested what I think is the most graphic way of stating the answer—namely, that what the realist adds on is a desk-thumping, foot-stamping shout of 'Really!' So, when the realist and the anti-realist agree, say, that there really are electrons and that they really carry a unit negative charge and really do have a small mass (of about 9.1×10^{-28} grams), what the realist wants to add is the emphasis that all this is really so. 'There really are electrons, really!' This typical realist emphasis serves both a negative and a positive function. Negatively, it is meant to deny the additions that the anti-realist would make to that core acceptance which both parties share. The realist wants to deny, for example, the phenomenalistic reduction of concepts or the pragmatic conception of truth. The realist thinks that these addenda take away from the substantiality of the accepted claims to truth or existence. 'No,' says he, 'they *really* exist, and not in just your diminished anti-realist sense.' Positively, the realist wants to explain the robust sense in which *he* takes these claims to truth or existence; namely, as claims about reality—what is really, really the case. The full-blown version of this involves the conception of truth as correspondence with the world, and the surrogate use of approximate truth as near-correspondence. We have already seen how these ideas of correspondence and approximate truth are supposed to explain what *makes* the truth *true*, whereas, in fact, they function as mere trappings, that is, as superficial decorations that may well attract our attention but do not compel rational belief. Like the extra 'really', they are an arresting foot thump and, logically speaking, of no more force.

It seems to me that when we contrast the realist and the anti-realist in terms of what they each want to add to the core position, a third alternative emerges—and an attractive one at that. It is the core position itself, *and all by itself.* If I am correct in thinking that, at heart, the grip of realism only extends to the homely connection of everyday truths with scientific truths, and that good sense dictates our acceptance of the one on the same basis as our acceptance of the other, then the homely line makes the core position, all by itself, a compelling one, one that we ought to take to heart. Let us try to do so and see whether it constitutes a philosophy, and an attitude toward science, that we can live by.

The core position is neither realist nor anti-realist; it mediates between the two. It would be nice to have a name for this position, but it would be a shame to appropriate another 'ism' on its behalf, for then it would appear to be just one of the many contenders for ontological allegiance. I think it is not just one of that crowd, but rather, as the homely line behind it suggests, it is for common-sense epistemology—the natural ontological attitude. Thus, let me introduce the acronym NOA (pronounced as in 'Noah'), for *natural ontological attitude*, and, henceforth, refer to the core position under that designation.

To begin showing how NOA makes for an adequate philosophical stance toward science, let us see what it has to say about ontology. When NOA counsels us to accept the results of science as true, I take it that we are to treat truth in the usual referential way, so that a sentence (or statement) is true just in case the entities referred to stand in the referred-to relations. Thus, NOA sanctions ordinary referential semantics, and commits us, via truth, to the existence of the individuals, properties, relations, processes, and so forth referred to by the scientific statements that we accept as true. Our belief in their existence will be just as strong (or weak) as our belief in the truth of the bit of science involved, and degrees of belief here, presumably, will be tutored by ordinary relations of confirmation and evidential support, subject to the usual scientific canons. In taking this referential stance, NOA is not committed to the progressivism that seems inherent in realism. For the realist, as an article of faith, sees scientific success, over the long run, as bringing us closer to the truth. His whole explanatory enterprise, using approximate truth, forces his hand in this way. But, a 'NOAer' (pronounced as 'knower') is not so committed. As a scientist, say, within the context of the tradition in which he works, the NOAer, of course will believe in the existence of those entities to which his theories refer. But should the tradition change, say, in the manner of the conceptual revolutions that Kuhn dubs 'paradigm shifts', then nothing in NOA dictates that the change be assimilated as being progressive, that is, as a change where we learn more accurately about *the same things*. NOA is perfectly consistent with the Kuhnian alternative, which counts such changes as wholesale changes of reference. Unlike the realist, adherents to NOA are free to examine the facts in cases of paradigm shift, and to see whether or not a convincing case for stability of reference across paradigms can be made without superimposing on these facts a realist-pro-gressivist superstructure. I have argued elsewhere (Fine 1975) that if one makes oneself free, as NOA enables one to do, then the facts of the matter will not usually settle the case; and that this is a good reason for thinking that cases of so-called incommensurability are, in fact, genuine cases where the question of stability of reference is indeterminate. NOA, I think, is the right philosophical position for such conclusions. It sanctions reference and existence claims, but it does not force the history of science into pre-fit moulds.

So far I have managed to avoid what, for the realist, is the essential point: what of the 'external world'? How can I talk of reference and of existence claims unless I am talking about referring to things right out there in the world? And here, of course, the realist, again, wants to stamp his feet.[20] I think the problem that makes the realist want to stamp his feet, shouting 'Really!' (and invoking the external world), has to do with the stance the realist tries to take *vis-à-vis* the game of science. The realist, as it were, tries to stand outside the arena watching the ongoing game, and then tries to judge (from this external point of view) what the point is. It is, he says, *about* some area external to the game. The realist, I think, is fooling himself. For he cannot (really!) stand outside the arena; nor can he survey some area off the playing field and mark it out as what the game is about.

Let me try to address these two points. How are we to arrive at the judgement that, in addition to, say, having a rather small mass, electrons are objects 'out there in the external world'? Certainly, we can stand off from the electron game and survey its claims, methods, predictive success, and so forth. But what stance could we take that would enable us to judge what the theory of electrons is *about*, other than agreeing that it is about electrons? It is not like matching a blueprint to a house being built, or a map route to a country road. For we are in the world, both physically and conceptually.[21] That is, *we* are among the objects of science, and the

---

[20] In his remarks at the Greensboro conference, my commentator, John King, suggested a compelling reason to prefer NOA over realism: viz. because NOA is less percussive! My thanks to John for this nifty idea, as well as for other comments.

[21] 'There is, I think, no theory-independent way to reconstruct phrases like "really there"; the notion of a match between the ontology of a theory and its "real" counterpart in nature now seems to me illusive in principle' (Kuhn 1970: 206). The same passage is cited for rebuttal by Newton-Smith (1981). But the 'rebuttal' sketched there in ch. 8, sects. 4 and 5, not only runs afoul of the objections stated here in my first section, it also fails to provide for the required theory-independence. For Newton-Smith's explication of verisimilitude (p. 204) makes explicit reference to some unspecified background theory. (He offers either current science or the Peircean limit as candidates.) But this is not to rebut Kuhn's challenge (and mine); it is to concede its force.

concepts and procedures that we use to make judgements of subject-
matter and correct application are themselves part of that same scientific
world. Epistemologically, the situation is very much like the situation with
regard to the justification of induction. For the problem of the external
world (so-called) is how to satisfy the realist's demand that we justify the
existence claims sanctioned by science (and, therefore, by NOA) as claims
to the existence of entities 'out there'. In the case of induction, it is clear
that only an inductive justification will do, and it is equally clear that no
inductive justification will do at all. So too with the external world, for only
ordinary scientific inferences to existence will do, and yet none of them
satisfies the demand for showing that the existent is really 'out there'. I
think we ought to follow Hume's prescription on induction with regard to
the external world. There is no possibility for justifying the kind of externality that realism requires, yet it may well be that, in fact, we cannot help
yearning for just such a comforting grip on reality.

If I am right, then the realist is chasing a phantom, and we cannot
actually do more, with regard to existence claims, than follow scientific
practice, just as NOA suggests. What, then, of the other challenges raised
by realism? Can we find in NOA the resources for understanding scientific
practice? In particular (since it was the topic of the first part of this
chapter), does NOA help us to understand the scientific method, say, the
problems of the small handful or of conjunctions? The sticking point with
the small handful was to account for why the few and narrow alternatives
that we can come up with result in successful novel predictions, and the
like. The background was to keep in mind that most such narrow alternatives are not successful. I think that NOA has only this to say. If you
believe that guessing based on some truths is more likely to succeed than
guessing pure and simple, then if our earlier theories were in large part
true and if our refinements of them conserve the true parts, then guessing
on this basis has some relative likelihood of success. I think this is a weak
account, but then I think the phenomenon here does not allow for anything
much stronger since, for the most part, such guesswork fails. In the same
way, NOA can help with the problem of conjunctions (and, more generally, with problems of logical combinations). For if two consistent theories
in fact have overlapping domains (a fact, as I have just suggested, that is
not so often decidable), and if the theories also have true things to say
about members in the overlap, then conjoining the theories just adds to the
truths of each, and thus *may*, in conjunction, yield new truths. Where one
finds other successful methodological rules, I think we will find NOA's grip
on the truth sufficient to account for the utility of the rules.

Unlike the realist, however, I would not tout NOA's success at making science fairly intelligible as an argument in its favour, *vis-à-vis* realism or various anti-realisms. For NOA's accounts are available to the realist and the anti-realist, too, provided what they add to NOA does not negate its appeal to the truth, as does a verificationist account of truth or the realist's longing for approximate truth. Moreover, as I made plain enough in the first section of this chapter, I am sensitive to the possibility that explanatory efficacy can be achieved without the explanatory hypothesis being true. NOA may well make science seem fairly intelligible and even rational, but NOA could be quite the wrong view of science for all that. If we posit as a constraint on philosophizing about science that the scientific enterprise should come out in our philosophy as not too unintelligible or irrational, then, perhaps, we can say that NOA passes a minimal standard for a philosophy of science.

Indeed, perhaps the greatest virtue of NOA is to call attention to just how minimal an adequate philosophy of science can be. (In this respect, NOA might be compared to the minimalist movement in art.) For example, NOA helps us to see that realism differs from various anti-realisms in this way: realism adds an outer direction to NOA, that is, the external world and the correspondence relation of approximate truth; anti-realisms (typically) add an inner direction, that is, human-oriented reductions of truth, or concepts, or explanations (as in my opening citation from Hume). NOA suggests that the legitimate features of these additions are already contained in the presumed equal status of everyday truths with scientific ones, and in our accepting them both as *truths*. No other additions are legitimate, and none are required.

It will be apparent by now that a distinctive feature of NOA, one that separates it from similar views currently in the air, is NOA's stubborn refusal to amplify the concept of truth by providing a theory or analysis (or even a metaphorical picture). Rather, NOA recognizes in 'truth' a concept already in use and agrees to abide by the standard rules of usage. These rules involve a Davidsonian–Tarskian referential semantics, and they support a thoroughly classical logic of inference. Thus NOA respects the customary 'grammar' of 'truth' (and its cognates). Likewise, NOA respects the customary epistemology, which grounds judgements of truth in perceptual judgements and various confirmation relations. As with the use of other concepts, disagreements are bound to arise over what is true (for instance, as to whether inference to the best explanation is always truth-conferring). NOA pretends to no resources for settling these disputes, for NOA takes to heart the great lesson of twentieth-century analytic and

Continental philosophy, namely, that there *are* no general methodological or philosophical resources for deciding such things. The mistake common to realism and all the anti-realisms alike is their commitment to the existence of such non-existent resources. If pressed to answer the question of what, then, does it *mean* to say that something is true (or to what does the truth of so-and-so commit one), NOA will reply by pointing out the logical relations engendered by the specific claim and by focusing, then, on the concrete historical circumstances that ground that particular judgement of truth. For, after all, there *is* nothing more to say.[22]

Because of its parsimony, I think the minimalist stance represented by NOA marks a revolutionary approach to understanding science. It is, I would suggest, as profound in its own way as was the revolution in our conception of morality, when we came to see that founding morality on God and his order was *also* neither legitimate nor necessary. Just as the typical theological moralist of the eighteenth century would feel bereft to read, say, the pages of *Ethics*, so, I think, the realist must feel similarly when NOA removes that 'correspondence to the external world' for which he so longs. I too have regret for that lost paradise, and too often slip into the realist fantasy. I use my understanding of twentieth-century physics to help me firm up my convictions about NOA, and I recall some words of Mach, which I offer as a comfort and as a closing. With reference to realism, Mach writes:

It has arisen in the process of immeasurable time without the intentional assistance of man. It is a product of nature, and preserved by nature. Everything that philosophy has accomplished . . . is, as compared with it, but an insignificant and ephemeral product of art. The fact is, every thinker, every philosopher, the moment he is forced to abandon his one-sided intellectual occupation . . . , immediately returns [to realism].

Nor is it the purpose of these 'introductory remarks' to discredit the standpoint [of realism]. The task which we have set ourselves is simply to show why and for what purpose we hold that standpoint during most of our lives, and why and for what purpose we are . . . obliged to abandon it.

These lines are taken from Mach's *The Analysis of Sensations* (sect. 14). I recommend that book as effective realism therapy, a therapy that works best (as Mach suggests) when accompanied by historico-physical investi-

---

[22] No doubt I am optimistic, for one can always think of more to say. In particular, one could try to fashion a general, descriptive framework for codifying and classifying such answers. Perhaps there would be something to be learned from such a descriptive, semantic framework. But what I am afraid of is that this enterprise, once launched, would lead to a proliferation of frameworks not so carefully descriptive. These would take on a life of their own, each pretending to ways (better than its rivals) to settle disputes over truth-claims, or their import. What we need, however, is less bad philosophy, not more. So here, I believe, silence is indeed golden.

gations (real versions of the breakneck history of my second section). For a better philosophy, however, I recommend NOA.[23]

REFERENCES

Boyd, R. (1981). 'Scientific Realism and Naturalistic Epistemology.' In P. Asquith and R. Giere, (eds.), *PSA 1980*, ii. 613–62. East Lansing, Mich.: Philosophy of Science Association.
—— (1984). 'The Current Status of Scientific Realism.' *Erkenntnis*, 19 (1983): 45–90. Repr. in Leplin (ed.) 1984, ch. 3.
Cartwright, N. (1983). *How the Laws of Physics Lie*. Oxford: Clarendon Press.
Earman, J., and Glymour, C. (1978). 'Lost in the Tensors.' *Studies in History and Philosophy of Science*, 9: 251–78.
—— *et al.* (1983). 'On Writing the History of Special Relativity.' In. P Asquith and T. Nichols (eds.), *PSA 1982*, ii. 403–16. East Lansing, Mich.: Philosophy of Science Association.
Einstein, A., *et al.* (1952). *The Principle of Relativity*, trans. W. Perrett and G. B. Jeffrey. New York: Dover.
Fine, A. (1975). 'How to Compare Theories: Reference and Change.' *Noûs*, 9: 17–32.
—— (1981). 'Conceptual Change in Mathematics and Science: Lakatos' Stretching Refined.' In P. Asquith and I. Hacking (eds.), *PSA 1978*, ii. 328–41. East Lansing, Mich.: Philosophy of Science Association.
—— (1982). 'Antinomies of Entanglement: The Puzzling Case of the Tangled Statistics.' *Jourual of Philosophy*, 79: 733–47.
—— (1986a). *The Shaky Game*. Chicago: University of Chicago Press.
—— (1986b).'Unnatural Attitudes: Realist and Instrumentalist Attachments to Science.' *Mind*, 95: 149–79.
Heisenberg, W. (1925). 'Uber den anschaulichen Inhalt der quantentheoretischen Kinematik und Mechanik.' *Zeitschrift für Physik*, 33: 879–93.
Hoffman, B. (1972). *Albert Einstein, Creator and Rebel*. New York: Viking Press.
Holton, G. (1973). *Thematic Origins of Scientific Thought*. Cambridge, Mass.: Harvard University Press.
Horwich, P. (1982). 'Three Forms of Realism.' *Synthese*, 51: 181–201.
Jones, R. (1991). 'Realism about What?' *Philosophy of Science*, 58: 185–202.
Kuhn, T. S. (1970). *The Structure of Scientific Revolutions*, 2nd edn. Chicago: University of Chicago Press.

[23] My thanks to Charles Chastain, Gerald Dworkin, and Paul Teller for useful preliminary conversations about realism and its rivals, but especially to Charles—for only he, then, (mostly) agreed with me, and surely that deserves special mention. This paper was written by me, but co-thought by Micky Forbes. I don't know any longer whose ideas are whose. That means that the responsibility for errors and confusions is at least half Micky's (and she is two-thirds responsible for 'NOA'). Finally, I am grateful to the many people who offered comments and criticisms at the conference on realism sponsored by the Department of Philosophy, University of North Carolina at Greensboro, in March 1982, where an earlier version of this paper was first presented under the title 'Pluralism and Scientific Progress'. I am also grateful to the National Science Foundation for a grant in support of this research.

Lande, A. (1974). 'Albert Einstein and the Quantum Riddle.' *American Journal of Physics*, 42: 459–64.
Laudan, L. (1984). 'A Confutation of Convergent Realism. *Philosophy of Science*, 48 (1981): 19–48. Repr. in Leplin (ed.) 1984, Ch. II and as Ch. VI this volume.
Leplin, J. (ed.) (1984). *Scientific Realism*. Berkeley: University of California Press.
Lycan, W. (1985). 'Epistemic Value.' *Synthese*, 64: 137–64.
——(1988). *Judgment and Justification*. New York: Cambridge University Press.
Miller, A. (1981). *Albert Einstein's Special Theory of Relativity*. Reading, Mass.: Addison-Wesley.
Newton-Smith, W. H. (1981). *The Rationality of Science*. London: Routledge and Kegan Paul.
Niiniluoto, I. (1982). 'What shall we do with Verisimilitude?' *Philosophy of Science*, 49: 181–97.
Pitt, J. (1981). *Pictures, Images and Conceptual Change*. Dordrecht: Reidel.
Popper, K. (1972). *Conjectures and Refutations*. London: Routledge and Kegan Paul.
Putnam, H. (1975). 'The Meaning of "Meaning."' In K. Gunderson (ed.), *Language, Mind and Knowledge*, 131–93. Minneapolis: University of Minnesota Press.
——(1981). *Reason Truth and History*. Cambridge: Cambridge University Press.
Rorty, R. (1979). *Philosophy and the Mirror of Nature*. Princeton: Princeton University Press.
Stachel, J. (1979). 'The Genesis of General Relativity.' In H. Nelkowski (ed.), *Einstein Symposium Berlin*, 428–42. Berlin: Springer Verlag.
Teller, P. (1981). 'The Projection Postulate and Bohr's Interpretation of Quantum Mechanics.' In P. Asquith and R. Giere (eds.), *PSA: 1980*, ii. 201–23. East Lansing, Mich.: Philosophy of Science Association.
van Fraassen, B. (1980). *The Scientific Image*. Oxford: Clarendon Press.
——(1982). 'The Charybdis of Realism: Epistemological Implications of Bell's Inequality.' *Synthese*, 52: 25–38.
Weinberg, S. (1972). *Gravitation and Cosmology: Principles and Applications of the General Theory of Relativity*. New York: Wiley.
Wessels, L. (1979). 'Schrödinger's Route to Wave Mechanics.' *Studies in History and Philosophy of Science*, 10: 311–40.
Woolf, H. (ed.) (1980). *Some Strangeness in the Proportion*. Reading, Mass.: Addison-Wesley.

II

NOA'S ARK—FINE FOR REALISM

ALAN MUSGRAVE

Arthur Fine says that scientific realism is dead, drowned by floods of criticism. In its place he puts the *natural ontological attitude*, NOA, pronounced 'Noah'. Fine thinks that NOA is a minimalist view which is neither realist nor anti-realist. I think that NOA is a thoroughly realist view: in NOA's Ark the realist can sail happily above the floods of criticism.

NOA stems from the 'homely line' that we should accept the results of science as true in the same way that we accept the evidence of the senses as true. Fine writes:

Let us say, then, that both realist and anti-realist accept the results of scientific investigations as 'true', on a par with more homely truths. . . . And call this acceptance of scientific truths the 'core position'. What distinguishes realists from anti-realists, then, is what they add on to this core position. (Fine 1984b: 96; p. 36, this volume)

NOA is the core position all by itself, California-pure, without additives.

This is mysterious. As usually understood, the realism–anti-realism issue centres precisely on the question of truth. As usually understood, realists can accept Fine's core position, but anti-realists cannot. Positivists deny the existence of the 'theoretical entities' of science, and think that any theory which asserts the existence of such entities is *false*. Instrumentalists think that scientific theories are tools or rules which are *neither true nor false*. Epistemological anti-realists like van Fraassen or Laudan concede that theories have truth-values, even that some of them might be true, but insist that no theory should be *accepted as true*. None of these anti-realist positions, as usually understood, is consistent with Fine's core position.

The mystery unravels, it seems, when Fine says that anti-realists often go in for some peculiar theory of truth:

Reprinted from *Philosophical Quarterly*, 39 (1989): 383–98, by permission of Basil Blackwell Ltd.

> The anti-realist may add on to the core position a particular analysis of the concept
> of truth, as in the pragmatic and instrumentalist and conventionalist conceptions of
> truth. . . . These addenda will then issue in a special meaning, say, for existence
> statements. (Fine 1984*b*: 97; p. 37, this volume)

So the positivist is seen as saying, not that theoretical science is all false
because there are no 'theoretical entities', but that some theoretical science is true because in its application to science 'true' *means useful*. The
instrumentalist is seen as saying something similar. Van Fraassen is seen as
saying, not that we should never accept a theory about the 'unobservable'
as true, but that we sometimes may because in its application to such
theories 'true' *means empirically adequate*. Laudan is seen as saying something similar.

It is certainly possible to see the anti-realisms this way. But it is not the
way the anti-realists see themselves, nor is it the clearest way to see them.
Adding to the core position a peculiar truth-theory for science actually
demolishes that position. The results of science are not being accepted as
true on a par with more homely truths. Homely statements are accepted
as true in the homely sense of the term 'true'—bits of science are
accepted as true in some esoteric sense of the term 'true'. The latter
acceptances are not on a par with the former at all. Only the equivocation
on the term 'true' could make us think otherwise.

Fine might object that there is no such equivocation: esoteric anti-realist
truth-theories are meant to apply both to homely truths and to scientific
ones. I doubt that this will work. Some scientific theories are true, meaning
useful for 'saving the phenomena'—and some statements of the phenomena are true, meaning useful for saving . . . *what*? Some statements about
unobservables are true, meaning they yield nothing but truths about
observables—and some statements about observables are true, meaning
they yield nothing but truths about . . . *what*? The esoteric truth-theories
developed for science seem to be parasitic upon the homely conception of
truth being applied to homely statements about the phenomena or about
observables.

But even if an esoteric truth-theory can be applied across the board, this
does not help. Now realists and anti-realists cannot both accept the core
position, because there is no one core position for them both to accept.
Now we have *several different core positions*, depending upon the meaning
to be attached (across the board) to the term 'true'. Confusingly, these
different positions are all expressed using the same *words*. Realists and
anti-realists can both assent to the *words*—but only because they mean
quite different things by them.

I wrote a paragraph back about 'the homely conception of truth being applied to homely statements'. It may be objected that there is no such 'homely conception of truth', that the core position leaves it open what sense is to be attached to the term 'true', that anti-realists attach esoteric senses to the term, *and so do realists*. This does not help either—nor does it seem to be Fine's own view of the matter.

It does not help because now we do not have different core positions confusingly expressed by the same words, but *no core position at all*. With no sense yet attached to the term 'true', neither the realist nor the anti-realist knows what it is to assent to the so-called core position. If each assents to it by tacitly reading 'true' a different way, then we are back to the plethora of core positions.

Nor, it seems, does Fine think that the core position (NOA) leaves it open what sense is to be attached to the term 'true'. On the contrary, a very definite conception of truth is built into it:

When NOA counsels us to accept the results of science as true, I take it that we are to treat truth in the usual referential way, so that a sentence (or statement) is true just in case the entities referred to stand in the referred-to relations. Thus, NOA sanctions ordinary referential semantics, and commits us, via truth, to the existence of the individuals, properties, relations, processes, and so forth referred to by the scientific statements that we accept as true. (Fine 1984*b*: 98; p. 38, this volume)

NOA recognizes in 'truth' a concept already in use and agrees to abide by the standard rules of usage. These rules involve a Davidsonian–Tarskian referential semantics, and they support a thoroughly classical logic of inference. Thus NOA respects the customary 'grammar' of 'truth' (and its cognates). (Fine 1984*b*: 101; p. 41, this volume)

Now 'Davidsonian–Tarskian referential semantics' is already a philosophical theory or analysis of truth. It is hardly part of the 'standard rules of usage' for the term 'true', though it yields or explains many of those rules. No matter. The key point is that referential semantics yields precisely the notion of truth that *realists* want to apply across the board, both to homely truths and to scientific ones. Anti-realists of any ilk could not accept the core position once they realized that this committed them to accepting some scientific theories as true *in the usual referential way*. NOA, the core position all by itself, is already a thoroughly realist position.

Fine thinks otherwise. But he has some difficulty in explaining what the realist adds to the core position, what distinguishes the NOA (pronounced 'knower') from the realist. It transpires that the realist shouts while the NOA speaks quietly, that the realist traffics in certain slogans that the NOA avoids, and that the realist has a certain metaphysical picture that the NOA does not have. Let us consider these one by one, beginning with the shouting. Fine writes:

What, then, of the realist; what does he add to his core acceptance of the results of science as really true? . . . what the realist adds on is a desk-thumping, foot-stamping shout of 'Really!' (Fine 1984a: 97; p. 37, this volume)

So the NOA is a realist who avoids desk-thumping, foot-stamping, and shouting. Whereas older realists shouted and stamped in opposition to anti-realist conceptions of truth, on NOA's Ark realists content themselves with a 'stubborn refusal to amplify' their referential semantic concept of truth. As NOA's Ark sails into the sunset, it carries only polite and restrained realists. I promise to shout no more, so that I can join this happy ship.

The slogans will take longer to dispose of. The realist is supposed to traffic in slogans, like 'Truth is correspondence with Reality', which the NOA avoids. Now for me the content of this slogan is exhausted by something the NOA does say: 'A sentence (or statement) is true just in case the entities referred to stand in the referred-to relations.' I have always thought (with Tarski himself) that the semantic conception of truth is a version of the common-sense correspondence theory of truth. Earlier correspondence theorists gave partial accounts (like Aristotle's) or trafficked in general slogans (like the one under discussion). Tarski (1944) showed how to give a complete account (for each well-defined language) which explains what the slogans mean. In this way Tarski rehabilitates the common-sense correspondence theory.

Most philosophers (including Arthur Fine) think that the correspondence theory of truth is one thing and Tarski's theory another. What does the correspondence theorist provide (or seek to provide) which Tarski does not? Fine thinks he provides (or seeks to provide) a general *account* of correspondence, a *theory* of the way in which language and reality can 'match up' so that truth results. Such a theory would 'explain what *makes* the truth true' (Fine 1984b: 97; p. 37, this volume). Armed with an account of *the* relation which all truths bear to reality, the correspondence theorist will know what all truths have in common, their 'essence', what makes them a 'natural kind'. And here Fine is sceptical:

Of course we are all committed to there being some kind of truth. But need we take that to be something like a 'natural' kind? (Fine 1984a: 56)

I share the scepticism. One thing Tarski taught us is that an *essentialist* correspondence theory is out, that the *essence* of truth is a chimera. (If we must talk of 'essences' here, let us talk thus: Tarski gives us the essence of the correspondence theory *without* giving truth an essence.) Truths are many and various, and so are the ways they correspond to facts. Tarskian referential semantics captures 'the' correspondence relation as well as it *can* be captured. Michael Levin puts it well:

Tarski tells us that all true conjunctions have in common the truth of each conjunct, that each true existential generalisation is such that its matrix is satisfied by at least one sequence, and so on. To be sure, at the level of the basis clauses, the definition goes strongly extensional, but that is the way it ought to go. The truths 'Ron Reagan is a man' and 'This tulip is red' are not shown to have much in common: each is a matter of a different object . . . satisfying a different primitive open sentence. But *do* they have more in common than being a man has in common with being red, which is to say, very little? I cannot see that they do, or at least more than Tarski gives them. (Levin 1984: 126)

When you think about it, Tarski's work shows that the idea that truths might all share an essence is quite absurd. It is meaningful *linguistic* items that are true or false in Tarski's sense. Languages are largely conventional human inventions, suited to different human purposes. Why suppose that the bewildering variety of truths languages contain form a *natural kind*? It is surely better to suppose that there is no more to the 'correspondence relation' than Tarski gives us. Nor need it be any part of scientific realism to think that there is more.

Fine disagrees. He thinks realists must add to Tarskian truth a hankering after truth's essence. In this connection he mentions Hilary Putnam. Now with friends like Putnam, realism needs no enemies. Perhaps Putnam did hanker after truth's essence—but he did not find it. Instead, he found his model-theoretical argument against realism, and abandoned the realist cause. (By the way, the model-theoretic argument is invalid—but that would be another paper.) Never mind Putnam. Realists *need* not hanker after truth's essence (proof: *I don't*). Realists can think that Tarski gives them as much of a correspondence theory as they need. But perhaps, in view of all the dust philosophers have raised with the word 'correspond-ence' down the ages (only to complain afterwards that they cannot see), realists would do better to drop the word. Perhaps realists would do better to drop the correspondence slogan too in favour of that of the NOA: 'A sentence (or statement) is true just in case the entities referred to stand in the referred-to relations.'

So much for one philosophical worry about Tarski's theory of truth: whether or to what extent it is a *correspondence* theory of truth. There are many other philosophical worries, and I want to digress into one of them, for it will reveal another possible way to drive a wedge between NOA and scientific realism.

I have been arguing that Tarski's theory gives the scientific realist all he needs from a theory of truth. But since Tarski's theory can be applied across the board, so to speak, will it not give all sorts of *suspect* realists all that they need also? That the theory applies across the board can be seen from the following list of instances of Tarski's Convention T:

(1) The statement 'There is a full moon tonight' is true if and only if there is a full moon tonight.

(2) The statement 'Electrons are negatively charged' is true if and only if electrons are negatively charged.

(3) The statement 'Two plus two equals four' is true if and only if two plus two equals four.

(4) The statement 'Eating people is wrong' is true if and only if eating people is wrong.

(5) The statement 'Ronald Reagan gives me the creeps' is true if and only if Ronald Reagan gives me the creeps.

The worry is this. Suppose that (1) yields common-sense realism about the moon, and (2) yields scientific realism about electrons. Then will not (3) yield *Platonic realism* about natural numbers, (4) *moral realism* about wrongness and rightness, and (5) realism regarding a mysterious entity (*the creeps*) which Ronald Reagan gives to me (and simultaneously, no doubt, to others too)? Since creeps-realism is absurd, and moral realism and Platonism philosophically suspect, so is the Tarskian theory which yields them.

This worry is quite groundless. Tarski's Convention T (often called Tarski's 'disquotational scheme', or a disquotational, deflationary, or re-dundancy theory of truth) *by itself yields none of these realisms*. We all avoid creeps-realism by saying that 'Ronald Reagan gives me the creeps', though true (which it is), is an idiom which is not to be taken at face value for logico-philosophical purposes. We replace it with a non-idiom (say, 'Ronald Reagan makes me nervous') and avoid ontological commitment to *the creeps*. Similarly, one sceptical of moral realism might refuse to take 'Eating people is wrong' at face value for logico-philosophical purposes— which is just what emotivists, prescriptivists, and the like do. Alternatively, one might go in morals the way Hartry Field (1982) goes in mathematics. Field accepts Tarski's scheme *and* takes 'Two plus two equals four' at face value, but eschews arithmetical Platonism by saying that it is false *because* there are no numbers for the numerals 'two' and 'four' to be names of. Moral realism might be eschewed similarly: take 'Eating people is wrong' at face value, apply Tarski's scheme to it, and say that it is false *because* there is no property for 'wrong' to refer to.

Similarly again, those sceptical of scientific realism can *either* refuse to take 'Electrons are negatively charged' at face value (as instrumentalists do in their different ways), *or* take it at face value and say that it is false because there are no theoretical entities for the term 'electron' to refer to (as positivists do), *or* take it at face value and concede that it might be true but insist that we should not accept it as true (as epistemological anti-realists do). Finally, those sceptical of common-sense realism can *either* refuse to take 'There is a full moon tonight' at face value, *or* take it at face value and say that it is false because there is no external object for the word 'moon' to refer to (as Bishop Berkeley and the phenomenalists do). The disjunctions here are not, of course, exclusive ones. Indeed, anti-realists typically adopt a mixed strategy, first saying that a certain kind of statement is false taken at face value (or taken literally), then trying to soften that conclusion by telling us what such statements 'really mean', that is, how they are to be taken for logico-philosophical purposes. (Think of phenomenalist translations, so called, of external-object statements, or of emotivist translations, so called, of moral statements.) The idea that Tarski's Convention T *by itself* begs all kinds of metaphysical questions is simply mistaken. Tarski himself said as much long ago.

By the way, calling Convention T a 'disquotational theory of truth' is highly misleading. Tarski's theory of truth is not exhausted by Convention T, though most of the philosophical worries about it are. Instances of Convention T will not be 'disquotational' if something other than a quotation-mark name of a statement appears on the left-hand side, or if the statement talked about comes from a different language from the one in which we talk. It is cases where neither of these conditions obtains, cases like my (1)–(5), that are also responsible for the mistaken view that Convention T is trivial *because circular*. There are similar reservations, for similar reasons, about calling the theory a 'deflationary' theory or a 'redundancy' theory.

If Tarski's Convention T does not by itself yield realism about the moon, electrons, natural numbers, moral properties, the creeps, or anything else, *what is its importance for realism*? The answer is obvious: it makes realism about all of these things *possible*. This is its importance for realism; this is why it is the theory of truth that realists need. Anti-realist theories of truth identify it with some *internal* feature of our beliefs (their coherence, their usefulness, their self-evidence, their ultimate undisbelievability, or whatever).[1] Such theories make realism impossible; they leave no room for it.

---

[1] Down the ages the chief motive for anti-realist truth-theories has been anti-sceptical, to make the truth accessible to us. By identifying truth with some internal features of beliefs, they make it something the believer is an authority on and can know for certain. Hence they are all aptly called *subjective* truth-theories.

Their immediate consequence is *relativism* about truth. If A's belief that P possesses the

So realists need Tarski's theory of truth—but they also need more. To be a *realist about Xs* (whatever Xs may be) you must:

(a) take statements about Xs at face value for logico-philosophical purposes;
(b) apply Tarski's Convention T to those statements;
(c) accept some of those statements (*appropriate* ones: 'There are no Xs' will not do) as true.

Let us end the digression and return to Fine's NOA. He does all of these three things when it comes to the theoretical statements of science. He insists that such statements are to be taken at face value (unlike the instrumentalists). He applies Tarski's scheme to them (unlike those who traffic in unrealistic truth-theories). And he accepts appropriate theoretical statements as true (unlike the positivists or epistemological anti-realists). All this places Fine's NOA squarely in the realist camp.

But wait! The digression may have a Fine point after all. It reveals a possible way to drive a wedge between the NOA and the realist. Suppose, contrary to what was just said, that the NOA does *not* take at face value any body of statements about X's, whatever X might be. Suppose that to do this is already to *add* to the California-pure core position. The NOA is philosophically neutral between realist and instrumentalist interpretations of theoretical scientific statements like (2). The NOA is also neutral between realist and phenomenalist interpretations of common-sense statements like (1). Bishop Berkeley counts as a NOA—even a solipsist counts as a NOA. The NOA accepts homely statements and scientific statements as true. The NOA reads 'true' Tarski-style, with all this brings in the way of commitment to the entities referred to in accepted statements which are taken at face value. *But* the NOA leaves it open what those commitments actually are, because he leaves it open which statements are to be taken at face value (that is, realistically) and which are not.

internal feature and *B*'s belief that *P* does not, then *P* is true for *A* and false for *B*. (Alternatively put, the 'laws of truth', contradiction, and excluded middle fail.)

To avoid relativism (and preserve the laws of truth), subjective truth-theorists tend to go *ideal* and *to the long run*. For example, coherence theorists will say that something is true (by definition) if it 'coheres', not with your or my beliefs, but with the beliefs an *ideally rational inquirer* would have *in the long run*.

Of course, such moves immediately threaten the anti-sceptical virtues (if virtues they be) of subjective truth-theories. What God will be coherently believing at the end of time is just as inaccessible to you or me as is truth-as-correspondence. (I think it is *more* inaccessible.)

To solve these problems, subjectivists tend to adopt a policy of *flipping back and forth*. When epistemological concerns are paramount, they stay subjective; when semantic concerns are paramount, they go ideal and to the long run. The resulting fandango is one of the least edifying sights in philosophy. Let us view it no more. Fine (1984*a*) has some excellent arguments against anti-realist truth-theories.

This is a possible position. It is consistent (just) with Fine's accounts of NOA. It attributes to the NOA (pronounced 'knower', remember) *a complete philosophical know-nothing-ism*. The NOA is not committed to electrons, the moon, tables and chairs, physical objects, other people, his self, *anything at all*. I first overlooked this possible interpretation of Fine's NOA. It was suggested to me by discussions with Arthur Fine in Indiana in 1987 (discussions for which I am grateful). I now think it might be the correct interpretation—as our unfinished business will show.

The NOA was said to differ from the realist on three counts. We have dealt with two of these: realists need not shout or stamp; and they can confine themselves to slogans acceptable to the NOA, like 'A sentence (or statement) is true just in case the entities referred to stand in the referred-to relations'. I cite this slogan for the third time because it would seem to give the realist all he needs by way of *metaphysics*. Fine disagrees, which brings us to the third alleged point of difference: the realist has a metaphysical picture which the NOA lacks.

Fine describes the realist metaphysic thus:

For realism, science is *about* something; something *out there*, 'external' and (largely) independent of us. The traditional conjunction of externality and independence leads to the realist picture of an objective, external world; what I shall call the *World*. According to realism, science is about *that*. Being about the *World* is what gives significance to science. (Fine 1986b: 150)

What exactly is wrong with this realist metaphysical picture? Is not the NOA committed to precisely the same picture?

Fine talks about 'the obscurity of the correspondence relation and the inscrutability of realist-style reference'. He elaborates:

The problem is one of access. The correspondence relation would map true statements (let us say) to states of affairs (let us say). But if we want to compare a statement with its corresponding state of affairs, how do we proceed? How do we get at a state of affairs when that is to be understood, realist-style, as a feature of the *World*? (Fine 1986b: 151)

What exactly is the problem here? Somebody says 'There is a full moon tonight', and I look up into the night sky and ascertain that the statement is true. (I use humdrum commonsensical examples, rather than esoteric scientific ones, because if there is a problem, it will be a quite general one, which afflicts the common-sense realist metaphysic as much as the scientific one.) I have access to both terms of the 'correspondence relation': my linguistic competence gives me access to what was said, my eyes give me access to the full moon out there in the world (or if you prefer, the *World*). Of course, *explaining* linguistic competence or sensory awareness gives rise to a host of scientific and philosophical problems—but to explain them is not to explain them away, to show that they do not exist at all.

Perhaps the worry is that if I were to report pedantically (similarity to (1) is intended) 'The statement "There is a full moon tonight" is true since there is a full moon tonight', then I would still be trapped inside language, would not have got 'at a state of affairs . . . understood, realist-style, as a feature of the *World*'. Perhaps the worry is that instances of Convention T, such as (1), do not relate language to the World, but rather one language (that talked about) to another (that in which we talk).

If this is the worry, it is a queer one. (1) speaks about a bit of language *and* about the World. True, to speak about the way in which language relates to the World, one must *use language*. But this is no deep truth; rather, it is a pallid truism. Sweeney had it right:

> . . . I gotta use words when I talk to you
> But if you understand or if you don't
> That's nothing to me and nothing to you
> We gotta do what we gotta do . . .

We are not trapped inside language in the *serious* sense that all we ever talk *about* is language. To think otherwise is to ignore the hard-won distinction between use and mention.[2]

I hesitate to attribute this worry to Arthur Fine. So let us see how he continues:

A similar question comes up if we move to reference and try to establish truth-conditions compositionally, for there again, what the realist needs by way of the referent for a term is some entity in the *World*. The difficulty is that whatever we observe, or, more generously, whatever we causally interact with, is certainly not independent of us. This is the problem of *reciprocity*. Moreover, whatever information we retrieve from such interaction is, prima facie, information about interacted-with things. This is the problem of *contamination*. How then, faced with reciprocity and contamination, can one get entities both independent and objective? Clearly the realist has no direct access to his *World*. (Fine 1986*b*: 151)

Return to my humdrum example (once again, any problems here will afflict humdrum examples as much as esoteric ones). There the term 'moon' referred to the moon, which is an entity out there in the World if anything is. Do the 'problems' of reciprocity and contamination show that

---

[2] Wittgenstein ignored this distinction in his *Tractatus*, and was led to the following 'deep' thoughts: the way language relates to the world cannot be said, it can only be shown; the limits of my language are the limits of my world; what the solipsist means is quite correct, only it cannot be said. Wittgenstein's 'logocentric predicament' is simply old psychologistic wine poured into new linguistic bottles. The British empiricists thought that thinking consists in having a stream of 'ideas', and concluded mistakenly that all we ever think *about* are our own ideas.

this is incorrect? Reciprocity is supposed to show that the moon is not independent of us because we can see it or otherwise causally interact with it. But implicit in this is a silly account of independence: an entity is independent of us if we cannot causally interact with it. The only independent entities in this sense will be Platonic entities, which do not exist in space and time, and which have no relations causal or otherwise with beings (like us) which do exist in space and time. No scientific realist should accept an account of independence which means that the only independent entities are Platonic entities and the only independent reality the Platonic realm of abstract entities. When a scientific realist says that the moon is (largely) independent of us, he obviously means that it is non-mental, it exists outside of us, we did not create it, it existed long before we did, it continues to exist when we are not looking at it, and so forth.

What of the 'problem' of *contamination*? It is supposed to show that when we see that the moon is full, we gain information not about an objective moon out there in the World, but rather about an *interacted-with-moon*. This hyphenated entity is presumably not the same entity as the moon (or, if hyphens thrill you, as the moon-in-itself). Fine suggests that, unlike the moon-in-itself, the interacted-with-moon is not objective, not out there in the World. Where is it then, subjective and inside our heads? This smacks of the long-discarded view that we do not see external objects like the moon at all, but rather moonish-sense-data inside our heads. I doubt that Fine wants to return to that view. I know that Fine's NOA ought not to attach that bit of bad philosophy to his core acceptance of the results of science.

Perhaps the thought is that the interacted-with-moon, the moon-as-observed-by-us, is not 'objective' in the sense that it is somehow partly constituted by the moon-concept which is our invention. After all, in order to see that there is a full moon, I must possess the moon-concept, and in order to say that there is a full moon, I must possess the word 'moon'. What I can see (or say) depends partly upon the concepts (words) that I possess.[3]

---

[3] Between this sentence and its predecessor there is a subtle slide which I have not disrupted the text to remark upon. It is trivially true that a being lacking the typewriter-concept (the word 'typewriter') cannot *see that* there is a typewriter on the table. It is trivially false that a being lacking the typewriter-concept (the word 'typewriter') cannot *see* the typewriter on the table. The Kalahari bushman may see the typewriter perfectly well, as evidenced by his response to the request (couched in Kalahari-bushman-ese) to pass him that thing over there. The cat might see the typewriter perfectly well, as evidenced by her not bumping into it when the mouse she is chasing hides under it. The slide results from conflating *seeing-that* with *seeing*. (In between these there is *seeing-as*: the cat or the Kalahari bushman may see the typewriter as non-food, and the mouse as food, even though they lack the typewriter-concept or the mouse-concept.)

This slide is responsible for the view that beings possessed of different 'conceptual schemes', different languages, even different theories, literally see different worlds. The Aristotelian and

The world (the world-in-itself, Fine's World-with-a-capital-'W') is not carved up according to any conceptual or linguistic scheme. It is we who carve things up according to such schemes. Having carved, we cannot partake of the world-in-itself, the world-as-it-is-independently-of-any-conceptual-scheme, the World-with-a-capital-'W'. That the moon is full is not a fact about the world-in-itself, since it traffics in the moon-concept. The statement 'The moon is full tonight' does not state a truth about the World-with-a-capital-'W', since it traffics in the word 'moon'. The world we experience and talk about is not the world-as-it-is-independently-of-any-conceptual-scheme. Rather, it is a world partly of our own conceptual or linguistic making, a world-as-conceived-by-us or a world-as-talked-about-by-us. This is conceptual (or linguistic) *idealism*.

It quickly turns into conceptual (or linguistic) *relativism*. Our concepts vary and change. There is no *one* world-as-conceived-by-us at all. The world-as-conceived-by-the-Aristotelian differs radically from the world-as-conceived-by-the-Newtonian. The world-of-the-Eskimo is not at all the world-of-the-Kalahari-bushman. This gets really exciting once we cease to be human chauvinists and consider non-human animals too. They experience the world too, but differently from us. The world-of-the-chimpanzee is not at all the world-of-Albert-Einstein, and both are *worlds apart* from the world-of-the-honeybee or the world-of-the three-spined-stickleback. And so on and so forth—tediously.

Kant is, of course, the philosopher who started the rot here. Kant stopped the rot from spreading, blocked the slide from idealism into relativism, by assuming that humans all have the same immutable set of basic concepts. Contemporary philosophical wisdom has outgrown that assumption. Even if contemporary wisdom is misguided, we might still ask Kant and the Kantians about non-human animals. Do they *experience* the world at all? If so, do they possess all those Kantian categories of the understanding deployment of which is a condition of the possibility of all experience? Do chimps, honey-bees, and flatworms structure incoming stimuli the way humans do? One alternative is to say that humans are the only animals that have experiences. This must be deemed implausible by anyone who takes Darwinism seriously. Another alternative is to say that the Kantian categories are only conditions of the possibility of *human* experience and that

the Copernican, watching a sunrise, *see different things*. This is, of course, nonsense. What might be true is that the Aristotelian says of the sunrise 'I see that the sun is still orbiting the earth', while the Copernican says 'I see that the earth is still rotating on its axis'. The profundity 'The limits of my language are the limits of my world' is false. What is true is the triviality that the limits of my language limit what I can say of the world.

other creatures can do without them. But if chimps can do without them, why not us?

Of course, all this talk of different worlds-as-experienced (conceived, talked about)-by-*X*s *need not be taken seriously.* We can see it just as a fancy way of drawing attention to the great diversity of experience, concepts, and talk *of the world.* On this view, all such entities as the moon-as-experienced-by-us are ersatz entities. (After all, is not 'ersatz' German for 'hyphenated', and 'hyphenated' philosopher's English for 'artificial' or 'unnatural' or 'unreal'?) The moon-as-experienced-by-us is just the moon—and similarly for all other hyphenated entities (*including* the Kantian moon-in-itself). On this view, 'The moon-as-conceived-by-Aristotelians was perfectly spherical' is just philosopher's gobbledy-gook for 'Aristotelians thought the moon is perfectly spherical'.

If we do take talk of different worlds-as-experienced (conceived, talked about)-by-*X*s seriously, we become experiential (conceptual, linguistic) idealists. And we come to inhabit a strange world indeed. Consider the moon-as-observed-by-us (*moon$_0$* for short) and the moon-in-itself (*moon$_1$* for short). Is *moon$_0$* identical with *moon$_1$*? Presumably not: if it were, the distinction would have no point, and we could rest content just with the *moon* (without subscript, unhyphenated). But if *moon$_0$* is distinct from *moon$_1$*, then there is presumably some property which the one lacks and the other possesses. But to know this is to know something about *moon$_1$*, when our knowledge was supposed to be confined to *moon$_0$*! Certainly, there could be no empirical evidence that *moon$_0$* is different from *moon$_1$*— it is just a piece of idealist metaphysics. Kantians object that we can know, not through evidence but through transcendental argument (whatever that is), that *moon$_1$* not only *exists*, but also lacks various properties that *moon$_0$* possesses. For example, *moon$_1$* has no position in space and time, these being 'forms of sensibility' in which only *moon$_0$* is located. Nor does *moon$_1$* cause (or help cause) moon-visions down on earth, causality being a category of the understanding which applies only in the world-of-appearance, not in the world-of-things-in-themselves. No wonder that some of Kant's immediate followers, realizing that *moon$_1$* is nowhere, at no time, and does nothing, decided that it was an *idle* metaphysical posit, did away with it altogether, and became fully-fledged idealists. At this point my Kantian friends (and I still have one or two) tell me that Kant was an 'empirical realist' and only a 'transcendental idealist'. I do not understand these Kantian slogans. I am reminded how fond Berkeley was of presenting himself as a defender of common-sense realism. The Kantian metaphysic, seen as it really is, is a form of *idealism*, as is Berkeley's metaphysic. Modern idealism is just Kantian idealism 'gone linguistic' or 'gone conceptual' and generalized.[4]

We have come a long way from Fine's remark that when we observe or otherwise interact with things, the information we retrieve is information about interacted-with-things. The remark may have been a perfectly innocent one. I do not know whether Fine is an idealist of this kind (or rather, of these kinds). I do think that Fine's NOA should have nothing to do with these idealisms. NOA stands for *natural* ontological attitude. The ersatz hyphenated entities involved in these idealisms are artificial and unnatural entities. These idealisms are dubious, perhaps in the end unintelligible, philosophical theories which no NOA should attach to his core acceptance of the results of science. Indeed, some pretty mundane and well-entrenched results of science tell us that the moon (not some hyphenated moon, not even the Kantian moon-in-itself, just the moon) is objective and independent of us: it exists outside of our heads, it was not created by us, it existed long before we did, and so forth. The NOA who accepts these bits of science as true has precisely the same metaphysical picture as the realist. Fine rejects the realist's metaphysical picture, not as unproved, but as false. Its falsity follows from what might be called an unnatural idealist attachment to science. But the unphilosophical NOA ought not to be trafficking in the 'problems' of reciprocity and contamination—for that traffic is *philosophy*. The unphilosophical NOA should do no more than accept homely truths and scientific ones. Will that not provide the NOA with the same 'metaphysical picture' as the realist?

An affirmative answer to this question overlooks the possibility of the NOA who knows *nothing* philosophical. That NOA's acceptance of bits of science as true implies nothing whatever about the objectivity and independence of the moon. For that NOA leaves it quite open how the accepted statements are to be 'interpreted', what they mean, what their ontological commitments are. Perhaps homely truths and scientific ones are to be taken at face value for logico-philosophical purposes, and *perhaps they are not*. Perhaps the homely truths are to be given a phenomenalist construal and the scientific truths an instrumentalist one. (Remember, Berkeley counts as a NOA in this minimalist sense.) The completely unphilosophical NOA leaves all this open.

---

[4] Permit me a true story. I was once told in all seriousness that when the concept 'person with an IQ two standard deviations above the mean' was invented, *new entities* were brought into being. It turns out (I replied) that there are two ways of making babies, the way we all know and love, namely love, and this new way, psychological theorizing! I was told that the new enities are not babies, indeed, are not persons with IQ's two standard deviations above the mean. I could gain no clear idea *what* they were.

In traditional discussions of scientific realism, common-sense realism regarding tables and chairs (or the moon) is accepted as unproblematic by both sides. Attention is focused on the difficulties of scientific realism regarding 'unobservables' like electrons. But Fine's discussion is not a traditional discussion. Fine's NOA, on this interpretation, begs *no* metaphysical question. That is why both realists and anti-realists of any ilk (even a solipsist) can accept NOA.

One special consideration suggests that this is the correct interpretation of Fine's position. In his acclaimed biography of Einstein, Abraham Pais tells how Einstein once turned to him and asked 'if I really believed that the moon exists only if I look at it' (Pais 1982: 5). Einstein's question concerned, of course, that interpretation of quantum mechanics according to which entities do not exist in well-defined states unless they are being observed. What if quantum mechanics, thus interpreted, should turn out to be correct? Will not science have turned out to overthrow even common-sense realism and to vindicate Bishop Berkeley? Science and common sense have often clashed—why should the clash not be as radical as this one? If this is possible, then we should not encumber our *philosophy* of science with metaphysical assumptions (those of common-sense realism) which science may outgrow. Hence the minimalist position called NOA is the only defensible position in *philosophy* of science—the rest is *up to science*.

Let me hasten to add that Fine himself is no friend of the interpretation of quantum mechanics just mentioned. Indeed, he has taken Einstein's side in his discussions of those debates (particularly in his 1986a). The point is that he wishes to define a philosophy of science (NOA) which will leave open all metaphysical questions for science to decide.

I doubt that this will work. The interpretation of quantum mechanics just mentioned is not a result of *science*—it is a philosophical interpretation of science inspired in some part by dubious philosophical theories like the verifiability theory of meaning. Science unaided by philosophy could not overthrow common-sense realism. Indeed, the quantum physicist presupposes common-sense realism every time he sets up some experimental apparatus (as the principle of complementarity acknowledges). Most of the alleged clashes between science and common sense (but not all of them) stem from Eddington's (1929) mistake (to *explain* the solidity of the table, in terms of the behaviour of things that are not solid, is to *explain solidity away*).

But these are difficult questions, not to be answered in a paragraph. Let us return from them to the minimalist position NOA, to see how minimalist it really is. The NOA is supposed to 'accept', say, 'There is a full

60 ALAN MUSGRAVE

moon tonight' and 'Electrons are negatively charged'. But does the NOA know what he has accepted? Remember, he leaves it open whether or not these statements are to be taken at face value, says nothing about how they are to be interpreted, what their ontological commitments are. The unphilosophical NOA does not just know nothing philosophical—he knows nothing at all.

We need to inject some content into NOA (pronounced 'knower'). One way to do this is clear: let the NOA give 'homely truths' a face-value realist construal. (Scientific realists and their opponents were agreed upon this much anyway.) But then the same will apply to the scientific statements which the NOA accepts 'on a par with' the homely ones. And NOA's Ark will after all be Fine (with a capital 'F') for realism (with a small 'r').

REFERENCES

Eddington, A. S. (1929). *The Nature of the Physical World*. Cambridge: Cambridge University Press.
Field, H. (1982). 'Realism and Anti-Realism about Mathematics.' *Philosophical Topics*, 13: 45–69.
Fine, A. (1984a). 'And Not Anti-Realism Either.' *Noûs*, 18: 51–65.
——(1984b). 'The Natural Ontological Attitude.' In Leplin (ed.) 1984: 83–107.
——(1986a). *The Shaky Game: Einstein, Reality and the Quantum Theory*. Chicago: University of Chicago Press.
——(1986b). 'Unnatural Attitudes: Realist and Instrumentalist Attachments to Science.' *Mind*, 95: 149–79.
Leplin, J. (ed.) (1984). *Scientific Realism*. Berkeley: University of California Press.
Levin M. (1984). 'What Kind of Explanation is Truth?' In Leplin (ed.) 1984: 124–39.
Pais, A. (1982). *'Subtle is the Lord...': The Science and the Life of Albert Einstein*. Oxford: Oxford University Press.
Tarski, A. (1944). 'The Semantic Conception of Truth.' *Philosophy and Phenomenological Research*, 4: 341–75.

III

SAVING THE NOUMENA

LAWRENCE SKLAR

I want to consider another defense which I find most unsatisfactory. It might be called the *positivist defense*; and some philosophers seem to think it is the only defense available to someone who wants to claim that the apparent conflict between two theories is merely verbal. According to the positivist defense, *whenever* we have two theories that have all the same observational consequences, any apparent disagreements between the two theories are merely verbal ones. Call this the *positivist principle....* For instance Sklar's discussion of conventionalism about geometry seems to presuppose this view. In all the standard cases of alternate geometries (plus compensating adjustments elsewhere in the physical theories), the geometric objects of one theory are definable out of the geometric objects of the other. Sklar obscures this fact by comparing the conflict between alternative geometries to the conflict between the normal world-view and Descartes' 'evil demon' hypothesis; but this latter example is one where the objects of one theory are clearly not definable in terms of the objects of the other, so only by some form of the positivist principle could one claim that the conflict between those theories was purely verbal.

Field, 'Conventionalism and Instrumentalism in Semantics'

I

Most, but not all, philosophers would agree that surface differences between two theory presentations might mask underlying identity of theory. Sometimes the verbal presentations of a theory may differ in a way which does not even suggest incompatibility, for example, in the case of the presentation of a theory in two different natural languages. Few would want to argue that Newton's *Principia* in Latin does not present the same theory as that work in English. More interesting are those cases where only one natural language is involved, and where, at least on the surface, some

Reprinted from *Philosophical Topics*, 13 (1982): 49–72, by permission of the University of Arkausas Press.

sort of incompatibility or other seems present, but where many, if not all, would agree that this surface incompatibility is deceptive.

Are our theories of the world really incompatible if they differ simply in that you talk of temporally extended physical objects whereas I deal with time-slices as my basic particulars? More radically, even if it is a case of a thing theory versus a theory which takes spatio-temporal regions as primitive and treats things as substantival predicates of such regions, many would certainly argue that we have here a case of theoretical equivalence. Hardly anyone would deny that the Schrödinger and Heisenberg 'representations' are, indeed, representations of one and the same theory, despite the fact that in the former the state function varies with time and the operators do not, and in the latter the reverse is the case.

A more controversial case from physics is the famous pair, curved space-time versus flat space-time with compensating metric and force fields. Here some (Eddington, Schlick, Reichenbach) assure us that we have a classic case of theoretical equivalence, while others (at various times Putnam and Glymour, for example) think, rather, that the 'theories' are not equivalent at all and that, indeed, the former is much preferable to the latter as a plausible account of the data. We will have more to say about this example later. Finally, we have the positivist philosophers' paradigm of equivalence, material world versus malevolent demon or brain-in-a-vat, an alleged equivalence which, as a consequence of positivism, is taken by many (including Field, above) as a *reductio ad absurdum* of the positivist position.

The positivist position on theoretical equivalence, whatever one thinks of its adequacy or ultimate tenability, is at least clear. It is especially designed to handle those cases where we want to claim that surface incompatibility masks underlying commonality of import. The line is familiar. One discriminates among the consequences of the theory a special class of sentences which allegedly contain among them all the consequences of the theory open to empirical determination of truth independently of assuming the truth of the theory in question. If two theories agree on this observational set then they are alleged equivalent, no matter how radical the apparent incompatibility which resides in the remaining, theoretical, level of the two theories. The vices of the positivist programme are well known, but here I would like to remind us of some of its virtues.

However fraught with untenable dualisms, incoherent foundationalist notions, etc. positivism may be, it at least offers an account of theoretical equivalence which brings into a harmonious whole a theory of meaning, of evidence, of ontology, of truth, of explanation, and of equivalence itself. Observationally equivalent theories do not differ in meaning, since the

total meaning of the theory is exhausted by its set of observational consequences. Since the consequences of the theory exhaust the totality of
possible evidence for or against a theory, evidentially equivalent theories
are automatically genuinely equivalent, blocking any opening for scepticism. One need not take the apparently conflicting ontological claims of
the theories seriously, since the ontology we are committed to by apparently referring terms at the theoretical level is only *façon de parler* in any
case. Nor need we be concerned about which theory is true, since, once
again, all the theories really say is what their observational consequences
say, and at that level they say the same thing. Finally, we are offered an
account of explanation, basically the subsumption of particular correlated
observable events under general rules of constant conjunction, which
makes it clear that, despite apparent differences, theories equivalent in the
positivist sense offer the same explanations of events, if explanation too is
positivistically understood. I am, of course, laying all this out here gro-
tesquely sketchily, and without anything like the necessary attention to
either detail or to variance and controversy within the positivist camp, but
it will have to do for the time being.

Nor will I be concerned here with the manifold, familiar objections to
the positivist programme, replete with the well-known proofs of its alleged
impossibility and absurdity. I will focus, rather, on one particular kind of
realist alternative to positivism. In particular, I will explore one way a
realist might try to offer an alternative to the positivistic account of theoretical equivalence, and will explore the problems encountered in such a
programme which an anti-realist might put forward as reasons for
scepticism regarding the viability of a notion of equivalence differing
from his own. I make no claim that the realist account of equivalence I
will offer is the only one a realist might offer, nor that this particular
account of equivalence could not possibly overcome the difficulties I
will raise in its path. I do think, however, that the account I offer does
capture the instincts many realists have had regarding theoretical equivalence, and the problems I lay out do capture the sceptical doubts regarding such a realist programme latent in many recalcitrant positivists'
sceptical remarks.

What is wrong with the positivist notion of equivalence according to the
realist? And how is the central deficiency of this account of equivalence to
be remedied? What is wrong with the positivistic notion, according to the
realist I have in mind, is this: designed to handle plausible cases of theoretical equivalence, say the Schrödinger and Heisenberg formulations of
quantum mechanics, the positivistic notion of equivalence misconstrues
just what it is about these two theory presentations that makes them

merely two expressions of one and the same theory. His misconstrual leads him, fallaciously, to believe that any reasonable commitment to the equivalence of theories must force one by an irresistible slippery slope argument to such absurd conclusions as the equivalence of a realistic account of the world with an extended-dream account or a brain-in-a-vat account. The source of this misconstrual, says my realist, is the positivist's failure to take sufficient account of the interrelationships theories can bear to one another at the theoretical level. Once one sees that in all the plausible cases of theoretical equivalence there is a far stronger relationship between the two theory presentations than merely a common ability to save the phenomena, a relationship at the theoretical level, one will realize that it is this stronger relationship which warrants the justified claim of theoretical equivalence. But then we will realize that we are by no means forced from the acceptance of such genuine cases of equivalence into any assertions of equivalence of the kind devastating to the realist programme.

So what I wish to explore is this: what is the structural relationship between theories sufficient for a realist to attribute equivalence to them? In particular, how does this additional requirement go beyond the weak positivist constraint of merely having all observational consequences in common? I will argue that the realist will find, unless he is very careful, that his programme all too often seems to lead him into assuming far more in common with the positivist than he might at first think. And, I will argue, it is incumbent upon the realist to offer us not just a notion of equivalence more stringent than that of the positivist, but an integrated theory of meaning, truth, ontology, confirmation, and explanation into which his notion of equivalence will naturally fit. I certainly will not maintain that this is impossible for the realist, but only insist that profound difficulties must be faced up to by any realist who takes this programme seriously enough to counter the positivist's claim that, dubious as his account may be, it at least offers us a fully integrated, coherent account of all these metatheoretical aspects of theories.

II

The realist I have in mind argues that it is not sufficient for theoretical equivalence that the two theories in question save all the same phenomena. A more profound and tighter structural interrelationship between the theories, an interrelationship which takes account of the relationship of the structure of one theory at the theoretical level to the structure of the other theory at its theoretical level, is required for genuine equivalence. What could this additional structural component be?

One thing we can be sure of. Whatever this structural 'isomorphism' is to be, it cannot be a purely formal notion. It cannot be, that is, an interrelationship which can be determined to hold solely on the basis of the logical form of the theories in question. Why not? We will consider the strongest such possible formal interrelationship one can imagine. Let us suppose that the theories are term by term 'intertranslatable', that is, that each can be obtained from the other merely by a substitution of terms of one theory for terms of the other. Would that be enough to show the theories equivalent?

Surely not. Let the two theories be 'All lions have stripes' and 'All tigers have stripes', with all the words in both theories taking on their usual meanings. The theories are intertranslatable in the purely formal sense. They are exactly alike in logical form, and one can be obtained from the other by a simple term-for-term substitution. But they are most assuredly not equivalent. I am not denying that if we found a speaker who persistently asserted that lions had stripes and tigers didn't that we would probably take him to mean by 'tiger' what we mean by 'lion' and vice versa. Nor am I denying that it is just the question of the meaning of the theoretical terms of a theory that is at issue in many questions about the necessary and sufficient conditions for theoretical equivalence. We will return to that issue in detail. All I am claiming here is that mere commonality of logical form, even of a total theory when compared with another total theory, is certainly not by itself sufficient for theoretical equivalence. The meanings of the terms in the theories, however construed, are crucial to questions of equivalence.

But 'translatability' of one theory to another, in some sense, is just the additional constraint, over and above saving the phenomena, that some realists want to demand as necessary for theoretical equivalence. So if that isn't the purely formal notion of commonality of logical form, what is it?

Two options suggest themselves. If we have some grasp of the meanings of the terms of the theories which comes to us from outside the role these terms play in the theories, and if on the basis of this knowledge of meanings we can affirm the logical equivalence of the theories, then, of course, the theories are equivalent. Straightforward translations of a theory phrased in one language to a version of the theory in some other language, where the theories are small fragments of the totality of beliefs of speakers of the languages and where the full vocabulary of the theories appears embedded in a far broader context than the particular theory in question and in such a way that we are inclined to say that we grasp the meanings of the terms independently of the role they play in the theories in question, are of this sort. Thus we won't have any trouble affirming that the theory that 'Salt is white' is equivalent to the theory that 'Salz ist weiss'. But this notion of equivalence will hardly be of help to us in the most interesting and crucial cases. Our conviction that Schrödinger quantum mechanics is equivalent to Heisenberg quantum mechanics hardly comes from some outside grasp of the meanings of the respective psi-functions which informs us that one simply means the time transform of the other.

So what does assure us over and above mere formal similarity, which we saw was never enough, that in this and similar cases the theories are genuinely equivalent to one another? I think the answer is clear. We believe that we have no antecedent understanding of the meanings of the theoretical terms (psi-functions and operators in the example of quantum mechanics) other than the role they play in the theories in question. We have no external grasp on the meanings of these terms that we could go to to determine whether the existence of a formal mapping from one theory to the other did or did not constitute a genuine translation, and hence a genuine demonstration of theoretical equivalence. Rather, all we have is this: relative to whatever understanding we have of the notion of observational consequence of a theory, we are satisfied that the two theories in question really do have all their observational consequences in common. Over and above this we can demonstrate the existence of a formal mapping of some appropriate kind between the theories at the theoretical level. For the purpose of our argument it really won't matter very much just exactly what this mapping, necessary to show appropriate commonality of theoretical structure, amounts to. Together, these *two* features of the theory, commonality of observational consequences and the existence of the appropriate structural mapping at the theoretical level, are taken to be enough to establish theoretical equivalence.

The intuition behind the position is clear. The meaning of the theoretical terms is fixed entirely by the role they play in their respective theories, in the holistic account of the meaning of theoretical terms now so familiar to us. The combined force of observational commonality and structural correspondence at the theoretical level is taken to be sufficient to establish commonality of meaning for the theoretical terms, their having, as they do, their meaning entirely determined by the place they occupy in the theoretical structure which generates the observational consequences. The details of the argument will depend on one's detailed account of the holistic structure of theories and on what one takes to be a sufficient structural interrelationship for theoretical equivalence. But consideration of such simple cases as theories which have common observational consequences and which can be obtained from one another by such simple operations as interchanging of terms for terms ('You just are using "positive charge" to mean what we meant by "negative charge" and vice versa') will be enough to give us the general idea of the motivations and arguments of the programme.

Seen from this point of view, a realist conception of theoretical equivalence should not be viewed as a total rejection of the positivist account. Rather, it amounts to the claim that the positivist constraint of commonality in saving the phenomena, while necessary for theoretical equivalence, is simply not sufficient. Again the intuition is clear. Two theories might have all the same observational predictions but be so radically different in their structure at the theoretical level that one ought to take them as attributing (realistically) quite different explanatory structures to the world. Only commonality of structure at the level of the theoretical ontology introduced by the theories to explain the commonly predicted observational results is enough for us to say that the two accounts are genuinely, realistically, equivalent to one another.

Notice how much of the positivist account the realist, construed this way, must first accept. The project of characterizing a notion of theoretical equivalence in this way at least presupposes that the notion of commonality of observational consequences is a coherent one. The argument which takes Schrödinger and Heisenberg to be offering us the same theory must at least assume that some coherent limits to the notion of observationality can be given. If the psi-function itself could be considered, under any circumstances, an observational quantity, then the existence of the appropriate mathematical transformation from one representation to another would no more constitute a demonstration of equivalence than does the trivial formal transformation from 'Lions have stripes' to 'Tigers have stripes' demonstrate the genuine equivalence of those assertions. Of course the realist need not lay down where observationality ends and theoreticity begins. He need only be assured that the consequences common to the two theories (in the example chosen probabilities of outcomes of measurements in the quantum-mechanical sense) *include* anything which could be called observational. He must, that is, be sure that the apparent incompatibilities between the theories, incompatibilities he will demonstrate to be only apparent by means of his equivalence-establishing interrelationship shown to hold among them, are all firmly 'trapped' at the non-observational level. At least that much of the presuppositions of positivism is presupposed by this kind of realist as well.

To summarize, what my realist asserts is this: the point of theories is to introduce theoretically posited structures to explain the observable phenomena. Merely predicting the same phenomena is not enough for two theories to be equivalent, for they may explain these phenomena in radically different ways. But if the theories share both commonality of predicted phenomena *and* an appropriately characterized commonality of structure at the theoretical level, as demonstrated by the existence of the appropriate structural mapping between them at the theoretical level, then the theories plainly are equivalent. Any residual appearance of incompatibility must be due to merely verbal equivocation at the theoretical level. But realizing as we do, on the basis of the theory of the meaning of theoretical terms, which tells us that the totality of their meaning is the place they play in the theoretical structure in question, that the apparent disagreement between the two theories is a mere superficiality of alternative verbal designations for common structural elements, we will not be deceived, but will realize that, properly speaking, we have only one theory expressed in two misleadingly different ways. But it is essential for genuine equivalence that two conditions be met. On the observational level the theories must be identical. On the theoretical level they must bear the appropriate structural similarity.

III

What interrelationship at the theoretical level should the realist demand, over and above commonality in saving the phenomena, before he admits two theories to be equivalent? I hope I don't have to say, for each specific proposal, be it term-by-term translation, common definitional extensions, or whatever, would require its own careful analysis. I hope that I will be able to say all I wish to say without pinning the imagined realist or myself down to a specific proposal. But then how can we say anything interesting about potential pitfalls in his realist path?

One obvious problem for the realist is getting his principles to coincide with his intuitions. He may, in the so-familiar manner, find that his proposal is too weak or too strong in its demands to coincide with what he takes intuitively to be genuine cases of equivalent theories and what he takes to be cases of non-equivalent theories which happen to save the same phenomena.

Reichenbach, for example, wanted to hold, along with Eddington and Schlick, that curved space-time and flat space-time with universal forces were equivalent theories. But on a model of equivalence suggested by *some* of his remarks, remarks to the effect that equivalent theories simply called the same theoretical entities by alternative names, the model of equivalence he should have in mind would require a strict term-by-term interdefinability of the two theories. This won't be the case, however, since the flat space-time theory, being 'ontologically otiose' compared to the curved space-time theory, can't have its parameters defined from those of the ontologically more parsimonious theory. That is, given a flat space-time metric and the 'universal forces', we can determine space-time curvature in the curved space-time theory. But a full specification of curvature is not enough to *uniquely* specify the proper flat space-time and universal forces. Crudely, this is related to the nineteenth-century observation that uniform gravitational fields are empirically undetectable. (I will have more to say about this case later.) So one will either have to give up his stringent model of equivalence or renege on his intuition that these two theories are, indeed, equivalent. In the case of Reichenbach, matters are more complex, of course. Actually, what he does is to suggest in other, dominating places that commonality of observational consequences is sufficient for equivalence, which certainly saves the equivalence of curved space-time and flat space-time plus universal forces. Naturally, though, adopting this notion of equivalence plays havoc with his alleged scientific realism.[1]

One can see this problem of a conflict of intuition with principles arise in the other direction as well. Even quite stringent realist notions of equivalence may not be strong enough to exclude as equivalent pairs of theories the realist does not wish to think of as equivalent.

Surely the realist, although he wants Schrödinger and Heisenberg to say the same thing, does not want the usual wild Cartesian possibilities to count as theories equivalent to our ordinary scheme of the external material world. Now some Cartesian alternatives, say the one that tells us that there is simply nothing but our private experience and nothing 'out there' to explain it, will be rejected as inequivalent to our ordinary theory because of a lack of structural similarity to our ordinary theory on the theoretical level. But other Cartesian fantasies, say the one in which all our private experience is caused by an appropriate signal fed into our tank-immersed brain along a cord (or caused by the appropriate state of the cybernetic machine at the other end of the cord), look as though

---

[1] Reichenbach's allegation of the equivalence of curved space-time with flat space-time plus universal forces can be found in Reichenbach 1958: sects. 3–8. The contrasting 'realist' attitude toward theories is in Reichenbach 1938. Specifically useful on the problem in Reichenbach, and generally useful as background to this essay throughout, is Glymour 1971. See also Quine 1975.

they have the possibility of being made to look as structurally similar to our ordinary 'external world' theory as we like. Indeed, what seems to differentiate the real cases of discovered scientific equivalence from these spurious Cartesian cases is not interdefinability in the former cases and not in the latter. Rather, what makes the scientific cases interesting scientifically, and the Cartesian cases not, is that it is a significant scientific task to demonstrate interdefinability in the scientific cases, whereas interdefinability is just too trivial to be interesting in the kind of Cartesian cases I have in mind.

Here quite a bit of caution is called for, though. There is the kind of realist who rejects the holistic-role-in-a-theory account of the meaning of theoretical terms. For him there is no problem in asserting that the Cartesian alternatives, no matter how structurally similar to our ordinary external-world account, are not equivalent to the ordinary account. But then just what account of theoretical equivalence he will offer, at least what account which will handle the cases we are interested in, where we think equivalence is clear despite surface incompatibility, isn't clear. The realist we have in mind, espousing as he does the doctrine of meaning which gives to theoretical terms only the meaning they possess from their holistic role in the theories in question, will find it difficult, I think, to avoid attributing equivalence of the suitably constructed Cartesian alternatives to our ordinary account. But then his account of theoretical equivalence is sliding perilously close to the positivist account. For if the brain-in-a-vat account of the world is really equivalent to the ordinary material-object world account, so long as the brain-in-a-vat account is suitably formally structured, then have we really gotten very far from merely saving the phenomena as sufficient for theoretical equivalence? Once again the tension between an 'instrumentalistic' account of the meaning of theoretical terms and a realist account of theoretical ontology is clear. But our realist can just bite the bullet and affirm that, appearances to the contrary, such cleverly tidied-up Cartesian alternatives are simply equivalent to our ordinary world-view. Contrary to the view expressed in the quote of Field at the beginning of this paper, all 'evil demon' hypotheses would not then be equal. Some would be genuine alternatives to our ordinary world-view, presumably to be rejected as unacceptable on some epistemic ground or other. Other 'evil demon' hypotheses would just be our ordinary world-view dressed out in peculiarly misleading terminology. It all depends on whether or not our hypothesis about the doings of the evil demon has him failing or succeeding in structurally duplicating our ordinary world in his alternative causal structure of our private experiences.

IV

Let us suppose that our realist is satisfied that his formal notion of theoretical equivalence is consonant with his intuitions about just what pairs are pairs of genuinely equivalent theories. What further concerns must then arise?

One is fairly obvious given a principle component of the motivation behind the positivist position on equivalence to which our realist is opposed. Surely one of the major thrusts behind the positivistic notion is its ability to limit the possibilities of scepticism. Faced with alternative theories all equally compatible with all possible empirical data, we are at a loss as to how to decide which of these we ought to believe. Taking the positivist line and adjudicating all of these theories as saying the same thing, there is no longer a decision to be made. Hence there is less room for scepticism to intrude. We still are faced with the inference problem from observed empirical phenomena to full generalizations over all possible empirical phenomena, of course the problem of induction in its most general form, but at least we need no longer fret about alternatives at the level of the in-principle unobservable haunting us.

Once we insist on stricter conditions for equivalence than mere commonality at the empirical level, the problem of rationalizing the choice of alternatives to be believed is reinstated. Nor, from a realistic point of view, with its insistence on truth as the goal of inquiry and on a correspondence theory of truth (whatever that means) even at the theoretical level, do such ways out as allowing rationality to be permissive ('You are rational if you believe any one of the non-equivalent theories which saves the phenomena equally well') appeal. What we have a right to expect from the realist is a systematic account of what rationalizes the choice of one theory as true over any of the non-equivalent but empirically indiscriminable alternatives to it.

The realist can, of course, simply refuse the challenge and accept the sceptical consequences. But suppose he doesn't. What we will be offered is an account of the rationality of theory choice which will allow us to go beyond the conformity of a theory with all possible empirical data in deciding its worthiness for our belief. One problem will be, though, that we will need to be assured that the powers of this model of confirmation are sufficient to bring us up to the boundaries of equivalence. To avoid the intrusion of scepticism, that is, we will need assurance that for any two non-equivalent theories one will be preferred epistemically to the other. In other words, there will have to be a close internal harmony between the notion of theoretical equivalence offered and the notion of the epistemic worthiness of theories.

Without a specific model of theoretical equivalence and a specific, allegedly harmonized, model of confirmation, I can only illustrate this problem by briefly considering in a qualitative way the issues I have in mind. Typically one can generate alternative theories saving the same phenomena by some process which introduces into a theory otiose elements whose place in the theory 'cancels out'. Most interesting, of course, are the historical cases where the theory with the otiose elements came first and where it was an important scientific discovery that one could eliminate them by a conceptual revision. For example, one could replace the ether theories used to account for the null round-trip experiment results with special relativity, thereby replacing a manifold of theories with a 'cancelled out' element (the velocity of the observer relative to the ether) with a theory without that vacuous element. Again, Einstein showed us that one could replace earlier gravitational theories with their otiose elements of 'real inertial frames' and 'real gravitational fields' by curved space-time with its merely local inertial frames.

While the positivist seems to be committed to the view that a theory and its excessively otiose counterparts are all equivalent, the realist will, in general, claim that they are not. What we can expect from him is some attempt at a systematization of the notion of ontological simplicity which will credit the less otiose theory with greater simplicity and the simpler theory with greater warrant for belief. Obviously this will require a notion of confirmation which differs from that which associates confirmation with having the right observational consequences, and attempts in this direction have been made.[2]

But what if we have a pair of theories such that each member of the pair is preferable to the other in ontological simplicity in some respect? We can easily imagine a situation where A is preferable to B in terms of ontological parsimony for one aspect of theoretical structure, and B preferred to A along another component. Here one will probably hope to establish that relative to this pair of theories there is some third alternative, C, which is notologically more parsimonious than either A or B, and, hence, preferable to both of them. What we will expect from the realist, though, if he takes this approach, is some reason to believe that relative to the notion of theoretical equivalence which he espouses, in each possible case there will always result a single unique 'maximally parsimonious' theory. Or, rather, that relative to his notions of theoretical equivalence and ontological

---

[2] For an attempt at a theory of confirmation appropriate to this realist notion of equivalence, see Glymour, 1980, esp. chs. 5 and 9.

parsimony, any two theories which, relative to a given set of observational consequences, are both adjudged maximally parsimonious will turn out, on this criterion, to be theoretically equivalent. Unless he can show us that there is a most parsimonious theory, and that this is unique modulo his notion of theoretical equivalence, we will once again be faced with room for the intrusion of scepticism.

But there is a more profound problem of an epistemic sort than this for the realist. Taking equivalence to demand more than commonality of observational consequences, the realist is faced with the threat of scepticism which arises when he tolerates inequivalent theories having all their observational consequences in common. So his confirmation theory is supplemented with elements going beyond reflection on the observational consequences of a theory in evaluating its merit for belief. Whether this be ontological simplicity, as I have chosen for the example above, or conservatism with respect to antecedent theory, which has frequently been proposed, or some other possible feature of theories altogether, is immaterial. The first problem the realist faces, unless he is simply willing to tolerate states of irresolvable suspension of belief, is to try to demonstrate to us that his combination of a notion of theoretical equivalence and a notion of warranted believability always produces as most believable only a single theory or a class of theories all equivalent to one another relative to his criterion of theoretical equivalence.

But even if he does this, should we then be fully satisfied? I think not. For the question will always arise: why should we accept *this* notion of confirmation or epistemic preferability? Why should we believe that the simpler theory, or the theory which deviates minimally from previously accepted theory, is the theory which is, in any sense, most likely to be true? The point is hardly new. In so far as one places in one's structure for warranted belief any considerations that go beyond mere conformity with the observational data, it is hard to see why one ought to take these additional espoused 'marks of believability' as genuine 'marks of truth'.

Of course there are the familiar options open of adopting as a priori principles that simpler theories are more likely to be true, etc. I do not wish to pursue these here. We ought to note that one familiar approach is not open to the realist we have in mind. There is a familiar 'justification' for such principles of warranted belief of a pragmatist sort which claims that the normative principles of believability rest ultimately on our actual practice of belief, and that the connection between truth and believability rests not on showing that some process for belief leads (more likely than not) to the truth, but rather on understanding the very notion of truth as ideal warranted believability. However persuasive such a tack may be in trying to rationalize our actual reliance on principles of simplicity, conservatism, etc., in science, the realist we have in mind should be loath to undercut the possibility of scepticism by this means. For if there is anything realistic to his realism, it should include at least the rejection of such a pragmatist line on truth and on justification. Whatever realism about theories means, it should include a claim that truth is correspondence to objective reality (in the metaphysical sense in which this is a controversial thesis), and this brings with it strong pressure for an account of just *why* we ought to accept anything over and above conformity with the observational data as relevant to the believability of a theory. But some such invocation of principles over and above conformity with observational data is necessary for anyone who wishes to impose a constraint on equivalence which outruns commonality in saving the phenomena and who does not wish to accede to the possible sceptical consequences of that demand.

There is, of course, a familiar response the realist can give. Unfortunately, it is of the 'You're no better' form, with the usual unsatisfactoriness of the *tu quoque* reply. This response is the familiar one of the realist to the positivist that even eschewing postulation of unobservables still leaves us with the problem of leaping beyond present data to a fully general theory of the world. And, the reply will go on, the problem of purely inductive inference is as infected with the invocation of notions like 'simplest hypothesis' etc., as is the discussion of inference from the observable to the unobservable in principle. Perhaps so. Still we shall want some rationalization from the realist for his invocation of whatever principles he chooses to suggest limit the range of scepticism inevitably introduced by his belief in an ontology of in-principle unobservables, and by his invocation of a criterion of theoretical equivalence which demands that in at least some cases theories be inequivalent even when they save all the same phenomena in all possible worlds.

V

If the kind of realist notion of theoretical equivalence we have been discussing runs into problems in the upward direction of confirmation, there are difficulties it must deal with in the other, downward, relation of theory to data as well. We take it that our theories are explanatory. How must the realist's notions of explanation and of theoretical equivalence mesh?

First, there will be the obvious requirement that his account of explanation satisfy at least some version of an equivalence principle. If two theories are genuinely equivalent, then, one hopes, in at least some sense they ought to offer the 'same explanation' of the phenomena. And if the theories are inequivalent on the realist's criterion, then there ought to be a discrimination on the level of explanation which tells us why they are not, even if they are predictively equivalent, explaining the predicted phenomena in the same way. Actually, the case here is subtle, for there might be alleged to be a kind of intensionality which allows theories to be equivalent in the sense the realist has in mind yet not fully equivalent in some 'finer-grained' explanatory sense. The individuation of 'kind of explanation' might, for example, depend upon alleged modes of expression so that one and the same theory might in two equivalent versions be said in some sense to offer different explanations due solely to the mere difference in the manner in which the theory is expressed in the two versions. Someone who finds the notion of explanation loaded with 'pragmatic' aspects might indeed hold to this. In the same vein it would, I suppose, be possible to maintain that two equivalent versions of a theory received differential confirmation from the same data, despite the obvious persuasiveness of a principle of equivalence as a criterion of adequacy of an acceptable notion of confirmation.

What is certain, I think, is that an exponent of a realistic notion of equivalence must needs be quite wary of allowing equivalent theories to either offer inequivalent explanations of the data or to be inequivalently confirmed by the same evidential base. If he allows both of these, his notion of equivalence is beginning to drain into a bloodless ghost of that notion as pre-analytically understood. At the very least, we would expect from him coarse-grained notions of explanation and confirmation which would mesh with his notion of theoretical equivalence in the natural way, and an account as to why the finer-grained notions, which allowed equivalent theories to offer inequivalent explanations or to be inequivalently confirmed, did not vitiate the real sense of theoretical equivalence he originally had in mind.

Once again, a detailed critique of a realistic attempt to 'mesh' notions of theoretical equivalence and explanation will be impossible for us here, for that would require what I am trying to avoid, the presentation of a specific principle of equivalence and specific accounts of explanation and confirmation. Instead, I will focus on what to the positivist seem to be certain fundamental difficulties with any such realistic programme. What the positivist will allege is familiar. Suppose we have available that minimal, best-confirmed theory compatible with the data (and, of course, all of those equivalent members of the equivalence class of theories of which this particular one is representative). Presumably this theory is the best possible explanatory account of the data. Indeed, any reasonable notion of confirmation and explanation should lead us in at least some weak sense to take it that the best-confirmed, most plausible account is the best explanatory account, a relationship which becomes, of course, trivial if we adopt notions of confirmation as inference to the best explanation.

One thing had better be the case if the realist account is to hold on to even a skeletal version of its fundamentals. That theory which is the Craigian reduction of the theory in question had better not appear among the set of equivalent best alternative theories. For if it does, the notion of theoretical equivalence in mind would seem automatically to reduce to the positivist notion. But how is the realist to avoid the introduction of the Craigian reduction, if not as equivalent to his best theoretical account, then as a simpler (hence superior) alternative account to it?

Relying on the syntactical complexity (or, indeed, infinitistic nature) of the Craigian alternative to exclude, it seems rather 'unrealistic' in nature, a move more suitable to a frank pragmatist notion of theoretical preferability. Surely the right way out for the realist is to argue that the Craigian alternative is to be rejected for its explanatory inadequacy. According to the realist, the best realistic theory and its equivalent alternatives explain the observational phenomena in a way in which the Craigian reduction does not. Indeed, the Craigian surrogate is alleged to be devoid of explanatory power at all. Theoretical postulation, we are told, does not merely summarize observational generality or accommodate it in a compact syntactical form. Theoretical postulation is taken to *explain* the observational generality. Hence the mere statement of that generality is, in a fundamental sense, totally devoid of any real explanatory power.

But what is this notion of explanation which makes the best realist theory explanatory in a way that the Craigian reduction of it is not? Here we must once again focus on two quite distinct kinds of realism, one the kind of realism which takes a view of the meaning of theoretical terms as given by their holistic role in the theories in which they appear, and the other which allows the attribution of 'excess' meaning to theoretical terms, by semantic analogy or otherwise. Thus, for example, for the first kind of realist, molecules exist, but the meaning of 'molecule' is given solely by the complex role it plays in the sum total of molecular theories. For the other kind of realist, we understand 'molecule' by means of our understanding of such notions as 'particle', notions whose meaning originally accrued in their use to refer to observables, but whose meaning is preserved intact when the term is used in the quite different contexts of reference to 'particles too small to see' or even 'particles unobservable in principle'. The first kind of realist, the one with the idea of the meaning of theoretical terms as being solely their role in the holistic theory, is the one we have been emphasizing. Why? Because, I have claimed, it is only for such a realist that the notion of theoretical equivalence as being structural iso-morphism superadded to observational equivalence is plausible. Without that theory of the meaning of theoretical terms, it is hard to see why even the most strict commonality of structure on the theoretical level would be enough to guarantee equivalence. But, I want to argue here, it is from the point of view of the second kind of realist that the notion of realistic explanation as something over and above subsumption of observed facts into generalizations about observables makes its most plausible sense. In other words, I want to argue that there is a fundamental tension between the realist's desire to posit a notion of explanation over and above that adopted by the positivist, and his desire to maintain a theory of the mean-ing of theoretical terms which allows for the notion of theoretical equiv-alence we have been discussing.[3]

Suppose we have a generalization over observable phenomena. What does the realist demand over and above that? Presumably the postulation of a theoretical entity or property which explains the observational gen-erality. But what additional explanatory power does this theoretical pos-tulation give us? A frequent answer is 'unification'. Positing the theoretical structure provides, somehow, a *unified* account of the phenomena.[4]

Let us look at some cases to see why I think there are difficulties for the realist here. The observational facts of dynamics require us to single out a set of preferred states of motion, the local inertial motions. Additionally, optical phenomena pick out as distinguished these same reference frames. The realist accounts for all of this by positing the space-time structure itself as existing over and above the (for this purpose) observational moving systems and light waves. The dynamic and optical distinctions between inertial and non-inertial states are then accounted for by the relation of the physical systems to the underlying space-time structure. But, replies the positivist, what kind of additional explanatory force does positing 'space-time itself' give us? Whereas before we had the irreducibly inexplicable basic distinction between inertial systems and non-inertial, now we have a new, and trivially introduced, 'deeper' explanation in terms of the under-lying space-time and the relations of the physical systems to it. But have we done anything more than add an otiose layer to what was already a per-

---

[3] For a discussion of that notion of the meaning of theoretical terms which allows meaning to accrue to them over and above the role they play in the theory, see Sklar 1980.

[4] On realistic theories as explaining by unifying, see Friedman 1974, and 1983, esp. ch. 6.

fectly adequate theory, a layer which only gives the spurious appearance of further 'unification' or further 'explanation' to the phenomena?

Many positivists would argue like this: the appearance of additional explanatory value in the substantivalist space-time account, over the purely relationist theory, is due to the fact that while the reference to space-time itself is really introduced only in the holistic-place-in-the-theory manner of the realist we have been emphasizing, the naïve picture of space-time as a kind of 'ghostly' substance, sort of like a thin 'rigid' extended material thing, gives us the impression that we are getting an explanation like that which the other kind of realist would offer. If we really understood what we were talking about when we posit the theoretical apparatus, independently of positing its explanatory role in the situation in question, then we could attribute further explanatory force to the substantival theory over the relational in the manner familiar from some older realists (reducing the unfamiliar to the familiar, demonstrating 'mechanism' in the Newtonian sense, etc.). But we have no such grasp of what the theoretical apparatus is over and above its place in the theory in question. As a consequence, its putative explanatory force is void.

To see this, the positivist continues, consider another case. Study of the symmetries of the interactions of elementary particles gives us a systematization of these particles in terms of various symmetry groups. One can capture this symmetry structure by positing the existence of various 'charges' for the particles (strangeness, charm, etc.) and associating with the posited charges various appropriate conservation rules. But in positing these charges and their conservation, are we *explaining* the symmetries involved? Most theoretical physicists would argue, I think rightly, that the invocation of the charges is simply another way of stating what the various symmetries are. There is a redundancy of the 'theoretical structure' on the 'observational data' to be explained. If anything would count as genuinely explanatory of the symmetries, they would argue, it would be the subsumption of these symmetries into some higher, more general symmetries naturally generated as the consequence of a more profound theory (say the generation of the symmetries out of some posited gauge field or other unified field account). The congruence of all of this with the positivistic notions of explanation is clear.

Now consider the alleged explanation of the symmetries of baryons and mesons given us by quark theory. Here we do seem to feel that a genuine explanatory account has been offered. Part of this intuition is, I think, perfectly acceptable positivistically, quark theory being more general and profound than the symmetry accounts derivable from it. But part of the intuition rests on our viewing quarks 'analogically' as tiny particles constituting larger particles in a manner understandable to us from the whole–

part relationship and its explanatory value in the realm of observables. Once again we see that the invocation of terms whose meaning rests solely on the role they play in the appropriate theory suggests an account of explanation which is wholly positivistic, whereas the other kind of realism, the one which allows terms to have 'excess meaning' attributable to them through semantic analogy seems to fit more harmoniously with the realistic notion that explanation by positing of theoretical structure is more than the subsumption of observational facts under broad generalities.

I do not, naturally, intend to pursue this problem of explanation from the realist point of view in any depth here. To repeat, my point is only this: a realist may take two attitudes toward the meaning of theoretical terms, either the holistic-role-in-the-theory attitude or the alternative which grants to theoretical terms meaning accrual over and above the role the terms play in the theory. From the latter point of view it is fairly clear how we will be offered a realist account of explanation which takes explanation to be something over and above hierarchical subsumption of observational possibilities into generalities. But, from this point of view, it is hard to see what the realist account of equivalence of theories will amount to. Basically it must come down to the line that theories are equivalent only when they 'say the same thing', and this notion will have to be explicated in terms of whatever theory of meaning (over and above role-in-theory) the realist offers us. From the alternative point of view regarding theoretical meaning, a much simpler notion of theoretical equivalence can be constructed. This is the notion which simply takes the positivistic notion of equivalence as necessary and superadds to it some notion of structural isomorphism between the theories at the theoretical level to obtain a sufficient condition for equivalence. But this notion of theoretical meaning, and of theoretical equivalence, besides borrowing many dubious presuppositions from the positivist position, suffers also from the difficulty of making it hard for us to see just what it is that is explanatory in a theory, or in any of its equivalents, which is not already there in the allegedly non-equivalent and non-explanatory Craigian reduction of the theory so beloved of positivists.

VI

To summarize: positivism, for all its defects, offers us a theory of theoretical equivalence neatly integrated with its theory of confirmation and its theory of explanation. The realist is obliged to do the same. If the realist adopts a theory of the meaning of theoretical terms which attributes to these terms a meaning over and above that which accrues to them by means of the role they play in the theory, then his notion of theoretical equivalence will be complex, and will depend in detail on just what theory of meaning for theoretical terms he offers us. If he adopts a theory of meaning for theoretical terms which takes their meaning to be fixed solely by the role they play in the theory, then his notion of theoretical equivalence is likely to be simpler. It will be the notion of two theories, first of all, sharing commonality of observational prediction and, secondly, having an appropriate structural isomorphism at the theoretical level.

One will first ask if the notion of equivalence offered corresponds with the realist's intuitions. Does it take as equivalent theories only those the realist wishes to count as equivalent, and will it count as equivalent all those pairs pre-analytically thought equivalent?

Next, one will want to examine the interaction of the notion of equivalence offered with the 'upward' notion of confirmation. Will equivalent theories always be equally confirmed by the same data or, if not, will the violation of the 'equivalence condition' for confirmation receive an adequate explanation? Will the theory of confirmation allow for differential confirmation of theories designated non-equivalent, or instead, will it leave room for scepticism. Even if the theory of confirmation does always select from a group of non-equivalent alternatives a unique 'best-confirmed' member, will there be an appropriate *realist* rationale for adopting that notion of confirmation?

Finally, will one want to look at the interaction of the notion of equivalence with the 'downward' notion of explanation? Will equivalent theories always offer 'the same explanation' of the phenomena, and non-equivalent theories 'different' explanations? More important, what notion of explanation, over and above the positivist notion, does the realist have in mind? Will it be such as to allow us to understand why a theory is not, according to the realist, equivalent to its Craigian reduction, and why the theory with its realistic posits is explanatory in a way in which the Craigian surrogate is not?

Far be it from me to claim that these questions cannot be answered by the realist. But until they are, we should be reluctant to dismiss the positivist notions with a sneering reference to 'outmoded verificationism'. It is sometimes said (wrongly I believe) that scientists do not reject one theory until a better one is available to take its place. Just exactly what is the realist notion of equivalence, and which associated realist notions of confirmation and explanation are supposed to take the place of the positivist's integrated if implausible accounts?[5]

[5] My thanks to Steve White for the title and to Tim McCarthy for many useful discussions.

REFERENCES

Friedman, M. (1974). 'Explanation and Scientific Understanding.' *Journal of Philosophy*, 71: 5–19.
——(1983). *Foundations of Space-Time Theories.* Princeton: Princeton University Press.
Glymour, C. (1971). 'Theoretical Realism and Theoretical Equivalence.' *Boston Studies in the Philosophy of Science*, 8: 275–88.
——(1980). *Theory and Evidence.* Princeton: Princeton University Press.
Quine, W. V. O. (1975). 'On Empirically Equivalent Systems of the World.' *Erkenntnis*, 9: 313–28.
Reichenbach, H. (1938). *Experience and Prediction.* Chicago: University of Chicago Press.
——(1958). *The Philosophy of Space and Time.* New York: Dover.
Sklar, L. (1980). 'Semantic Analogy.' *Philosophical Studies*, 38: 217–34.

IV

TO SAVE THE PHENOMENA

BAS C. VAN FRAASSEN

After the demise of logical positivism, scientific realism has once
more returned as a major philosophical position. I shall not try here
to criticize that position, but rather attempt to outline a comprehensive
alternative.[1]

I

What exactly is scientific realism? Naïvely stated, it is the view that the
picture science gives us of the world is true, and the entities postulated
really exist. (Historically, it added that there are real necessities in nature;
I shall ignore that aspect here.[2]) But that statement is too naïve; it attributes to the scientific realist the belief that today's scientific theories are
(essentially) right.

The correct statement, it seems to me, must indeed be in terms of
epistemic attitudes, but not so directly. The aim of science is to give us *a
literally true story of what the world is like*; and the proper form of acceptance of a theory is to believe that it is true. This is the statement of
scientific realism: 'To have good reason to accept a theory is to have good
reason to believe that the entities it postulates are real,' as Wilfrid Sellars
has expressed it. Accordingly, all anti-realism is a position according to
which the aims of science can well be served without giving such a literally
true story, and acceptance of a theory may properly involve something less
(or other) than belief that it is true.

The idea of a literally true account has two aspects: the language is to
be literally construed; and, so construed, the account is true. This divides
the anti-realists into two sorts. The first sort holds that science is or aims to

Reprinted from *Journal of Philosophy*, 73/18 (1976), 623–32, by permission of the author and
the *Journal of Philosophy*.
[1] For some criticisms, see van Fraassen 1974 and 1975.
[2] Cf. van Fraassen 1977.

be true, properly (but not literally) construed. The second holds that the language of science should be literally construed, but its theories need not be true to be good. The anti-realism I advocate belongs to the second sort.

II

When Newton wrote his *Mathematical Principles of Natural Philosophy* and *System of the World*, he carefully distinguished the phenomena to be saved from the reality he postulated. He distinguished the 'absolute magnitudes' that appear in his axioms from their 'sensible measures' which are determined experimentally. He discussed carefully the ways in which, and extent to which, 'the true motions of particular bodies [may be determined] from the apparent', via the assertion that 'the apparent motions . . . are the differences of true motions' (Cajori (ed.) 1960: 12).

The 'apparent motions' form relational structures defined by measuring relative distances, time intervals, and angles of separation. For brevity, let us call these relational structures *appearances*. In the mathematical model provided by Newton's theory, bodies are located in absolute space, in which they have real or absolute motions. But within these models we can define structures that are meant to be exact reflections of those appearances and are, as Newton says, identifiable as differences between true motions. These structures, defined in terms of the relevant relations between absolute locations and absolute times, which are the appropriate parts of Newton's models, I shall call *motions*, borrowing Simon's (1954) term.

When Newton claims empirical adequacy for his theory, he is claiming that his theory has some model such that *all actual appearances are identifiable with (isomorphic to) motions in that model.*

Newton's theory goes a great deal further than this. It is part of his theory that there is such a thing as absolute space, that absolute motion is motion relative to absolute space, that absolute acceleration causes certain stresses and strains, and thereby deformations in the appearances, and so on. He offered, in addition, the hypothesis (his term) that the centre of gravity of the solar system is at rest in absolute space. But, as he himself noted, the appearances would be no different if that centre were in any other state of constant absolute motion.

Let us call Newton's theory (mechanics and gravitation) *TN*, and *TN(v)* the theory *TN* plus the postulate that the centre of gravity of the solar system has constant absolute velocity. By Newton's own account, he claims empirical adequacy for $TN(0)$; and also claims that, if $TN(0)$ is empirically adequate, then so are all the theories $TN(v)$.

Recalling what it was to claim empirical adequacy, we see that all the theories $TN(v)$ are empirically equivalent exactly *if all the motions in a model of $TN(v)$ are isomorphic to motions in a model $TN(v + w)$*, for all constant velocities v and w. For now, let us agree that these theories are empirically equivalent, referring objections to a later section.

III

What exactly is the 'empirical import' of $TN(0)$? Let us focus on a fictitious and anachronistic philosopher Leibniz*, whose only quarrel with Newton's theory is that he does not believe in the existence of absolute space. As a corollary, of course, he can attach no 'physical significance' to statements about absolute motion. Leibniz* believes, like Newton, that $TN(0)$ is empirically adequate; but not that it is true. For the sake of brevity, let us say that Leibniz* *accepts* the theory, but that he does not *believe* it; when confusion threatens we may expand that idiom to say that he *accepts the theory as empirically adequate*, but does not *believe it to be true*. What does Leibniz* believe, then?

Leibniz* believes that $TN(0)$ is empirically adequate, and hence, equivalently, that all the theories $TN(v)$ are empirically adequate. Yet we cannot identify the theory that Leibniz* holds about the world—call it *TNE*—with the common part of all the theories $TN(v)$. For each of the theories $TN(v)$ has such consequences as that the earth has *some* absolute velocity, and that absolute space exists. In each model of each theory $TN(v)$ there is to be found something other than motions, and there is the rub.

To believe a theory is to believe that one of its models correctly represents the world. A theory may have isomorphic models; that redundancy is easily removed. If it has been removed, then to believe the theory is to believe that exactly one of its models correctly represents the world. Therefore, if we believe of a family of theories that all are empirically adequate, but each goes beyond the phenomena, then we are still free to believe that each is false, and hence their common part is false. For that common part is phrasable as: one of the models of one of those theories correctly represents the world.

IV

It may be objected that theories will seem empirically equivalent only so long as we do not consider their possible extensions.[3] The equivalence may generally, or always, disappear when we consider their implications for some further domain of application. The usual example is Brownian motion; but this it imperfect, for it was known that phenomenological and statistical thermodynamics disagreed even on macroscopic phenomena over sufficiently long periods of time. But there is a good, *fictional* example: the combination of electromagnetism with mechanics, if we ignore the unexpected null results that led to the replacement of classical mechanics.

Maxwell's theory was not developed as part of mechanics, but it did have mechanical models. This follows from a result of König, as detailed by Poincaré in the preface of his *Electricité et Optique* and elsewhere. But the theory had the strange new feature that velocity itself, not just its derivative, appears in the equations. A spate of thought-experiments was designed to measure absolute velocity, the simplest perhaps that of Poincaré:

Consider two electrified bodies; though they are both carried along by the motion of the earth; . . . therefore, equivalent to two parallel currents of the same sense and these two currents should attract each other. In measuring this attraction, we shall measure the velocity of the earth; not its velocity in relation to the sum or the fixed stars, but its absolute velocity. (1958: 98)

The null outcome of all experiments of this sort led to the replacement of classical by relativistic mechanics. But let us imagine that values *were* found for the absolute velocities; specifically for that of the centre of the solar system. Then, surely, one of the theories $TN(v)$ would be confirmed and the others falsified?

This reasoning is spurious. Newton made the distinction between true and apparent motions without presupposing more than the basic mechanics in which Maxwell's theories had models. Each motion in a model of $TN(v)$ is isomorphic to one in some model of $TN(v + w)$, for all constant velocities v and w. Could this assertion of empirical equivalence possibly be controverted by those nineteenth-century reflections? The answer is *no*. The thought-experiment, we may imagine, confirmed the theory that added to TN the hypothesis:

---

[3] See e.g. Boyd 1973.

HO. The centre of gravity of the solar system is at absolute rest.
EO. Two electrified bodies moving with absolute velocity v attract each other with force $F(v)$.

This theory has a consequence strictly about appearances:

CON. Two electrified bodies moving with velocity v relative to the centre of gravity of the solar system, attract each other with force $F(v)$.

However, that same consequence can be had by adding to TN the two alternative hypotheses:

Hw. The centre of gravity of the solar system has absolute velocity w.
Ew. Two electrified bodies moving with absolute velocity $v + w$ attract each other with force $F(v)$.

More generally, for each theory $TN(v)$ there is an electromagnetic theory $E(v)$ such that $E(0)$ is Maxwell's and all the combined theories $TN(v)$ plus $E(v)$ are empirically equivalent.

There is no originality in this observation, of which Poincaré discusses the equivalent immediately after the passage I cited above. Only familiar examples, but rightly stated, are needed, it seems, to show the feasibility of concepts of empirical adequacy and equivalence. In the remainder of this paper I shall try to generalize these considerations, while showing that the attempts to explicate those concepts *syntactically* had to reduce them to absurdity.

V

The idea that theories may have hidden virtues by allowing successful extensions to new kinds of phenomena, is too pretty to be left. Nor is it a very new idea. In the first lecture of his *Cours de philosophie positive*, Comte referred to Fourier's theory of heat as showing the emptiness of the debate between partisans of calorific matter and kinetic theory. The illustrations of empirical equivalence have that regrettable tendency to date; calorifics lost. Federico Enriques seemed to place his finger on the exact reason when he wrote: 'The hypotheses which are indifferent in the limited sphere of the actual theories acquire significance from the point of view of their possible extension' (1929: 230). To evaluate this suggestion, we must ask what exactly is an extension of a theory.

Suppose that experiments really had confirmed the combined theory $TN(0)$ plus $E(0)$. In that case mechanics would have won a *victory*. The claim that $TN(0)$ was empirically adequate would have been borne out by the facts. But such victorious extensions could never count for a theory as against one of its empirical equivalents.

Therefore, if Enriques's idea is to be correct, there must be another sort of extension, which is really a defeat—but qualified For a theory T may have an easy or obvious modification which is empirically adequate, while another theory empirically equivalent to T does not. One example may be the superiority of Newton's celestial mechanics over the variant produced by Brian Ellis; Ellis (1965) himself seems to be of this opinion. This is a *pragmatic* superiority, and cannot suggest that theories, empirically equivalent in the sense explained, can nevertheless have different empirical import.

VI

We still need a general account of empirical adequacy and equivalence. It is here that the syntactic approach has most conspicuously failed. A theory was conceived as identifiable with the set of its theorems in a specified language. This language has a vocabulary, divided into two classes—the observational and theoretical terms. Let the first class be E; then the empirical import of theory T was said to be its subtheory T/E—those theorems expressible in that subvocabulary. T and T' were declared empirically equivalent if T/E was the same as T'/E.

Obvious questions were raised and settled. Craig showed that, under suitable conditions, T/E is axiomatizable in the vocabulary E. Logicians attached importance to questions about restricted vocabularies, and this was apparently enough to make philosophers think them important too. The distinction between observational and theoretical terms was more debatable, and some changed the division into 'old' and 'newly introduced' terms.[4] But all this is mistaken. Empirical import cannot be isolated in this syntactic fashion. If that could be done, then T/E would say exactly what T says about what is observable, and nothing else. But consider: the quantum theory, Copenhagen version, says that there are things which sometimes have a position in space and sometimes do not. This consequence I have just stated without using theoretical terms. Newton's theory TN implies that there is something (to wit, absolute space) which neither has a position nor occupies volume. As long as unobservable entities differ systemically from observation entities with

---

[4] e.g. Lewis 1970. This paper is not subject to my criticisms here; on the contrary, it provides independent reasons to conclude that the empirical import of a theory cannot be syntactically isolated.

respect to observable characteristics, T/E will say that there are such if T does.

The reduced theory T/E is not a description of the observable part of the world of T; rather, it is a hobbled and hamstrung version of T's description of everything. Empirical equivalence fares as badly. In section II, $TN(0)$ and TNE *must* be empirically equivalent, but the above remark about TN shows that $TN(0)/E$ is not TNE/E. To eliminate such embarrassments, extensions of theories were considered in attempts to redefine empirical equivalence.[5] But these have similar absurd consequences.

The worst consequence of the syntactic approach was surely the way it focused philosophical attention on irrelevant technical questions. The expressions 'theoretical object' and 'observational predicate' mark category mistakes. Terms may be theoretical, but 'observable' classifies putative entities. Hence there cannot be a 'theoretical/observable distinction'. It is true, surely, that elimination of all theory-laden terms would leave no usable language; also all that 'observable' is as vague as 'bald'. But these facts imply not at all the 'observable' marks on unreal distinction. The latter refers quite clearly to our limitations, the limits of observation, which are not incapacitating, but also not negligible.

VII

The phenomena are saved when they are exhibited as fragments of a larger unity. For that very reason it would be strange if scientific theories described the phenomena, the observable part, in different terms from the rest of the world they describe. And so an attempt to draw the conceptual line between phenomena and the trans-phenomenal by means of a distinction of vocabulary must always have looked too simple to be good.

Not all philosophers who discussed unobservables, by any means, did so in terms of vocabulary. But there was a common assumption: that the distinction marked is philosophical. Hence it must be drawn, if at all, by philosophical analysis and, if attacked, by philosophical arguments. This attitude needs a grand reversal. If there are limits to observation, these are empirical, and must be described by empirical science. The classification marked by 'observable' must be of entities in the world of science. And science, in giving content to the distinction, will reveal how much we believe when we accept it as empirically adequate.

[5] See Boyd 1973. We could say that Boyd's paper, like Lewis's (1970), provides independent evidence that empirical import cannot be syntactically isolated. But Boyd concludes also that there can be no distinction between truth and empirical adequacy for scientific theories.

A future unified science may detail the limits of observation exactly; meanwhile, extant theories are not silent on them. We saw Newton's delineation; for relativity theory, we have two revealing studies by Clark Glymour. The first (1972) shows that local (hence, I should think, measurable) quantities do not uniquely determine global features of space-time.[6] The second shows that these features also are not uniquely determined by structures each lying wholly within some absolute past cone—hence, I should think, by observable structures. It is the theory of relativity itself, after all, that places an *absolute* limit on the information we can gather, through the limiting function of the speed of light.

In the foundations of quantum mechanics, much more attention has been given to measurement. Much of the discussion is about necessary limitations: the role of noise in amplification, the distinction between macro- and micro-observables.[7] Yet we have no such clarity as Glymour gave us for relativity theory, concerning the extent to which macro-structure determines micro-structure. The debate over scientific realism may at least have the virtue of directing attention to such questions.

Science itself distinguishes the observable that it postulates from the whole it postulates. The distinction, being in part a function of the limits science discloses on human observation, is anthropocentric. But, since science places human observers among the physical systems it means to describe, it also gives itself the task of describing anthropocentric distinctions. It is in this way that even the scientific realist must observe a distinction between the phenomena and the trans-phenomenal in the scientific world picture.

VIII

I have laid some philosophical misfortunes at the door of a mistaken orientation toward syntax. The alternative is to say that theories are presented directly by describing their models. But does this really introduce a new element? When you give the theorems of T, you give the set of models of T—namely, all those structures which satisfy the theorems. And, if you give the models, you give at least the set of theorems of T—namely all those sentences which are satisfied in all the models. Does it not follow that we can as advantageously identify T with its theorems as with its models?

---

[6] Glymour 1972 is discussed in my 1972 (referred to therein by an earlier title).
[7] See e.g. Cartwright 1974 and references therein.

But there is an ellipsis in the argument. It is being assumed that there is a specific language L which is the one language that belongs to T. And indeed, the theorems of T in L determine and are determined by the set of model structures of L (i.e. structures in which L is interpreted) in which those theorems are satisfied. However, the assumption that there is a language L which plays this role for T places important restrictions on what the set of models of T can be like.

A theory provides, among other things, a specification (more or less complete) of the parts of its models that are to be direct images of the structures described in measurement reports. In the case of Newton's mechanics, I called those parts *motions*; in general, let us call them *empirical substructures*. The structures described in measurement reports we may continue to call *appearances*. A theory is *empirically adequate* exactly if all appearances are isomorphic to empirical substructures in at least one of its models. Theory T is *empirically no stronger* than theory T' exactly if, for each model M of T, there is a model M' of T' such that all empirical substructures of M are isomorphic to empirical substructures of M'. Theories T and T' are *empirically equivalent* exactly if neither is empirically stronger than the other. In that case, as an easy corollary, each is empirically adequate if and only if the other is.

In section V, I distinguished two kinds of extensions, the first a sort of victory and the second a sort of defeat. Let us call the first a *proper extension*: this simply narrows the class of models. We may call a theory *empirically minimal* if it is not empirically equivalent to any of its proper extensions. Glymour has convincingly argued, in the work cited above, that general relativity is not empirically minimal. The reason is, in my present terms, that only local properties of space-time enter the descriptions of the appearances, but models may differ in global properties. This is a further non-trivial example of empirical equivalence.

The second sort of extension I shall not try to define precisely. The idea is that models of the theory may differ in structure other than that of the empirical substructures. In that case the theory is not empirically minimal, but this may put it in the advantageous position of offering modelling possibilities when radically new phenomena come to light. An example may yet be offered by hidden-variable theories in quantum mechanics.[8]

In terms of the concepts now at our disposal, and the examples given, we can conclude that there are indeed non-trivial cases of empirical equivalence, non-uniqueness, and extendability, both proper and improper.

---

[8] See Gudder 1968 and van Fraassen 1973, sect. 3.

Such cases are now seen to be quite possible *even if the formulation of the theory has not a single term that cannot be called observationial, in some way*. And now it should be possible to state the issue of scientific realism, which concerns our epistemic attitude toward theories rather than their internal structure.

All the results of measurements are not in; they are never all in. Therefore we cannot know what all the appearances are. We can say that a theory is empirically adequate, that all the appearances will fit (the empirical substructures of) its models. Though we cannot know this with certainty, we can reasonably believe it. All this is the case not only for empirical adequacy but for truth as well. Yet there are two distinct epistemic attitudes that can be taken: we can *accept* a theory (accept it as empirically adequate) or *believe* the theory (believe it to be true). We can take it to be the aim of science to produce a literally true story about the world, or simply to produce accounts that are empirically adequate. This is the issue of scientific realism versus its (divided) opposition. The intrascientific distinction between the observable and the unobservable is an anthropocentric distinction; but it is reasonable that the distinction should be drawn in terms of *us*, when it is a question of *our* attitudes toward theories.[9]

REFERENCES

Boyd, R. N. (1973). 'Realism, Underdetermination, and a Causal Theory of Evidence.' *Noûs*, 7/1: 1–12.
Cajori, F. (ed.) (1960). *Sir Isaac Newton's Mathematical Principles of Natural Philosophy and his System of the World*. Berkeley: University of California Press.
Cartwright, N. D. (1974). 'Superposition and Macroscopic Observation.' *Synthese*, 29: 229–42.
Ellis, B. (1965). 'The Origins and Nature of Newton's Laws of Motion.' In R. Colodny (ed.), *Beyond the Edge of Certainty*, 29–68. Englewood Cliffs, NJ: Prentice-Hall.
Enriques, F. (1929). *Historical Development of Logic*, tr. J. Rosenthal. New York: Holt.
Glymour, C. (1972). 'Cosmology, Convention, and the Closed Universe.' *Synthese*, 24/1–2: 195–218.
Gudder, S. (1968). 'Hidden Variables in Quantum Mechanics Reconsidered.' *Review of Modern Physics*, 11: 229–31.

[9] This paper was presented in an APA symposium on scientific realism, 28 Dec. 1976. The research for this paper was supported by Canada Council Grant S74-0590. An earlier version was presented at the Western Division of the Canadian Philosophical Association (Calgary Oct. 1975).
I want to acknowledge my debt to Clark Glymour, Princeton University, for the challenge of his critiques of conventionalism in his dissertation and unpublished manuscripts.

Lewis, D. (1970). 'How to Define Theoretical Terms.' *Journal of Philosophy*, 67/13: 427–46.

Poincaré, H. (1958). *The Value of Science*, tr. B. Halsted. New York: Dover.

Simon, H. A. (1954). 'The Axiomatization of Classical Mechanics.' *Philosophy of Science*, 21/4: 340–3.

Van Fraassen, B. C. (1972). 'Earman on the Causal Theory of Time.' *Synthese*, 24/ 1–2: 87–95.

——(1973). 'Semantic Analysis of Quantum Logic.' In C. A. Hooker (ed.), *Contemporary Research in the Foundations and Philosophy of Quantum Theory*, 80–113. Dordrecht: Reidel.

——(1974). 'Theoretical Entities: The Five Ways.' *Philosophia*, 4: 95–109.

——(1975). 'Wilfrid Sellars on Scientific Realism.' *Dialogue*, 14/4: 606–16.

——(1977). 'The Only Necessity is Verbal Necessity.' *Journal of Philosophy*, 74/2: 71–85.

V

IS THE BEST GOOD ENOUGH?

PETER LIPTON

I

Is it ever rational to believe that a scientific theory is even approximately true? The evidence, however extensive, will not entail the theory it supports: the grounds for belief always remain inductive. Consequently, the realist who holds that there can be rational grounds for belief remains hostage to wholesale Humean scepticism about induction. The Humean argument has yet to be conclusively turned, but that project is not my present concern. Instead, I propose to consider intermediate forms of scepticism which attempt to show that, even if we grant scientists considerable inductive powers, rational belief in theory remains impossible. I will argue that some of these intermediate forms of scepticism are unstable, leading either back to radical Humean doubt or towards a moderate realism.

I will focus especially on the argument from 'underconsideration'. This argument has two premisses. The *ranking* premiss states that the testing of theories yields only a comparative warrant. Scientists can rank the competing theories they have generated with respect to likelihood of truth. The premiss grants that this process is known to be highly reliable, so that the more probable theory is always ranked ahead of a less probable competitor, and the truth, if it is among the theories generated, is likely to be ranked first, but the warrant remains comparative. In short, testing enables scientists to say which of the competing theories they have generated is likeliest to be correct, but does not itself reveal how likely the likeliest theory is. The second premiss of the argument, the *no-privilege* premiss, states that scientists have no reason to suppose that the process by which they generate theories for testing makes it likely that a true theory will be among those generated. It always remains possible that the truth lies

Reprinted from *Proceedings of the Aristotelian Society*, 93/2 (1993): 89–104, by courtesy of the Editor of the Aristotelian Society. © 1993.

rather among those theories nobody has considered, and there is no way of judging how likely this is. The conclusion of the argument is that, while the best of the generated theories may be true, scientists can never have good reason to believe this. They know which of the competing theories they have tested is likeliest to be true, but they have no way of judging the likelihood that any of those theories is true. On this view, to believe that the best available theory is true would be rather like believing that Jones will win the Olympics when all one knows is that he is the fastest miler in Britain.

The argument from underconsideration is clearly different from the radical Humean problem. The upshot of Hume's argument is that all non-deductive evaluation is unjustifiable. By contrast, the argument from underconsideration concedes very substantial inductive powers, by granting scientists the ability to rank reliably whichever competing theories they generate. Indeed these powers are almost certainly stronger than any sensible scientific realist would wish to claim. This only seems to strengthen the underconsideration argument, however, since it appears to show that even these generous powers cannot warrant belief in any scientific theory.

The argument from underconsideration is much more similar to an argument from underdetermination. According to one version of this argument, scientists are never entitled to believe a theory true because, however much supporting evidence that theory enjoys, there must exist competing theories, generated or not, that would be as well supported by the same evidence. This is an argument from inductive ties. Like the argument from underconsideration, it is an intermediate form of scepticism, since it grants scientists considerable inductive powers, but the two arguments are not the same. The argument from underconsideration does not exploit the existence of inductive ties, though it may allow them. On the other side, the argument from underdetermination does not assume any limitations on the scientists' powers of theory generation. Roughly speaking, whereas the underdetermination argument depends on the claim that scientists' inductive powers are excessively coarse-grained, the underconsideration argument focuses instead on the claim that they are only comparative. Moreover, the argument from underdetermination is in one sense more extreme than the argument from underconsideration. The underdetermination problem would remain even if scientists knew all the possible competing hypotheses and all possible data, whereas the underconsideration problem would disappear if they only knew all the competitors. Nevertheless, the similarities between the two arguments are substantial. Towards the end of this essay I will suggest that some of the objections to the argument from underconsideration also threaten the argument from underdetermination.

II

Bas van Fraassen has recently deployed the argument from underconsideration as part of his attack on inference to the best explanation (van Fraassen 1989: 142–50). It may therefore be useful to clarify the connections between inference to the best explanation, van Fraassen's constructive empiricism, and the argument from underconsideration, before turning to a critical assessment of the argument itself. (In the same work, van Fraassen also mounts a quite different Dutch-book argument against inference to the best explanation: although this argument seems to me flawed, I will not criticize it here.)

Constructive empiricism is the view that an aim of science is not truth across the board, but only empirical adequacy, the truth about all observable entities and processes. Inference to the best explanation is an account of inductive inference. Its governing idea is that explanatory considerations are a guide to inference. In its simplest form, the account claims that scientists judge that the theory which would, if correct, provide the best explanation of the available evidence is also the theory that is likeliest to be correct. What, then, is the relationship between constructive empiricism and inference to the best explanation? They are widely supposed to be incompatible. Certainly champions of inference to the best explanation tend to be realists, and van Fraassen develops his case against inference to the best explanation as part of his argument for constructive empiricism. But the two views are in fact compatible, since one may have a constructive empiricist version of inference to the best explanation. To do this requires only that we construe 'correct' as empirically adequate rather than as true, and that we allow that false theories may explain. I see no special barrier to the former, and van Fraassen's own account of explanation allows the latter.

Is inference to the best explanation especially vulnerable to the argument from underconsideration, more vulnerable than other accounts of inference? Van Fraassen's discussion gives this impression, since he deploys the argument specifically against this account. Moreover, inference to the best explanation does seem particularly vulnerable, since it seems that 'best theory' can only mean 'best of those theories that have been generated'. Here too, however, the appearances may be deceptive. The governing idea of inference to the best explanation, as I have said, is simply that explanatory considerations are a guide to inference, and this need not be articulated in a way that makes the evaluation comparative. That is, inference to the best explanation might be more accurately if less memorably called 'inference to the best explanation if the best is sufficiently good'. The story one tells about the explanatory virtues may make them either comparative or absolute. In spite of my best efforts, inference to the best explanation still remains at such an early stage of articulation that we cannot yet say with any confidence which version is the more promising (Lipton 1991).

Finally, what is the relationship between the argument from underconsideration and constructive empiricism? Once again, van Fraassen's discussion may give a false impression, since one might suppose that the argument forms part of his general case for favouring constructive empiricism over realism. Yet the argument seems clearly to work against the constructive empiricist too, if it works at all. The ranking premiss is no less plausible for evaluation with respect to empirical adequacy than it is with respect to truth, and so far as I can tell, van Fraassen himself accepts it. Similarly, we have a constructive empiricist version of the no-privilege premiss, to the effect that scientists have no reason to suppose that the means by which they generate theories for testing in itself makes it likely that an empirically adequate theory will be among those generated. Recalling that empirically adequate means adequate to everything observable and not just everything observed, this too will seem plausible to someone who endorses the realist version of the premiss, and, once again, van Fraassen appears to accept it. Constructive empiricism can itself be seen as being based in part on a kind of intermediate scepticism, to the effect that our inductive powers extend only to the limits of the observable, but this form of scepticism is orthogonal to the one articulated by the argument from underconsideration. The argument from underconsideration is thus especially salient neither as part of an argument for constructive empiricism nor as an argument against inference to the best explanation. Its interest is more general, since it applies to many models of theory evaluation and views of the proper aims of science.

III

Let us now consider the argument from underconsideration in its own right. There are several quick replies that immediately suggest themselves. We may simply deny either or both of the premisses. That is, we may insist either that scientists are capable of absolute and not only comparative evaluation, or that their methods of theory generation do sometimes provide them with good reason to believe that the truth lies somewhere among the theories they have generated. These responses may well be correct but, baldly asserted, they lead to an unsatisfying stand-off between those who believe in absolute evaluation or privilege and those who do not. Moreover, it seems undeniable that scientists' actual evaluative practices do include a strong comparative element, and one that is reflected in the most popular accounts of confirmation. Examples of this include the use of 'crucial' experiments and the distribution of prior probabilities between the available hypotheses (cf. Sklar 1985: 151–3).

Another obvious reply would be to concede some force to the sceptical argument but to deny that it undermines the rationality of science. As we have seen, the ranking assumption grants to the scientist considerable inductive powers. In particular, it allows that theory change is a truth-tropic process, so that later theories are always likelier to be correct than those they replace. Thus we might maintain that science is a progressive activity with respect to the aim of truth, even if scientists are never in a position rationally to assert that the best theory of the moment is actually true. (This view would be a kind of inductively boosted Popperianism.) More ambitiously, it might be argued that this truth-tropism even justifies scientific belief, by appealing to the scientist's desire to avoid ignorance as well as error. But the cost of these truth-tropic approaches is high, since there are various aspects of scientific activity that appear to require absolute evaluations. The most obvious of these is the practical application of science. In order to decide whether to administer a drug with known and serious side-effects, one needs to know how likely it is that the drug will effect a cure, not merely that it is likelier to do so than any other drug. Absolute evaluations also seem indispensable to 'pure' research, for example, to the decision whether it is better to develop the best available theory or to search for a better alternative.

IV

The quick replies I have mentioned are not to be disdained, but they concede too much to the argument from underconsideration. The nub of the argument is the claim that there is an unbridgeable gap between comparative and absolute evaluation. This gap is, however, only a plausible illusion.

The most straightforward way to eliminate a gap between comparative and absolute evaluation would be by exhaustion. If the scientist could generate all possible competitors in the relevant domain, and he knew this, then he would know that the truth is among them. Given the reliability that the ranking premiss grants, he would also know that the best of them is probably true. This brute-force solution, however, seems hopeless, since it takes a wildly exaggerated view of the scientist's abilities. Even granting that we can make sense of the notion of all possible competitors, how could the scientists possibly generate them all?

But collapsing the distinction between relative and absolute evaluation does not require exhaustion. The scientist does not have to know that he has considered all the competitors, only that one of those he has considered must be true, and for this he needs only a pair of contradictories, not the full set of contraries. It is enough that the scientist consider a theory and its negation, or the claim that a theory has a probability greater than one-half and the claim that it does not, or the claim that X is a cause of some phenomenon and the claim that it isn't, or the claim that an entity or process with specified properties exists or it doesn't. Since scientists are plainly capable of considering contradictories, and the ranking premiss entails that, when they do, they will be able to determine which is true, the argument from underconsideration fails.

The sceptic has two natural replies to this objection from contradictories. The first is to modify and restrict the ranking premiss, so that it concedes only the ability to rank contraries, not contradictories. But while the original ranking premiss is epistemically over-generous, it is not clearly over-generous in this way. Scientists do, for example, compare the likelihood of the existence and non-existence of entities, causes, and processes. So the sceptic would owe us some argument for denying that these comparisons yield reliable rankings while accepting the reliability of the comparisons of contraries. Moreover, it is not clear that the sceptic can even produce a coherent version of this restricted doctrine. The problem is that a pair of contraries entails a pair of contradictories. To give a trivial example, (P&Q) and ~P are contraries, but the first entails P, which is the contradictory of ~P. Indeed, all pairs of contraries entail a pair of contradictories, since one member of such a pair always entails the negation of the other. Suppose, then, that we wish to rank the contradictories T1 and ~T1. If we find a contrary to T1 (say T2) that is ranked ahead of T1, then ~T1 is ranked ahead of T1, since T2 entails ~T1. Alternatively, if we find a contrary to ~T1 (say T3) that is ranked ahead of ~T1, then T1 is ranked ahead of ~T1, since T3 entails T1. So it is not clear how to ban the ranking of contradictories while allowing the ranking of contraries.

The second natural reply the sceptic might make to the objection from contradictories would concede contradictory ranking. For in most cases, only one of a pair of contradictories would mark a significant scientific discovery. Not to put too fine a point on it, usually one member of the pair will be interesting, the other boring. Thus, if the pair consists of the claim that all planets move in ellipses and the claim that some don't, only the former claim is interesting. Consequently, the sceptic may concede contradictory ranking but maintain that the result will almost always be that the boring hypothesis is ranked above the interesting one. In short, he will claim that the best theory is almost always boring, so the scientist will almost never be in a position rationally to believe an interesting theory.

This concession substantially changes the character of the argument from underconsideration, however, and it is a change for the worse. Like most important sceptical arguments, what made the original argument from underconsideration interesting was the idea that it might rule out reasons for belief, even in cases where the belief is in fact true. (Compare Hume's general argument against induction: he does not argue that the future will not resemble the past, but that, even if it will, this is unknowable.) With the concession, however, the argument from underconsideration reduces to the claim that scientists are unlikely to think of the truth. The idea that scientists are only capable of relative evaluation no longer plays any role in the argument, since ranking of contradictory theories has collapsed the distinction between relative and absolute evaluation, and the argument reduces to the observation that scientists are unlikely to think of interesting truths, since they are hidden behind so many interesting falsehoods.

So the revised argument is substantially less interesting than the original. But the situation is worse than this. For scientists do in fact often rank interesting claims ahead of their boring contradictories. The revised argument thus faces a dilemma. If it continues to grant that scientists are reliable rankers, then the fact that interesting claims often come out ahead refutes the claim that scientists do not generate interesting truths. If, on the other hand, reliable ranking is now denied, we have lost all sense of the original strategy of showing how even granting scientists substantial inductive powers would be insufficient for rational belief.

V

The argument from underconsideration depends on a gap between relative and absolute evaluation. I have suggested that contradictory ranking closes that gap, and that the argument cannot be modified to reopen it without substantial loss of interest or force. What I will argue now is that the original argument is fundamentally flawed, even if we restrict our attention to the ranking of contraries. Given an uncontroversial feature of the way scientists rank theories, the two premisses of the argument from underconsideration are incompatible.

Faced with the problem of justifying scientists' methods of evaluation, one may forget how difficult it is even to describe them. This is exacerbated by the general tendency of epistemologists to focus on normative issues at the expense of descriptive ones. In any event, the descriptive project has turned out to be enormously challenging. As the paradox of the ravens and the new riddle of induction illustrate, most standard accounts are remarkably crude, leading to the absurd consequence that almost anything is evidence for anything else. Moreover, as one might expect for any inquiry at such a primitive stage, there is little consensus about even the most basic features that a correct account should include. At least one feature of theory evaluation, however, is almost universally acknowledged, not least among those eager to cast doubt on the possibility of rational belief in science. This is the essential role played by background theories: theories already accepted, if only tentatively, at the time when a new theory is tested. These theories influence the scientists' understanding of the instruments they use in their tests, the way the data themselves are to be characterized, the prior plausibility of the theory under test, and bearing of the data on the theory. (The importance of background theories and their bearing on realism have been emphasized by Richard Boyd in many articles (e.g. Boyd 1985).)

Scientists rank new theories with the help of background theories. According to the ranking premiss of the argument from underconsideration, this ranking is highly reliable. For this to be the case, however, it is not enough that the scientists have any old background theories on the books with which to make the evaluation: these theories must be *probably true*, or at least probably approximately true. If most of the background theories were not even approximately true, they would skew the ranking, leading in some cases to placing an improbable theory ahead of a probable competitor, and perhaps leading generally to true theories, when generated, being ranked below falsehoods. The ranking premiss would be violated. So the ranking premiss entails that the background is probably (approximately) true. The problem for the argument from underconsideration then appears on iteration. These background theories are themselves the result of prior generation and ranking, and the best of the theories now being ranked will form part of tomorrow's background. Hence, if scientists are highly reliable rankers, as the ranking premiss asserts, the highest-ranked theories have to be absolutely probable, not just more probable than the competition. This is only possible if the truth tends to lie among the candidate theories the scientists generate, which contradicts the no-privilege premiss. Hence, if the ranking premiss is true, the no-privilege assumption must be false, and the argument from underconsideration self-destructs.

Given the role of background in theory evaluation, the truth of the ranking premiss entails the falsity of the no-privilege premiss. Moreover, since the ranking premiss allows not only that scientists are reliable rankers, but also that they know this, the situation is even worse. If a scientist knows that her method of ranking is reliable, then she is also in a position to know that her background is probably true, which entails that she is capable of absolute evaluation. Thus, knowing that she is capable of comparative evaluation (and, perhaps, reading this essay) enables the scientist to know that she is capable of absolute evaluation, and the claim of the ranking premiss that the scientist knows that she is only capable of reliable comparative evaluation must be false.

So the initially plausible idea that scientists might be completely reliable rankers yet arbitrarily far from the truth is an illusion. Might the sceptic salvage his case by weakening the ranking premiss, as he was tempted to do in response to the objection from contradictories? I do not think this will help. Of course, if ranking were completely unreliable, the sceptic would have his conclusion, but this just takes us back to Hume. The point of the argument from underconsideration was rather to show that the sceptical conclusion follows, even if we grant scientists considerable inductive powers. So the sceptic needs to argue that, if scientists were moderately but not completely reliable rankers, the connection between the best theory and the truth would be severed. Our sceptic has not, however, provided us with such an argument, and there is good reason to believe that no sound argument of this sort exists. For the levels of reliability seems to depend, not just on the degree of reliability of the prior ranking of background theories, but on their verisimilitude.

To see this, suppose that reliability did depend only on the reliability of the prior ranking process by which the background theories were selected. Consider now two isolated scientific communities that are equally reliable rankers, but who in the past generated quite different ranges of candidate theories and so come to have quite different backgrounds. One community was lucky enough to generate true theories, while the other was uninspired enough to generate only wildly erroneous ones. If present reliability depended only on prior ranking, we would have to suppose that these two communities are now equally reliable rankers of new theories, which is clearly incorrect. The general point is that the level of reliability a background confers depends on its *content*, not just on the method by which it was generated, and that what matters about the content is, among other things, how close it is to the truth. Consequently, even though scientists are in fact only moderately reliable rankers, this does not sever the connection between relative and absolute evaluation. Even moderately reliable ranking is not compatible with the claim that scientists' methods may leave them with theories that are arbitrarily far from the truth. In other words, even moderately reliable ranking requires moderate privilege.

VI

The moral of the story is that certain kinds of intermediate scepticism, of which the argument from underconsideration is one example, are incoherent. Because of the role of background beliefs in theory evaluation, what we cannot have are inductive powers without inductive achievements. At the beginning of this essay, I distinguished the argument from underconsideration from the better-known argument from underdetermination. Having seen what is wrong with the former, however, it appears that a similar objection applies to the latter, and I want now briefly to suggest why this may be so.

The central claim of the argument from underdetermination is sometimes expressed by saying that, however much evidence is available, there will always be many theories that are incompatible with each other but compatible with the evidence. This version of underdetermination, however, ought not to bother the realist, since it amounts only to the truism that the connection between data and theory is and always will be inductive. Like the argument from underconsideration, an interesting version of the underdetermination argument is an intermediate scepticism which attempts to show that rational belief is impossible, even granting the scientist considerable inductive powers. Such a version of the underdetermination argument is an argument from inductive ties. The central claim is that, although some theories are better supported by the evidence than others, for any theory there must exist a competitor (which scientists may not have generated) that is equally well-supported, and this situation remains, however much evidence the scientist has. The argument thus allows that scientists are reliable rankers, but insists that the ranking will not discriminate between every pair of competing theories. In particular, it is claimed that this 'coarse' ranking is such that, however much evidence a scientist has, there exist competitors to the highest-ranked theory, which, if they were considered, would do just as well. Consequently, even if one of the theories the scientist has actually generated is ranked ahead of all the others, he has no reason to believe that this one is true, since he has only avoided a tie through lack of imagination.

Coarse ranking is not quite the same as moderately reliable ranking; the difference is roughly that between a degree of ignorance and a degree of error. Nevertheless, the objection from the background seems to apply here too. Even coarse ranking requires that most of the background theories be close to the truth. If they were not, we would have more than a failure of discrimination; we would have misranking. In other words, even if the underdeterminationist is correct in claiming that there will always in principle be ties for the best theory, this does not support the conclusion that the theories we accept may none the less be arbitrarily far from the truth. To get that conclusion would require abandoning the concession that coarse ranking is reliable, as far as it goes, and we are back to an undiscriminating Humean scepticism about non-demonstrative inference.

The underdeterminationist might respond to the objection from the background by 'going global'. He could take the unit of evaluation to be the full set of candidate beliefs a scientist might endorse at one time, rather than a particular theory. The point then would be that there are always ties for the best total set of beliefs. By in effect moving the background into the foreground, the objection from the background appears to be blocked, since what is evaluated now always includes the background and so cannot be relative to it. At the same time, the argument appears able to grant the scientist considerable inductive powers, since it can allow that not all consistent sets are equally likely or equally ranked, and that the higher-ranked sets are more likely to be correct than those ranked below them.

I do not think this response is successful. One difficulty is that the global version of the underdetermination argument does not respect the fact that scientists' actual methods of evaluation are local and relative to a (revisable) background. Consequently, although the argument makes a show of granting scientists some sort of inductive powers, it does not grant reliability to the methods scientists actually employ. The reliability of the actual practice of local ranking relative to background cannot be accommodated within this global version without ruining the argument, since, as we have seen, local reliability requires that the background be approximately true, the consequence the underdeterminationist is trying to avoid.

A further and related difficulty with the global argument is that it appears tacitly to rely on an untenable distinction between methodological principles and substantive belief. The argument suggests a picture in which the principles of evaluation float somehow above the varying global sets of candidate beliefs, permitting a common scheme of ranking to be applied to them all. Since beliefs about inductive support (such as what is evidence for what) are themselves part of the scientists' total set of beliefs, however, this picture is untenable. What are we to put in its place? If we could say that all the sets shared the same principles, this would perhaps suffice for the argument, but we cannot say this. The problem is not simply that these principles will in fact vary, but that the very notion of a division of the elements of a global set into those that are methodological principles and those that are substantive beliefs is suspect.

There are two reasons for this suspicion. Notice first that, unlike the principles of deductive inference, reliable principles of induction are *contingent*. (This is the source of the Humean problem.) A pattern of non-demonstrative inference that generally takes us from truth to truth in this world would perhaps not do so in some other possible worlds. Moreover, although this is perhaps somewhat more controversial, the principles also appear to be a posteriori. Given all this, it is difficult to see why they are not tantamount to substantive claims about our world. A second reason for treating the distinction between principle and belief with suspicion pushes from the other side, and appeals to the main theme of this essay: the role of the background in evaluation. Given this role, it is unclear on what basis one is to deny that the substantive theories in a global set are themselves also principles of evaluation.

The intermixture of methodological principle and substantive belief, in part a consequence of the essential role of background belief in theory evaluation, makes it unclear how even to formulate the global argument, and in what sense the argument grants the scientist reliable inductive powers. The intermixture of principle and belief is also perhaps the root cause of the failure of the two forms of intermediate scepticism I have considered in this essay: it explains why it proves so difficult to grant the reliability of evaluation without also admitting the correctness of theory.

VII

'Of course! Why didn't I think of that!' The distinction between being able to generate the right answer and seeing that an answer is correct once someone else has proposed it is depressingly familiar. Meno's slave-boy (or the reader of the dialogue) might never have thought of doubling the square by building on its diagonal, but he has no trouble seeing that the answer must be correct, once Socrates suggests it. And it is apparently no great leap from this truism that there is a distinction between generation and evaluation, between the context of discovery and the context of justification, to the thought that powers of evaluation are quite distinct from powers of generation, that we might be good at evaluating the answers we generate, yet bad at generating correct answers. Hence the thought that scientists might be reliable rankers of the conjectures they generate, yet hopeless at generating conjectures that are true, or close to the truth. Yet this thought turns out to be mistaken, falling to the elementary observation that the scientists' methods of evaluation work relative to a set of background beliefs, and that these methods cannot be even moderately reliable unless that background is close to the truth—hence the failure of the argument from underconsideration and of at least some versions of the argument from underdetermination. Of course, in particular cases scientists fail to generate answers that are even approximately correct, but the idea that they might always so fail even though their methods of evaluation are reliable is incoherent. Scientists who did not regularly generate approximately true theories could not be reliable rankers.

What is the bearing of these considerations on scientific realism? Both the arguments from underconsideration and from underdetermination threaten the view that scientists may have rational grounds for believing that a theory is at least approximately true; in so far as these arguments have been turned, the realist who believes in the existence of such grounds will be comforted. It is important, however, to emphasize what has not been shown. I have argued against certain intermediate scepticisms, but have suggested no answer here to wholesale inductive scepticism. Moreover, I have not tried to show that all intermediate arguments are untenable. In particular, I have not argued against van Fraassen's own intermediate position, which depends in part on the claim that scientists' inductive powers extend only to statements about the observable. On this view, what the scientist is entitled to believe is not that theories are true, but only that they are empirically adequate, that their observable consequences are true. The objection from the background would gain purchase here if it could be shown that, in order for scientists reliably to judge the empirical adequacy of their theories, their background theories must themselves be true, not just empirically adequate. I suspect that this is the case, but I have not attempted to argue it here.

The role of the background in theory evaluation is something of a two-edged sword. It defeats some sceptical arguments, but it also shows both that the realist must take care not to exaggerate scientists' inductive powers and how much even a modest realism entails. Even the most fervent realist cannot afford to claim that scientists are completely reliable rankers, since this would require that all their background beliefs be true, a hopelessly optimistic view, and one that is incompatible with the way the scientific background changes over time. The objection from the background drives home the point that realists must also be thoroughgoing fallibilists, allowing the possibility of error not just about theory and the data which support it, but also about the assessment of support itself. The argument of this paper also shows that the realist cannot maintain that scientists are good at evaluation while remaining agnostic about their ability to generate true theories. Reliable evaluation entails privilege, so the realist must say that scientists do have the knack of thinking of the truth. This ability is, from a certain point of view, somewhat surprising, but it remains in my view far more plausible than the extreme ignorance, substantive and methodological, that a coherent critic must embrace.[1]

REFERENCES

Boyd, R. (1985). 'Lex Orandi est Lex Credendi.' In P. Churchland and C. Hooker (eds.), Images of Science, 3–34. Chicago: University of Chicago Press.
Lipton, P. (1991). Inference to the Best Explanation. London: Routledge.
Sklar, L. (1985). Philosophy and Spacetime Physics. Berkeley: University of California Press.
Van Fraassen, B. (1989). Laws and Symmertry. Oxford: Oxford University Press.

[1] Paper originally presented at a meeting of the Aristotelian Society held in the Senior Common Room, Birkbeck College, London, on 18 Jan. 1993. I am grateful for the very helpful comments of Jeremy Butterfield, Gavin Ferris, Chris Daly, Michael Gaylard, Mary Hesse, Alex Oliver, and Tim Williamson.

VI

A CONFUTATION OF CONVERGENT REALISM

LARRY LAUDAN

> The positive argument for realism is that it is the only philosophy that doesn't make the success of science a miracle.
>
> Putnam, *Mathematics, Matter, and Method*

THE PROBLEM

It is becoming increasingly common to suggest that epistemological realism is an empirical hypothesis, grounded in, and to be authenticated by, its ability to explain the workings of science. A growing number of philosophers (including Boyd, W. Newton-Smith, A. Shimony, Putnam, and I. Niiniluoto) have argued that the theses of epistemic realism are open to empirical test.[1] The suggestion that epistemological doctrines have much the same empirical status as the sciences is a welcome one; for, whether it stands up to detailed scrutiny or not, this suggestion marks a significant facing-up by the philosophical community to one of the most neglected (and most notorious) problems of philosophy: the status of epistemological claims.

There are, however, potential hazards as well as advantages associated with the 'scientizing' of epistemology. Specifically, once one concedes that epistemic doctrines are to be tested in the court of experience, it is possible that one's favourite epistemic theories may be refuted rather than confirmed. It is the thesis of this paper that precisely such a fate afflicts a form of realism advocated by those who have been in the vanguard of the move to show that realism is supported by an empirical study of the development of science. Specifically, I will show that epistemic realism, at least in certain of its extant forms, is neither supported by, nor has it made sense of, much of the available historical evidence.

Reprinted from *Philosophy of Science*, 48 (1981): 19–48, by permission of the author and the Philosophy of Science Association.
[1] See Boyd 1973, Newton-Smith 1978, Putnam 1975, Niiniluoto 1977.

CONVERGENT REALISM

Like other philosophical 'isms', the term 'realism' covers a variety of sins. Many of these will not be at issue here. For instance, 'semantic realism' (in brief, the claim that all theories have truth-values and that some theories are true, although we know not which) is not in dispute. Nor shall I discuss what one might call 'intentional realism' (i.e. the view that theories are generally intended by their proponents to assert the existence of entities corresponding to the terms in those theories). What I will focus on instead are certain forms of epistemological realism. As Hilary Putnam has pointed out, although such realism has become increasingly fashionable, 'very little is said about what realism is'. The lack of specificity about what realism asserts makes it difficult to evaluate its claims, since many formulations are too vague and sketchy to get a grip on. At the same time, any efforts to formulate the realist position with greater precision lay the critic open to charges of attacking a straw man. In the course of this paper, I shall attribute several theses to the realists. Although there is probably no realist who subscribes to all of them, most of them have been defended by some self-avowed realist or other; taken together, they are perhaps closest to that version of realism advocated by Putnam, Boyd, and Newton-Smith. Although I believe the views I shall be discussing can be legitimately attributed to certain contemporary philosophers (and I will cite the textual evidence for such attributions), it is not crucial to my case that such attributions can be made. Nor will I claim to do justice to the complex epistemologies of those whose work I will criticize. Rather, my aim is to explore certain epistemic claims which those who are realists might be tempted (and in some cases have been tempted) to embrace. If my arguments are sound, we will discover that some of the most intuitively tempting versions of realism prove to be chimeras.

The form of realism I shall discuss involves variants of the following claims:

(R1) Scientific theories (at least in the 'mature' sciences) are typically approximately true, and more recent theories are closer to the truth than older theories in the same domain.

(R2) The observational and theoretical terms within the theories of a mature science genuinely refer (roughly, there are substances in the world that correspond to the ontologies presumed by our best theories).

(R3) Successive theories in any mature science will be such that they preserve the theoretical relations and the apparent referents of earlier theories; that is, earlier theories will be limiting cases of later theories.[2]

(R4) Acceptable new theories do and should explain why their predecessors were successful in so far as they were successful.

To these semantic, methodological, and epistemic theses is conjoined an important metaphilosophical claim about how realism is to be evaluated and assessed. Specifically, it is maintained that:

(R5) Theses (R1) to (R4) entail that ('mature') scientific theories should be successful; indeed, these theses constitute the best, if not the only, explanation for the success of science. The empirical success of science (in the sense of giving detailed explanations and accurate predictions) accordingly provides striking empirical confirmation for realism.

I shall call the position delineated by (R1) to (R5) *convergent epistemological realism*, or CER for short. Many recent proponents of CER maintain that (R1), (R2), (R3), and (R4) are empirical hypotheses that, via the linkages postulated in (R5), can be tested by an investigation of science itself. They propose two elaborate abductive arguments. The structure of the first (argument 1) which is germane to (R1), is something like this:

1. If scientific theories are approximately true, then they typically will be empirically successful.
2. If the central terms in scientific theories genuinely refer, then those theories generally will be empirically successful.
3. Scientific theories are empirically successful.
4. (Probably) theories are approximately true and their terms genuinely refer.

The structure of the second abductive argument (argument 2), which is relevant to (R3), is of slightly different form, specifically:

1. If the earlier theories in a 'mature' science are approximately true, and if the central terms of those theories genuinely refer, then later, more successful theories in the same science will preserve the earlier theories as limiting cases.

---

[2] Putnam (1978: 20–1), evidently following Boyd, sums up theses (R1) and (R3) in these words:

(1) Terms in a mature science typically *refer*.
(2) The laws of a theory belonging to a mature science are typically approximately true . . . I will only consider [new] theories . . . which have this property—[they] contain the [theoretical] laws of [their predecessors] as a limiting case.

2. Scientists seek to preserve earlier theories as limiting cases and generally succeed in doing so.
3. (Probably) earlier theories in a 'mature' science are approximately true and genuinely referential.

Taking the success of present and past theories as givens, proponents of CER claim that *if* CER were true, it would follow, as a matter of course, that science would be successful and progressive. Equally, they allege that if CER were false, the success of science would be 'miraculous' and without explanation.[3] Because (on their view) CER explains the fact that science is successful, the theses of CER are thereby confirmed by the success of science, and non-realist epistemologies are discredited by the latter's alleged inability to explain both the success of current theories and the progress which science historically exhibits.

As Putnam and certain others (e.g. Newton-Smith) see it, the fact that statements about reference (R2, R3) or about approximate truth (R1, R3) function in the explanation of a contingent state of affairs, establishes that 'the notions of "truth" and "reference" have a causal explanatory role in epistemology'.[4] In one fell swoop, both epistemology and semantics are 'naturalized' and, to top it all off, we get an explanation of the success of science thrown into the bargain!

The central question before us is whether the realist's assertions about the interrelations between truth, reference, and success are sound. It will be the burden of this paper to raise doubts about both arguments 1 and 2. Specifically, I will argue that four of the five premises of those abductions are either false or too ambiguous to be acceptable. I will also seek to show that, even if the premises were true, they would not warrant the conclusions that realists draw from them. The next three sections of this essay deal with the first abductive argument. Then I turn to the second.

REFERENCE AND SUCCESS

The specifically referential side of the empirical argument for realism has been developed chiefly by Putnam, who talks explicitly of reference rather more than most realists. However, reference is usually implicitly smuggled

---

[3] Putnam (1975) insists, for instance, that if the realist is wrong about theories being referential, then 'the success of science is a miracle' (i. 69).
[4] Boyd remarks: 'Scientific realism offers an *explanation* for the legitimacy of ontological commitment to theoretical entities' (Putnam 1978: 2). It allegedly does so by explaining why theories containing theoretical entities work so well: because such entities genuinely exist.

in, since most realists subscribe to the (ultimately referential) thesis that 'the world probably contains entities very like those postulated by our most successful theories'.

If (R2) is to fulfil Putnam's ambition that reference can explain the success of science, and that the success of science establishes the presumptive truth of (R2), it seems he must subscribe to claims similar to these:

(S1) The theories in the advanced or mature sciences are successful.

(S2) A theory whose central terms genuinely refer will be a successful theory.

(S3) If a theory is successful, we can reasonably infer that its central terms genuinely refer.

(S4) All the central terms in theories in the mature sciences do refer.

There are complex interconnections here. (S2) and (S4) explain (S1), while (S1) and (S3) provide the warrant for (S4). Reference explains success, and success warrants a presumption of reference. The arguments are plausible, given the premises. But there is the rub, for with the possible exception of (S1), none of the premises is acceptable.

The first and toughest problem involves getting clearer about the nature of that 'success' which realists are concerned to explain. Although Putnam, W. Sellars, and Boyd all take the success of certain sciences as a given, they say little about what this success amounts to. So far as I can see, they are working with a largely *pragmatic* notion to be couched in terms of a theory's workability or applicability. On this account, we would say that a theory is successful if it makes substantially correct predictions, if it leads to efficacious interventions in the natural order, and if it passes a battery of standard tests. One would like to be able to be more specific about what success amounts to, but the lack of a coherent theory of confirmation makes further specificity very difficult.

Moreover, the realist must be wary, at least for these purposes, of adopting too strict a notion of success, for a highly robust and stringent construal of 'success' would defeat the realist's purposes. What he wants to explain, after all, is why science in general has worked so well. If he were to adopt a very demanding characterization of success (such as those advocated by inductive logicians or Popperians), then it would probably turn out that science has been largely 'unsuccessful' (because it does not have high confirmation), and the realist's avowed explanandum would thus be a non-problem. Accordingly, I will assume that a theory is successful so long as it has worked well; that is, so long as it has functioned in a variety of explanatory contexts, has led to confirmed predictions, and has been of broad explanatory scope. As I understand the realist's position, his concern is to explain why certain theories have enjoyed this kind of success.

If we construe 'success' in this way, (S1) can be conceded. Whether one's criterion of success is broad explanatory scope, possession of a large number of confirming instances, or conferring manipulative or predictive control, it is clear that science, by and large, is a successful activity.

What about (S2)? I am not certain that any realist would or should endorse it, although it is a perfectly natural construal of the realist's claim that 'reference explains success'. The notion of reference that is involved here is highly complex and unsatisfactory in significant respects. Without endorsing it, I shall use it frequently in the ensuing discussion. The realist sense of reference is a rather liberal one, according to which the terms in a theory may be genuinely referring even if many of the claims the theory makes about the entities to which it refers are false. Provided that there are entities that 'approximately fit' a theory's description of them, Putnam's charitable account of reference allows us to say that the terms of a theory genuinely refer.[5] On this account (and these are Putnam's examples), Bohr's 'electron', Newton's 'mass', Mendel's 'gene', and Dalton's 'atom' are all referring terms, while 'phlogiston' and 'ether' are not (Putnam 1978: 20–2).

Are genuinely referential theories (i.e. theories whose central terms genuinely refer) invariably or even generally successful at the empirical level, as (S2) states? There is ample evidence that they are not. The chemical atomic theory in the eighteenth century was so remarkably unsuccessful that most chemists abandoned it in favour of a more phenomenological, elective affinity chemistry. The Proutian theory that the atoms of heavy elements are composed of hydrogen atoms had, through most of the nineteenth century, a strikingly unsuccessful career, confronted by a long string of apparent refutations. The Wegenerian theory that the continents are carried by large subterranean objects moving laterally across the earth's surface was, for some thirty years in the recent history of geology, a strikingly unsuccessful theory until, after major modifications, it became the geological orthodoxy of the 1960s and 1970s. Yet all of these theories postulated basic entities which (according to Putnam's 'principle of charity') genuinely exist.

The realist's claim that we should expect referring theories to be empirically successful is simply false. And, with a little reflection, we can see good reasons why it should be. To have a genuinely referring theory is to have

---

[5] Whether one utilizes Putnam's earlier or later versions of realism is irrelevant for the central arguments of this eassy.

a theory that 'cuts the world at its joints', a theory that postulates entities of a kind that really exist. But a genuinely referring theory need not be such that all—or even most—of the specific claims it makes about the properties of those entities and their modes of interaction are true. Thus, Dalton's theory makes many false claims about atoms; Bohr's early theory of the electron was similarly flawed in important respects. *Contra* (S2), genuinely referential theories need not be strikingly successful, since such theories may be 'massively false' (i.e. have far greater falsity content than truth content).

(S2) is so patently false that it is difficult to imagine that the realist need be committed to it. But what else will do? The (Putnamian) realist wants attributions of reference to a theory's terms to function in an explanation of that theory's success. The simplest and crudest way of doing that involves a claim like (S2). A less outrageous way of achieving the same end would involve the weaker:

(S2′) A theory whose terms refer will usually (but not always) be successful.

Isolated instances of referring but unsuccessful theories, sufficient to refute (S2), leave (S2′) unscathed. But, if we were to find a broad range of referring but unsuccessful theories, that would be evidence against (S2′). Such theories can be generated at will. For instance, take any set of terms which one believes to be genuinely referring. In any language rich enough to contain negation, it will be possible to construct indefinitely many unsuccessful theories, all of whose substantive terms are genuinely referring. Now, it is always open to the realist to claim that such 'theories' are not really theories at all, but mere conjunctions of isolated statements— lacking that sort of conceptual integration we associate with 'real' theories. Sadly, a parallel argument can be made for genuine theories. Consider, for instance, how many inadequate versions of the atomic theory there were in the 2,000 years of atomic speculating, before a genuinely successful theory emerged. Consider how many unsuccessful versions there were of the wave theory of light before the 1820s, when a successful wave theory first emerged. Kinetic theories of heat in the seventeenth and eighteenth centuries, and developmental theories of embryology before the late nineteenth century, sustain a similar story. (S2′), every bit as much as (S2), seems hard to reconcile with the historical record.

As Richard Burian has pointed out to me (personal communication), a realist might attempt to dispense with both of those theses and simply rest content with (S3) alone. Unlike (S2) and (S2′), (S3) is not open to the objection that referring theories are often unsuccessful, for it makes no claim that referring theories are always or generally successful. But (S3) has difficulties of its own. In the first place, it seems hard to square with the fact that the central terms of many relatively successful theories (e.g. ether theories or phlogistic theories) are evidently non-referring. I shall discuss this tension in detail below. More crucial for our purposes here is that (S3) is *not strong enough* to permit the realist to utilize reference to explain success. Unless genuineness of reference entails that all or most referring theories will be successful, then the fact that a theory's terms refer scarcely provides a convincing explanation of that theory's success. If, as (S3) allows, many (or even most) referring theories can be unsuccessful, how can the fact that a successful theory's terms refer be taken to explain why it is successful? (S3) may or may not be true; but in either case it arguably gives the realist no explanatory access to scientific success.

A more plausible construal of Putnam's claim that reference plays a role in explaining the success of science involves a rather more indirect argument. It might be said (and Putnam does say this much) that we can explain why a theory is successful by assuming that the theory is true or approximately true. Since a theory can only be true or nearly true (in any sense of those terms open to the realist) if its terms genuinely refer, it might be argued that reference gets into the act willy-nilly when we explain a theory's success in terms of its truth-like status. On this account, reference is piggybacked on approximate truth. The viability of this indirect approach is treated at length in the next section, so I will not discuss it here except to observe that if the only contact point between reference and success is provided through the medium of approximate truth, then the link between reference and success is extremely tenuous.

What about (S3), the realist's claim that success creates a rational presumption of reference? We have already seen that (S3) provides no explanation of the success of science, but does it have independent merits? The question specifically is whether the success of a theory provides a warrant for concluding that its central terms refer. In so far as this is, as certain realists suggest, an empirical question, it requires us to inquire whether past theories which have been successful are ones whose central terms genuinely referred (according to the realist's own account of reference).

A proper empirical test of this hypothesis would require an extensive sifting of the historical record that is not possible to perform here. What I can do is to mention a range of once successful, but (by present lights) non-referring, theories. A fuller list will come later, but for now we will focus on a whole family of related theories, namely, the subtle fluids and ethers of eighteenth- and nineteenth-century physics and chemistry.

Consider specifically the state of etherial theories in the 1830s and 1840s. The electrical fluid, a substance that was generally assumed to accumulate on the surface rather than to permeate the interstices of bodies, had been utilized to explain *inter alia* the attraction of oppositely charged bodies, the behaviour of the Leyden jar, the similarities between atmospheric and static electricity, and many phenomena of current electricity. Within chemistry and heat theory, the caloric ether had been widely utilized since H. Boerhaave (by, among others, A. L. Lavoisier, P. S. Laplace, J. Black, Count Rumford, J. Hutton, and H. Cavendish) to explain everything from the role of heat in chemical reactions to the conduction and radiation of heat and several standard problems of thermometry. Within the theory of light, the optical ether functioned centrally in explanations of reflection, refraction, interference, double refraction, diffraction, and polarization. (Of more than passing interest, optical ether theories had also made some very startling predictions, e.g. A. Fresnel's prediction of a bright spot at the centre of the shadow of a circular disc, a surprising prediction which, when tested, proved correct. If that does not count as empirical success, nothing does!) There were also gravitational (e.g. G. LeSage's) and physiological (e.g. D. Hartley's) ethers which enjoyed some measure of empirical success. It would be difficult to find a family of theories in this period as successful as ether theories; compared with them, nineteenth-century atomism (for instance), a genuinely referring theory (on realist accounts), was a dismal failure. Indeed, on any account of empirical success which I can conceive of, non-referring nineteenth-century ether theories were more successful than contemporary, referring atomic theories. In this connection, it is worth recalling the remark of the great theoretical physicist J. C. Maxwell to the effect that the ether was better confirmed than any other theoretical entity in natural philosophy.

What we are confronted with in nineteenth-century ether theories, then, is a wide variety of once successful theories, whose central explanatory concept Putnam (1978: 22) singles out as a prime example of a non-referring one. What are (referential) realists to make of this historical case? On the face of it, this case poses two rather different kinds of challenges to realism: first, it suggests that (S3) is a dubious piece of advice in that *there can be* (and have been) *highly successful theories some central terms of which are non-referring*; and second, it suggests that *the realist's claim that he can explain why science is successful is false at least in so far as a part of the historical success of science has been success exhibited by theories whose central terms did not refer.*

But perhaps I am being less than fair when I suggest that the realist is committed to the claim that *all* the central terms in a successful theory refer. It is possible that when Putnam, for instance, says that 'terms in a mature [or successful] science typically refer' (1978: 20), he only means to suggest that *some* terms in a successful theory or science genuinely refer. Such a claim is fully consistent with the fact that certain other terms (e.g. 'ether') in certain successful, mature sciences (e.g. nineteenth-century physics) are none the less non-referring. Put differently, the realist might argue that the success of a theory warrants the claim that at least some (but not necessarily all) of its central concepts refer.

Unfortunately, such a weakening of (S3) entails a theory of evidential support which can scarcely give comfort to the realist. After all, part of what separates the realist from the positivist is the former's belief that the evidence for a theory is evidence for *everything* the theory asserts. Where the stereotypical positivist argues that the evidence selectively confirms only the more 'observable' parts of a theory, the realist generally asserts (in the language of Boyd) that:

the sort of evidence which ordinarily counts in favor of the acceptance of a scientific law or theory is, ordinarily, evidence for the (at least approximate) truth of the law or theory as an account of the causal relations obtaining between the entities ['observational or theoretical'] quantified over in the law or theory in question. (Boyd 1973: 1)[6]

For realists such as Boyd, either all parts of a theory (both observational and non-observational) are confirmed by successful tests, or none are. In general, realists have been able to utilize various holistic arguments to insist that it is not merely the lower-level claims of a well-tested theory that are confirmed but its deep-structural assumptions as well. This tactic has been used to good effect by realists in establishing that inductive support 'flows upward' so as to authenticate the most 'theoretical' parts of our theories. Certain latter-day realists (e.g. Glymour) want to break out of this holist web and argue that certain components of theories can be 'directly' tested. This approach runs the very grave risk of undercutting what the realist desires most: a rationale for taking our deepest-structure theories seriously, and a justification for linking reference and success. After all, if the tests to which we subject our theories only test *portions* of those theories, then even highly successful theories may well have central terms that are non-referring and central tenets that, because untested, we have no grounds for believing to be approximately true. Under those circumstances, a theory might be highly successful and yet contain important constituents that were patently false. Such a state of affairs would

---

[6] See also Boyd 1973: 3: 'Experimental evidence for a theory is evidence for the truth of even its non-observational laws.' See also Sellars 1963: 97.

wreak havoc with the realist's presumption (thesis R1) that success betokens approximate truth. In short, to be less than a holist about theory testing is to put at risk precisely that predilection for deep-structure claims which motivates much of the realist enterprise.

There is, however, a rather more serious obstacle to this weakening of referential realism. It is true that by weakening (S3) to only certain terms in a theory, one would immunize it from certain obvious counter-examples. But such a manœuvre has debilitating consequences for other central realist theses. Consider the realist's thesis (R3) about the retentive character of inter-theory relations (discussed below in detail). The realist both recommends as a matter of policy and claims as a matter of fact that successful theories are (and should be) rationally replaced only by theories that preserve reference for the central terms of their successful predecessors. The rationale for the normative version of this retentionist doctrine is that the terms in the earlier theory, *because it was successful, must have been referential*, and thus a constraint on any successor to that theory is that reference should be retained for such terms. This makes sense just in case success provides a blanket warrant for presumption of reference. But if (S3) were weakened so as to say merely that it is reasonable to assume that *some* of the terms in a successful theory genuinely refer, then the realist would have no rationale for his retentive theses (variants of R3), which have been a central pillar of realism for several decades.[7]

Something apparently has to give. A version of (S3) strong enough to license (R3) seems incompatible with the fact that many successful theories contain non-referring central terms. But any weakening of (S3) dilutes the force of, and removes the rationale for, the realist's claims about convergence, retention, and correspondence in inter-theory relations.[8] If the realist once concedes that some unspecified set of the terms of a successful theory may well not refer, then his proposals for restricting 'the class of candidate theories' to those that retain reference for the prima-facie referring terms in earlier theories is without foundation (Putnam 1973: 22).

More generally, we seem forced to say that such linkages as there are

---

[7] A caveat is in order here. *Even* if all the central terms in some theory refer, it is not obvious that every rational successor to that theory must preserve all the referring terms of its predecessor. One can easily imagine circumstances when the new theory is preferable to the old one even though the range of application of the new theory is less broad than the old. When the range is so restricted, it may well be entirely appropriate to drop reference to some of the entities that figured in the earlier theory.

[8] For Putnam and Boyd both, 'it will be a constraint on T_2 [i.e. any new theory in a domain] . . . that T_2 must have this property, the property that *from its standpoint* one can assign referents to the terms of T_1 [i.e. an earlier theory in the same domain]' (Putnam 1976: 181). For Boyd, see 1973: 8: 'New theories should, prima facie, resemble current theories with respect to their accounts of causal relations among theoretical entities.'

between reference and success are rather murkier than Putnam's and Boyd's discussions would lead us to believe. If the realist is going to make his case for CER, it seems that it will have to hinge on approximate truth, (R1), rather than reference, (R2).

APPROXIMATE TRUTH AND SUCCESS:
THE DOWNWARD PATH

Ignoring the referential turn among certain recent realists, most realists continue to argue that, at bottom, epistemic realism is committed to the view that successful scientific theories, even if strictly false, are none the less 'approximately true' or 'close to the truth' or 'verisimilar'.[9] The claim generally amounts to this pair:

(T1) If a theory is approximately true, then it will be explanatorily successful.

(T2) If a theory is explanatorily successful, then it is probably approximately true.

What the realist would *like* to be able to say, of course, is:

(T1′) If a theory is true, then it will be successful.

(T1′) is attractive because self-evident. Most realists, however, balk at invoking (T1′) because they are (rightly) reluctant to believe that we can reasonably presume of any given scientific theory that it is true. If all the realist could explain was the success of theories that were true *simpliciter*, his explanatory repertoire would be acutely limited. As an attractive move in the direction of broader explanatory scope, (T1) is rather more appealing. After all, presumably many theories which we believe to be false (e.g. Newtonian mechanics, thermodynamics, wave optics) were—and still are—highly successful across a broad range of applications.

Perhaps, the realist evidently conjectures, we can find an *epistemic* account of that pragmatic success by assuming such theories to be 'approximately true'. But we must be wary of this potential sleight of hand. It may be that there is a connection between success and approximate truth; *but*

<hr>

[9] For just a small sampling of this view, consider the following: 'The claim of a realist ontology of science is that the only way of explaining why the models of science function so successfully . . . is that they approximate in some way the structure of the object' (McMullin 1970: 63–4); 'The continued success of confirmed theories can be *explained* by the hypothesis that they are in fact close to the truth' (Niiniluoto 1980: 448); and the claim that 'the laws of a theory belonging to a mature science are typically approximately *true* . . . [provides] an explanation of the behavior of scientists and the success of science' (Putnam 1978: 20–1). J. J. Smart, W. Sellars, and Newton-Smith, among others, share a similar view.

if there is such a connection it must be independently argued for. The acknowledgedly uncontroversial character of (T1′) must not be surreptitiously invoked—as it sometimes seems to be—in order to establish T1. When the antecedent of (T1′) is appropriately weakened by speaking of approximate truth, it is by no means clear that (T1) is sound.

Virtually all the proponents of epistemic realism take it as unproblematic that if a theory were approximately true, it would deductively follow that the theory would be a relatively successful predictor and explainer of observable phenomena. Unfortunately, few of the writers of whom I am aware have defined what it means for a statement or theory to be 'approximately true'. Accordingly, it is impossible to say whether the alleged entailment is genuine. This reservation is more than perfunctory. Indeed, on the best-known account of what it means for a theory to be approximately true, it does *not* follow that an approximately true theory will be explanatorily successful.

Suppose, for instance, that we were to say in a Popperian vein that a theory, T_1, is approximately true if its truth content is greater than its falsity content; that is,

$$Ct_T(T_1) \gg Ct_F(T_1),^{10}$$

where $Ct_T(T_1)$ is the cardinality of the set of true sentences entailed by T_1, and $Ct_F(T_1)$ is the cardinality of the set of false sentences entailed by T_1. When approximate truth is so construed, it does *not* logically follow that an arbitrarily selected class of a theory's entailments (viz. some of its observable consequences) will be true. Indeed, it is entirely conceivable that a theory might be approximately true in the indicated sense and yet be such that *all* of its consequences tested thus far are *false*.[11]

Some realists concede their failure to articulate a coherent notion of

---

[10] Although Popper is generally careful not to assert that actual historical theories exhibit ever-increasing truth content (for an exception, see his 1963: 220), other writers have been more bold. Thus, Newton-Smith (1981) writes that 'the historically generated sequence of theories of a mature science is a sequence in which succeeding theories are increasing in truth content without increasing in falsity content'.

[11] On the more technical side, Niiniluoto (1977) has shown that a theory's degree of corroboration co-varies with its 'estimated verisimilitude'. Roughly speaking, 'estimated truthlikeness' is a measure of how closely (the content of) a theory corresponds to *what we take to be* the best conceptual systems that, so far, we have been able to find (Niiniluoto 1980: 443ff.). If Niiniluoto's measures work, it follows from the above-mentioned co-variance that an empirically successful theory will have a high degree of estimated truthlikeness. But because estimated truthlikeness and genuine verisimilitude are not necessarily related (the former being parasitic on existing evidence and available conceptual systems), it is an open question whether, as Niiniluoto asserts, the continued success of highly confirmed theories can be *explained* by the hypothesis that they in fact are so close to the truth, at least in the relevant respects. Unless I am mistaken, this remark of his betrays a confusion between 'true verisimilitude' (to which we have no epistemic access) and 'estimated verisimilitude' (which is accessible but non-epistemic).

approximate truth or verisimilitude but insist that this failure in no way compromises the viability of (T1). Newton-Smith (1981), for instance, grants that 'no one has given a satisfactory analysis of the notion of verisimilitude', but insists that the concept can be legitimately invoked 'even if one cannot at the time give a philosophically satisfactory analysis of it'. He quite rightly points out that many scientific concepts were explanatorily useful long before a philosophically coherent analysis was given for them. But the analogy is unseemly, for what is being challenged is not whether the concept of approximate truth is philosophically rigorous, but rather whether it is even clear enough for us to ascertain whether it entails what it purportedly explains. Until someone provides a clearer analysis of approximate truth than is now available, it is not even clear whether truthlikeness would explain success, let alone whether, as Newton-Smith insists,[12] 'the concept of verisimilitude is *required* in order to give a satisfactory theoretical explanation of an aspect of the scientific enterprise'. If the realist would demystify the 'miraculousness' (Putnam) or the 'mysteriousness' (Newton-Smith) of the success of science, he needs more than a promissory note that somehow, someday, someone will show that approximately true theories must be successful theories.[13]

It is not clear whether there is some definition of approximate truth that does indeed entail that approximately true theories will be predictively successful (and yet still probably false).[14] What can be said is that, promises to the contrary notwithstanding, none of the proponents of realism has yet articulated a coherent account of approximate truth which entails that approximately true theories will, across the range where we can test them, be successful predictors. Further difficulties abound. Even if the realist had a semantically adequate characterization of approximate or partial truth, and even if that semantics entailed that most of the consequences of an approximately true theory would be true, he would still be without any criterion that would *epistemically* warrant the ascription of approximate truth to a theory. As it is, the realist seems to be long

[12] Newton-Smith (1981) claims that the increasing predictive success of science through time 'would be totally mystifying . . . if it were not for the fact that theories are capturing more and more truth about the world'.

[13] I must stress again that I am *not* denying that there *may* be a connection between approximate truth and predictive success. I am only observing that until the realists show us what that connection is, they should be more reticent than they are about claiming that realism can explain the success of science.

[14] A *non-realist* might argue that a theory is approximately true just in case all its *observable* consequences are true or within a specified interval from the true value. Theories that were 'approximately true' in this sense would indeed be demonstrably successful. But, the realist's (otherwise commendable) commitment to taking seriously the theoretical claims of a theory precludes him from utilizing any such construal of approximate truth, since he wants to say that the theoretical as well as the observational consequences are approximately true.

on intuitions and short on either a semantics or an epistemology of approximate truth.

These should be urgent items on the realists' agenda since, until we have a coherent account of what approximate truth is, central realist theses such as (R1), (T1), and (T2) are just so much mumbo-jumbo.

APPROXIMATE TRUTH AND SUCCESS: THE UPWARD PATH

Despite the doubts voiced in the previous section, let us grant for the sake of argument that if a theory is approximately true, then it will be successful. Even granting (T1), is there any plausibility to the suggestion of (T2) that explanatory success can be taken as a rational warrant for a judgement of approximate truth? The answer seems to be 'no'.

To see why, we need to explore briefly one of the connections between 'genuinely referring' and being 'approximately true'. However the latter is understood, I take it that *a realist would never want to say that a theory was approximately true if its central terms failed to refer*. If there were nothing like genes, then a genetic theory, no matter how well confirmed it was, would not be approximately true. If there were no entities similar to atoms, no atomic theory could be approximately true; if there were no subatomic particles, then no quantum theory of chemistry could be approximately true. In short, a necessary condition, especially for a scientific realist, for a theory being close to the truth is that its central explanatory terms genuinely refer. (An *instrumentalist*, of course, could countenance the weaker claim that a theory was approximately true so long as its directly testable consequences were close to the observable values. But as I argued above, the realist must take claims about approximate truth to refer alike to the observable and the deep-structural dimensions of a theory.)

Now, what the history of science offers us is a plethora of theories that were both successful and (so far as we can judge) non-referential with respect to many of their central explanatory concepts. I discussed earlier one specific family of theories that fits this description. Let me add a few more prominent examples to the list:

the crystalline spheres of ancient and medieval astronomy;
the humoral theory of medicine;
the effluvial theory of static electricity;
'catastrophist' geology, with its commitment to a universal (Noachian) deluge;
the phlogiston theory of chemistry;

the caloric theory of heat;
the vibratory theory of heat;
the vital force theories of physiology;
the electromagnetic ether;
the optical ether;
the theory of circular inertia;
theories of spontaneous generation.

This list, which could be extended *ad nauseam*, involves in every case a theory that was once successful and well-confirmed, but which contained central terms that (we now believe) were non-referring. Anyone who imagines that the theories that have been successful in the history of science have also been, with respect to their central concepts, genuinely referring theories has studied only the more whiggish versions of the history of science (i.e. the ones which recount only those past theories that are referentially similar to currently prevailing ones).

It is true that proponents of CER sometimes hedge their bets by suggesting that their analysis applies exclusively to 'the mature sciences' (e.g. Putnam and W. Krajewski). This distinction between mature and immature sciences proves convenient to the realist, since he can use it to dismiss any prima-facie counter-example to the empirical claims of CER on the grounds that the example is drawn from a so-called immature science. But this insulating manœuvre is unsatisfactory in two respects. In the first place, it runs the risk of making CER vacuous, since these authors generally define a mature science as one in which correspondence or limiting-case relations obtain invariably between any successive theories in the science once it has passed 'the threshold of maturity'. Krajewski (1977: 91) grants the tautological character of this view when he notes that 'the thesis that there is [correspondence] among successive theories becomes, indeed, analytical'. None the less, he believes that there is a version of the maturity thesis which 'may be and must be tested by the history of science'. That version is that 'every branch of science crosses at some period the threshold of maturity'. But the testability of this hypothesis is dubious at best. There is no historical observation that could conceivably *refute* it, since, even if we discovered that no sciences yet possessed 'corresponding' theories, it could be maintained that eventually every science will become corresponding. It is equally difficult to *confirm* it, since, even if we found a science in which corresponding relations existed between the latest theory and its predecessor, we would have no way of knowing whether that relation will continue to apply to subsequent changes of theory in that science. In other words, the much-vaunted empirical testability of realism is seriously compromised by limiting it to the mature sciences.

But there is a second unsavoury dimension to the restriction of CER to the mature sciences. The realists' avowed aim, after all, is to explain why science is successful: that is the 'miracle' they allege the non-realists leave unaccounted for. The fact of the matter is that parts of science, including many immature sciences, have been successful for a very long time; indeed, many of the theories I alluded to above were empirically successful by any criterion I can conceive of (including fertility, intuitively high confirmation, successful prediction, etc.). If the realist restricts himself to explaining only how the mature sciences work (and recall that very few sciences indeed are yet mature as the realist sees it), then he will have completely failed in his ambition to explain why science in general is successful. Moreover, several of the examples I have cited above come from the history of mathematical physics in the last century (e.g. electromagnetic and optical ethers), and, as Putnam himself concedes (1978: 21), '*physics* surely counts as a "mature" science if any science does'. Since realists would presumably insist that many of the central terms of the theories enumerated above do not genuinely refer, it follows that none of those theories could be approximately true (recalling that the former is a necessary condition for the latter). Accordingly, cases of this kind cast very grave doubts on the plausibility of (T2), that is, the claim that nothing succeeds like approximate truth.

I dare say that for every highly successful theory in the history of science that we now believe to be a genuinely referring theory, one could find half a dozen once successful theories that we now regard as substantially non-referring. If the proponents of CER are the empiricists they profess to be about matters epistemological, cases of this kind and this frequency should give them pause about the well-foundedness of (T2).

But we need not limit our counter-examples to non-referring theories. There were many theories in the past that (so far as we can tell) were both genuinely referring and empirically successful which we are none the less loath to regard as approximately true. Consider, for instance, virtually all those geological theories prior to the 1960s which denied any lateral motion to the continents. Such theories were, by any standard, highly successful (and apparently referential); but would anyone today be prepared to say that their constituent theoretical claims—committed as they were to laterally stable continents—are almost true? Is it not the fact of the matter that structural geology was a successful science between (say) 1920 and 1960, even though geologists were fundamentally mistaken about many (perhaps even most) of the basic mechanisms of tectonic construction? Or what about the chemical theories of the 1920s which assumed that the atomic nucleus was structurally homogeneous? Or those chemical and physical theories of the late nineteenth century which explicitly assumed that matter was neither created nor destroyed? I am aware of no sense of approximate truth (available to the realist) according to which such highly successful, but evidently false, theoretical assumptions could be regarded as 'truth-like'.

More generally, the realist needs a riposte to the prima-facie plausible claim that there is no necessary connection between increasing the accuracy of our deep-structural characterizations of nature and improvements at the level of phenomenological explanations, predictions, and manipulations. It *seems* entirely conceivable intuitively that the theoretical mechanisms of a new theory, T_2, might be closer to the mark than those of a rival T_1, and yet T_1 might be more accurate at the level of testable predictions. In the absence of an argument that greater correspondence at the level of unobservable claims is more likely than not to reveal itself in greater accuracy at the experimental level, one is obliged to say that the realist's hunch that increasing deep-structural fidelity must manifest itself pragmatically in the form of heightened experimental accuracy has yet to be made cogent. (Equally problematic, of course, is the inverse argument to the effect that increasing experimental accuracy betokens greater truthlikeness at the level of theoretical, i.e. deep-structural, commitments.)

CONFUSIONS ABOUT CONVERGENCE AND RETENTION

Thus far, I have discussed only the static or synchronic versions of CER, versions that make absolute rather than relative judgements about truthlikeness. Of equal appeal have been those variants of CER that invoke a notion of what is variously called 'convergence', 'correspondence', or 'cumulation'. Proponents of the diachronic version of CER supplement the arguments discussed above ((S1)–(S4) and (T1)–(T2)) with an additional set. They tend to be of this form:

(C1) If earlier theories in a scientific domain are successful and thereby, according to realist principles (e.g. (S3) above), approximately true, then scientists should only accept later theories that retain appropriate portions of earlier theories.

(C2) As a matter of fact, scientists do adopt the strategy of (C1) and manage to produce new, more successful theories in the process.

(C3) The 'fact' that scientists succeed at retaining appropriate parts of earlier theories in more successful successors shows that the earlier theories did genuinely refer and that they are approximately true. And thus, the strategy propounded in (C1) is sound.[15]

Perhaps the prevailing view here is Putnam's and (implicitly) Popper's, according to which rationally warranted successor theories in a mature science must contain reference to the entities apparently referred to in the predecessor theory (since, by hypothesis, the terms in the earlier theory refer), and also contain the theoretical laws and mechanisms of the predecessor theory as limiting cases. As Putnam tells us (1978: 21), a realist should insist that *any* viable successor to an old theory T_1 must 'contain the laws of T as a limiting case'. John Watkins, a like-minded convergentist, puts the point this way (1978: 376–7):

It typically happens in the history of science that when some hitherto dominant theory T is superseded by T^1, T^1 is in the relation of correspondence to T [i.e. T is a 'limiting case' of T^1].

Numerous recent philosophers of science have subscribed to a similar view, including Popper, H. R. Post, Krajewski, and N. Koertge.[16]

This form of retention is not the only one to have been widely discussed. Indeed, realists have espoused a wide variety of claims about what is or should be retained in the transition from a once successful predecessor (T_1) to a successor theory (T_2). Among the more important forms of realist retention are the following cases: (1) T_2 entails T_1 (W. Whewell); (2) T_2 retains the true consequences or truth content of T_1 (Popper); (3) T_2 retains the 'confirmed' portions of T_1 (Post, Koertge); (4) T_2 preserves the theoretical laws and mechanisms of T_1 (Boyd, McMullin, Putnam); (5) T_2 preserves T_1 as a limiting case (J. Watkins, Putnam, Krajewski); (6) T_2 explains why T_1 succeeded in so far as T_1 succeeded (W. Sellars); and (7) T_2 retains reference for the central terrns of T_1 (Putnam, Boyd).

---

[15] If this argument, which I attribute to the realists, seems a bit murky, I challenge any reader to find a more clear-cut one in the literature! Overt formulations of this position can be found in Putnam, Boyd, and Newton-Smith.

[16] Popper: 'A theory which has been well-corroborated can only be superseded by one . . . [which] *contains* the old well-corroborated theory—or at least a good approximation to it' (1959: 276). Post: 'I shall even claim that, as a matter of empirical historical fact, [successor] theories [have] always explained the *whole* of [the well-confirmed part of their predecessors]' (1971: 229). Koertge: 'Nearly all parts of successive theories in the history of science stand in a correspondence relation and . . . where there is no correspondence to begin with, the new theory will be developed in such a way that it comes more nearly into correspondence with the old' (1973: 175–7).

Among other authors who have defended a similar view, one should mention Fine 1967: 231ff.; Kordig 1971; Margenau 1950; and Sklar 1967.

The question before us is whether, when retention is understood in *any* of these senses, the realist's theses about convergence and retention are correct.

Do Scientists Adopt the Retentionist Strategy of CER?

One part of the convergent realist's argument is a claim to the effect that scientists generally adopt the strategy of seeking to preserve earlier theories in later ones. As Putnam puts it: 'preserving the mechanisms of the earlier theory as often as possible, which is what scientists try to do. . . . That scientists try to do this . . . is a fact, and that this strategy has led to important discoveries . . . is also a fact' (p. 20).[17]

In a similar vein, I. Szumilewicz (although not stressing realism) insists that many eminent scientists made it a main heuristic requirement of their research programmes that a new theory stand in a relation of 'correspondence' with the theory it supersedes (1977: 348). If Putnam and the other retentionists are right about the strategy that most scientists have adopted, we should expect to find the historical literature of science abundantly provided with proofs that later theories do indeed contain earlier theories as limiting cases, or outright rejections of later theories that fail to contain earlier theories. Except on rare occasions (coming primarily from the history of mechanics), one finds neither of these concerns prominent in the literature of science. For instance, to the best of my knowledge, literally no one criticized the wave theory of light because it did not preserve the theoretical mechanisms of the earlier corpuscular theory; no one faulted C. Lyell's uniformitarian geology on the grounds that it dispensed with several causal processes prominent in catastrophist geology; Darwin's theory was not criticized by most geologists for its failure to retain many of the mechanisms of Lamarckian evolutionary theory.

For all the realist's confident claims about the prevalence of a retentionist strategy in the sciences, I am aware of *no* historical studies that would sustain as a *general* claim his hypothesis about the evaluative strategies utilized in science. Moreover, in so far as Putnam and Boyd claim to be offering 'an explanation of the retentionist behavior of scientists', they have the wrong explanandum (Putnam 1978: 21); for if there is any widespread strategy in science, it is one that says, 'accept an empirically successful theory, regardless of whether it contains the theoretical laws and

---

[17] Putnam fails to point out that it is also a fact that many scientists do *not* seek to preserve earlier theoretical mechanisms and that theories which have not preserved earlier theoretical mechanisms (whether the germ theory of disease, plate tectonics, or wave optics) have led to important discoveries is also a fact.

mechanisms of its predecessors'.[18] Indeed, one could take a leaf from the realist's (C2) and claim that the success of the strategy of assuming that earlier theories do not generally refer shows that it is true that earlier theories generally do not!

(One might note in passing how often, and on what evidence, realists imagine that they are speaking for the scientific majority. Putnam, for instance, claims that 'realism is, so to speak, "science's philosophy of science"' and that 'science taken at "face value" *implies* realism'.[19] C. A. Hooker (1974) insists that to be a realist is to take science 'seriously', as if to suggest that conventionalists, instrumentalists, and positivists such as Duhem, Poincaré, and Mach did not take science seriously. The willingness of some realists to attribute realist strategies to working scientists—on the strength of virtually no empirical research into the principles which *in fact* have governed scientific practice—raises doubts about the seriousness of their avowed commitment to the empirical character of epistemic claims.)

Do Later Theories Preserve the Mechanisms, Models, and Laws of Earlier Theories?

Regardless of the explicit strategies to which scientists have subscribed, are Putnam and several other retentionists right that later theories 'typically' entail earlier theories, and that 'earlier theories are, very often, limiting cases of later theories?' (Putnam 1978: 20, 123). Unfortunately, answering this question is difficult, since 'typically' is one of those weasel words that allows for much hedging. I shall assume that Putnam and Watkins mean that 'most of the time (or perhaps in most of the important cases) successor theories contain predecessor theories as limiting cases'. So construed, the claim is patently false. Copernican astronomy did not retain all the key mechanisms of Ptolemaic astronomy (e.g. motion along an equant); Newton's physics did not retain all (or even most of) the theoretical laws of Cartesian mechanics, astronomy, and optics; Franklin's electrical theory did not contain its predecessor (J. A. Nollet's) as a limiting case. Relativistic physics did not retain the ether, nor the mechanisms associated with it; statistical mechanics does not incorporate all the mechanisms of thermodynamics; modern genetics does not have Darwinian pangenesis as a limiting case; the wave theory of light did not appropriate the mecha-

---

[18] I have written a book about this strategy. See Laudan 1977.
[19] After the epistemological and methodological battles about science during the last 300 years, it should be fairly clear that science, taken at its face value, implies no particular epistemology.

nisms of corpuscular optics; modern embryology incorporates few of the mechanisms prominent in classical embryological theory. As I have shown elsewhere (Laudan 1976), loss occurs at virtually every level: the confirmed predictions of earlier theories are sometimes not explained by later ones; even the 'observable' laws explained by earlier theories are not always retained, not even as limiting cases; theoretical processes and mechanisms of earlier theories are, as frequently as not, treated as flotsam.

The point is that some of the most important theoretical innovations have been due to a willingness of scientists to violate the cumulationist or retentionist constraint which realists enjoin 'mature' scientists to follow.

There is a deep reason why the convergent realist is wrong about these matters. It has to do, in part, with the role of ontological frameworks in science and with the nature of limiting-case relations. As scientists use the term 'limiting case', T_1 can be a limiting case of T_2 only if *all* the variables (observable and theoretical) assigned a value in T_1 are assigned a value by T_2 and if the values assigned to every variable of T_1 are the same as, or very close to, the values T_2 assigns to the corresponding variable when certain initial and boundary conditions—consistent with T_2[20]—are specified. This seems to require that T_1 can be a limiting case of T_2 only if *all* the entities postulated by T_1 occur in the ontology of T_2. Whenever there is a change of ontology accompanying a theory transition such that T_2 (when conjoined with suitable initial and boundary conditions) fails to capture the ontology of T_1, then T_1 cannot be a limiting case of T_2. Even where the ontologies of T_1 and T_2 overlap appropriately (i.e. where T_2's ontology embraces all of T_1's), T_1 is a limiting case of T_2 only if *all* the laws of T_1 can be derived from T_2, given appropriate limiting conditions. It is important to stress that *both* these conditions (among others) must be satisfied before one theory can be a limiting case of another. Where 'closet' positivists might be content with capturing only the formal mathematical relations or only the observable consequences of T_1 within a successor T_2, any genuine realist must insist that T_1's underlying ontology is

---

[20] This matter of limiting conditions consistent with the 'reducing' theory is curious. Some of the best-known expositions of limiting-case relations depend (as Krajewski has observed) upon showing an earlier theory to be a limiting case of a later theory only by adopting limiting assumptions *explicitly denied by the later theory*. For instance, several standard textbook discussions present (a portion of) classical mechanics as a limiting case of special relativity, provided c approaches infinity. But special relativity is committed to the claim that c is a constant. Is there not something suspicious about a 'derivation' of T_1 from a T_2 which essentially involves an assumption inconsistent with T_2? If T_2 is correct, then it forbids the adoption of a premiss commonly used to derive T_1 as a limiting case. (It should be noted that most such proofs can be reformulated unobjectionably; e.g. in the relativity case, by letting $v \rightarrow 0$ rather than $v \rightarrow \infty$.)

preserved in T_2's, *for it is that ontology above all which he alleges to be approximately true.*

Too often, philosophers (and physicists) infer the existence of a limiting-case relation between T_1 and T_2 on substantially less than this. For instance, many writers have claimed one theory to be a limiting case of another when only some, but not all, of the laws of the former are derivable from the latter. In other cases, one theory has been said to be a limiting case of a successor when the mathematical laws of the former find homologies in the latter but where the former's ontology is not fully extractable from the latter's.

Consider one prominent example which has often been misdescribed: namely, the transition from the classical ether theory to relativistic and quantum mechanics. It can, of course, be shown that *some* laws of classical mechanics are limiting cases of relativistic mechanics. But there are other laws and general assertions made by the classical theory (e.g. claims about the density and fine structure of the ether, general laws about the character of the interaction between ether and matter, models and mechanisms detailing the compressibility of the ether) which could not conceivably be limiting cases of modern mechanics. The reason is a simple one: a theory cannot assign values to a variable that does not occur in that theory's language (or, more colloquially, it cannot assign properties to entities whose existence it does not countenance). Classical ether physics contained a number of postulated mechanisms for dealing *inter alia* with the transmission of light through the ether. Such mechanisms could not possibly appear in a successor theory like the special theory of relativity which denies the very existence of an etherial medium and which accomplishes the explanatory tasks performed by the ether via very different mechanisms.

Nineteenth-century mathematical physics is replete with similar examples of evidently successful mathematical theories which, because some of their variables refer to entities whose existence we now deny, cannot be shown to be limiting cases of our physics. As Adolf Grünbaum (1976) has cogently argued, when we are confronted with two incompatible theories, T_1 and T_2, such that T_2 does not 'contain' all of T_1's ontology, then not all the mechanisms and theoretical laws of T_1 that involve those entities of T_1 not postulated by T_2 can possibly be retained—not even as limiting cases—in T_2. This result is of some significance. What little plausibility convergent or retentive realism has enjoyed derives from the presumption that it correctly describes the relationship between classical and post-classical mechanics and gravitational theory. Once we see that even in this prima-facie most favourable case for the realist (where some of the

laws of the predecessor theory are genuinely limiting cases of the successor), changing ontologies or conceptual frameworks make it impossible to
capture many of the central theoretical laws and mechanisms postulated by
the earlier theory, then we can see how misleading Putnam's claim (1978:
20) is that 'what scientists try to do [is to preserve] the *mechanisms* of the
earlier theory as often as possible—or to show that they are "limiting
cases" of new mechanisms'. Where the mechanisms of the earlier theory
involve entities whose existence the later theory denies, no scientist does
(or should) feel any compunction about wholesale repudiation of the
earlier mechanisms.

But even where there is no change in basic ontology, many theories
(even in mature sciences such as physics) fail to retain all the explanatory
successes of their predecessors. It is well known that statistical mechanics
has yet to capture the irreversibility of macro-thermodynamics as a genuine limiting case. Classical continuum mechanics has not yet been reduced
to quantum mechanics or relativity. Contemporary field theory has yet to
replicate the classical thesis that physical laws are invariant under reflection in space. If scientists had accepted the realist's constraint (viz. that
new theories must have old theories as limiting cases), neither relativity
nor statistical mechanics would have been viewed as viable theories. It has
been said before, but it needs to be reiterated over and again: *a proof of the
existence of limiting relations between selected components of two theories is
a far cry from a systematic proof that one theory is a limiting case of the
other*. Even if classical and modern physics stood to one another in the
manner in which the convergent realist erroneously imagines they do, his
hasty generalization that theory successions in all the advanced sciences
show limiting-case relations is patently false.[21] But, as this discussion
shows, not even the realist's paradigm case will sustain the claims he is apt
to make about it.

What this analysis underscores is just how reactionary many forms of
convergent epistemological realism are. If one took seriously CER's advice to reject any new theory that did not capture existing mature theories
as referential and existing laws and mechanisms as approximately auth-
entic, then any prospect for deep-structure, ontological changes in our
theories would be foreclosed. Equally outlawed would be any significant
repudiation of our theoretical models. In spite of his commitment to the
growth of knowledge, the realist would unwittingly freeze science in its

---

[21] As Mario Bunge has cogently put it (1970: 309–10): 'The popular view on inter-theory
relations . . . that every new theory includes (as regards its extension) its predecessors . . . is
philosophically superficial . . . and it is false as a historical hypothesis concerning the advancement of science.'

present state by forcing all future theories to accommodate the ontology of contemporary (mature) science and by foreclosing the possibility that some future generation may come to the conclusion that some (or even most) of the central terms in our best theories are no more referential than was 'natural place', 'phlogiston', 'ether', or 'caloric'.

Could Theories Converge in Ways Required by the Realist?

These violations, within genuine science, of the sorts of continuity usually required by realists are by themselves sufficient to show that the form of scientific growth which the convergent realist takes as his explicandum is often absent, even in the mature sciences. But we can move beyond these specific cases to show in principle that the kind of cumulation demanded by the realist is unattainable. Specifically, by drawing on some results established by David Miller and others, the following can be shown:

1. The familiar requirement that a successor theory, T_2, must both preserve as true the true consequences of its predecessor, T_1, and explain T_1's anomalies is contradictory.
2. If a new theory, T_2, involves a change in the ontology or conceptual framework of a predecessor, T_1, then T_1 will have true and determinate consequences not possessed by T_2.
3. If two theories, T_1 and T_2, disagree, then each will have true and determinate consequences not exhibited by the other.

To establish these conclusions, one needs to utilize a 'syntactic' view of theories according to which a theory is a conjunction of statements and its consequences are defined à la Tarski in terms of content classes. Needless to say, this is neither the only, nor necessarily the best, way of thinking about theories; but it happens to be the way in which most philosophers who argue for convergence and retention (e.g. Popper, Watkins, Post, Krajewski, and Niiniluoto) tend to conceive of theories. What can be said is that if one utilizes the Tarskian conception of a theory's content and its consequences as they do, then the familiar convergentist theses alluded to in conclusions 1–3 make no sense.

The elementary but devastating consequences of Miller's analysis establish that virtually any effort to link scientific progress or growth to the wholesale retention of a predecessor theory's Tarskian content *or* logical consequences *or* true consequences *or* observed consequences *or* confirmed consequences is evidently doomed. Realists have not only got their history wrong is so far as they imagine that cumulative retention has prevailed in science, but we can also see that, given their views on what should be retained through theory change, history could not possibly have been the way their models require it to be. The realist's strictures on cumulativity are as ill-advised normatively as they are false historically.

Along with many other realists, Putnam has claimed (1978: 37) that 'the mature sciences do converge . . . and that convergence has great explanatory value for the theory of science'. As this section should show, Putnam and his fellow realists are arguably wrong on *both* counts. Popper once remarked that 'no theory of knowledge should attempt to explain why we are successful in our attempts to explain things'. Such a dogma is too strong. But what the foregoing analysis shows is that an occupational hazard of recent epistemology is imagining that convincing explanations of our success come easily or cheaply.

Should New Theories Explain Why their Predecessors were Successful?

An apparently more modest realism than that outlined above is familiar in the form of the requirement (R4) often attributed to Sellars—that every satisfactory new theory must be able to explain why its predecessor was successful in so far as it was successful. On this view, viable new theories need not preserve all the content of their predecessors, nor capture those predecessors as limiting cases. Rather, it is simply insisted that a viable new theory, T_N, must explain why, when we conceive of the world according to the old theory T_O, there is a range of cases where our T_O-guided expectations are correct or approximately correct.

What are we to make of this requirement? In the first place, it is clearly *gratuitous*. If T_N has more confirmed consequences (and greater conceptual simplicity) than T_O, then T_N is preferable to T_O even if T_N cannot explain why T_O is successful. Contrariwise, if T_N has fewer confirmed consequences than T_O, then T_N cannot be rationally preferred to T_O even if T_N explains why T_O is successful. In short, a theory's ability to explain why a rival is successful is neither a necessary nor a sufficient condition for saying that it is better than its rival.

Other difficulties likewise confront the claim that new theories should explain why their predecessors were successful. Chief among them is the ambiguity of the notion itself. One way to show that an older theory, T_O, was successful is to show that it shares many confirmed consequences with a newer theory, T_N, which is highly successful. But this is not an 'explanation' that a scientific realist could accept, since it makes no reference to, and thus does not depend upon, an epistemic assessment of either T_O or T_N. (After all, an instrumentalist could quite happily grant that if T_N 'saves the phenomena', then T_O—in so far as some of its observable consequences overlap with or are experimentally indistinguishable from those of T_N—should also succeed at saving the phenomena.)

The intuition being traded on in this persuasive account is that the pragmatic success of a new theory, combined with a partial comparison of the respective consequences of the new theory and its predecessor, will sometimes put us in a position to say when the older theory worked and when it failed. But such comparisons as can be made in this manner do not involve *epistemic* appraisals of either the new or the old theory *qua* theories. Accordingly, the possibility of such comparisons provides no argument for epistemic realism.

What the realist apparently needs is an epistemically robust sense of 'explaining the success of a predecessor'. Such an epistemic characterization would presumably begin with the claim that T_N, the new theory, was approximately true and would proceed to show that the 'observable' claims of its predecessor, T_O, deviated only slightly from (some of) the 'observable' consequences of T_N. It would then be alleged that the (presumed) approximate truth of T_N and the partially overlapping consequences of T_O and T_N jointly explained why T_O was successful in so far as it was successful. But this is a *non sequitur*. As I have shown above, the fact that T_N is approximately true does not even explain why it is successful; how, under those circumstances, can the approximate truth of T_N explain why some theory different from T_N is successful? Whatever the nature of the relations between T_N and T_O (entailment, limiting case, etc.), the epistemic ascription of approximate truth to either T_O or T_N (or both) apparently leaves untouched questions of how successful T_O or T_N are.

The idea that new theories should explain why older theories were successful (in so far as they were) originally arose as a rival to the 'levels' picture of explanation according to which new theories fully explained, because they entailed, their predecessors. It is clearly an improvement over the levels picture (for it does recognize that later theories generally do not entail their predecessors). But when it is formulated as a general thesis about inter-theory relations, designed to buttress a realist epistemology, it is difficult to see how this position avoids difficulties similar to those discussed in earlier sections.

The Realists' Ultimate Petitio Principii

It is time to step back a moment from the details of the realists' argument to look at its general strategy. Fundamentally, the realist is utilizing, as we have seen, an abductive inference which proceeds from the success of science to the conclusion that science is approximately true, verisimilar, or referential (or any combination of these). This argument is meant to show the sceptic that theories are not ill-gotten, the positivist that theories are not reducible to their observational consequences, and the pragmatist that classical epistemic categories (e.g. 'truth' and 'falsehood') are a relevant part of metascientific discourse.

It is little short of remarkable that realists would imagine that their critics would find the argument compelling. As I have shown elsewhere (Laudan 1978), ever since antiquity, critics of epistemic realism have based their scepticism upon a deep-rooted conviction that the fallacy of affirming the consequent is indeed fallacious. When E. Sextus or R. Bellarmine or Hume doubted that certain theories which saved the phenomena were warrantable as true, their doubts were based on a belief that the exhibition that a theory had some true consequences left entirely open the truth-status of the theory. Indeed, many non-realists have been non-realists precisely because they believed that false theories, as well as true ones, could have true consequences.

Now enters the new breed of realist (e.g. Putnam, Boyd, Newton-Smith) who wants to argue that epistemic realism can reasonably be presumed true by virtue of the fact that it has true consequences. But this is a monumental case of begging the question. The non-realist refuses to admit that a *scientific* theory can be warrantedly judged to be true simply because it has some true consequences. Such non-realists are not likely to be impressed by the claim that a philosophical theory such as realism can be warranted as true because it arguably has some true consequences. If non-realists are chary about first-order abductions to avowedly true conclusions, they are not likely to be impressed by second-order abductions, particularly when, as I have tried to show above, the premisses and conclusions are so indeterminate.

But, it might be argued, the realist is not out to convert the intransigent sceptic or the determined instrumentalist.[22] Perhaps, he is seeking to show that realism can be tested like any other scientific hypothesis, and that realism is at least as well confirmed as some of our best scientific theories. Such an analysis, however plausible initially, will not stand up to scrutiny. I am aware of no realist who is willing to say that a scientific theory can be reasonably presumed to be true or even regarded as well-confirmed just on the strength of the fact that its thus-far-tested consequences are true. Realists have long been in the forefront of those opposed to *ad hoc* and

---

[22] I owe the suggestion of this realist response to Andrew Lugg.

post hoc theories. Before a realist accepts a scientific hypothesis, he generally wants to know whether it has explained or predicted more than it was
devised to explain, whether it has been subjected to a battery of controlled
tests, whether it has successfully made novel predictions, and whether
there is independent evidence for it.

What, then, of realism itself as a 'scientific' hypothesis?[23] Even if we
grant (contrary to what I argued in the section on 'Approximate Truth and
Success') that realism entails and thus explains the success of science,
ought that (hypothetical) success warrant, by the realist's own construal of
scientific acceptability, the acceptance of realism? Since realism was devised to explain the success of science, it remains purely *ad hoc* with
respect to that success. If realism has made some novel predictions or has
been subjected to carefully controlled tests, one does not learn about it
from the literature of contemporary realism. At the risk of apparent inconsistency, the realist repudiates the instrumentalist's view that saving the
phenomena is a significant form of evidential support while endorsing
realism itself on the transparently instrumentalist grounds that it is confirmed by those very facts it was invented to explain. No proponent of
realism has sought to show that realism satisfies those stringent empirical
demands which the realist himself minimally insists on when appraising
scientific theories. The latter-day realist often calls realism a 'scientific'
or 'well-tested' hypothesis, but seems curiously reluctant to subject it
to those controls he otherwise takes to be a *sine qua non* for empirical
well-foundedness.

CONCLUSION

The arguments and cases discussed above seem to warrant the following
conclusions:

---

[23] I find Putnam's views on the 'empirical' or 'scientific' character of realism rather perplexing. At some points, he seems to suggest that realism is both empirical and scientific. Thus, he
writes (1976: 178): 'If realism is an explanation of this fact [namely, that science is successful],
realism must itself be an over-arching scientific *hypothesis*'. Since Putnam clearly maintains
the antecedent, he seems committed to the consequent. Elsewhere (1978: 37), he refers to
certain realist tenets as being 'our highest level empirical generalizations about knowledge'. He
says, moreover, that realism 'could be false', and that 'facts are relevant to its support (or to
criticize it)' (ibid. 78–9). None the less, for reasons he has not made clear, Putnam wants to deny
that realism is either scientific or a hypothesis (ibid. 79). How realism can consist of doctrines
which explain facts about the world, are empirical generalizations about knowledge, and can
be confirmed or falsified by evidence, and *yet* be neither scientific nor hypothetical, is left
opaque.

1. The fact that a theory's central terms refer does not entail that it will be successful, and a theory's success is no warrant for the claim that all or most of its central terms refer.
2. The notion of approximate truth is presently too vague to permit one to judge whether a theory consisting entirely of approximately true laws would be empirically successful. What is clear is that a theory may be empirically successful even if it is not approximately true.
3. Realists have no explanation whatever for the fact that many theories which are not approximately true and whose 'theoretical' terms seemingly do not refer are, none the less, often successful.
4. The convergentist's assertion that scientists in a 'mature' discipline usually preserve, or seek to preserve, the laws and mechanisms of earlier theories in later ones is probably false. His assertion that when such laws are preserved in a successful successor, we can explain the success of the latter by virtue of the truthlikeness of the preserved laws and mechanisms suffers from all the defects noted above confronting approximate truth.
5. Even if it could be shown that referring theories and approximately true theories would be successful, the realist's argument that successful theories are approximately true and genuinely referential takes for granted precisely what the non-realist denies, namely, that explanatory success betokens truth.
6. It is not clear that acceptable theories either *do* or *should* explain why their predecessors succeeded or failed. If a theory is better supported than its rivals and predecessors, then it is not epistemically decisive whether it explains why its rivals worked.
7. If a theory has once been falsified, it is unreasonable to expect that a successor should retain either all of its content *or* its confirmed consequences or its theoretical mechanisms.
8. Nowhere has the realist established, except by fiat, that non-realist epistemologists lack the resources to explain the success of science.

With these specific conclusions in mind, we can proceed to a more global one: it is not yet established—Putnam, Newton-Smith, and Boyd notwithstanding—that realism can explain any part of the successes of science. What is very clear is that realism *cannot*, even by its own lights, explain the success of those many theories whose central terms have evidently not referred and whose theoretical laws and mechanisms were not approximately true. The inescapable conclusion is that in so far as many realists are concerned with explaining how science works and with assessing the adequacy of their epistemology by that standard, they have, thus far, failed to explain very much. Their epistemology is confronted by anomalies that seem beyond its resources to grapple with.

It is important to guard against a possible misinterpretation of this essay. Nothing I have said here refutes the possibility, in principle, of a realistic epistemology of science. To conclude as much would be to fall prey to the same inferential prematurity with which many realists have rejected in principle the possibility of explaining science in a non-realist way. My task here is, rather, that of reminding ourselves that there is a difference between wanting to believe something and having good reasons for believing it. All of us would like realism to be true; we would like to think that science works because it has got a grip on how things really are. But such claims have yet to be made out. Given the present state of the art, it can only be wish fulfilment that gives rise to the claim that realism, and realism alone, explains why science works.

REFERENCES

Boyd, R. (1973). 'Realism, Underdetermination, and a Causal Theory of Evidence.' *Noûs*, 7: 1–12.
Bunge, M. (1970). 'Problems Concerning Intertheory Relations.' In P. Weingartner and G. Zecha (eds.), *Induction, Physics and Ethics*, 309–10. Dordrecht: Reidel.
Fine, A. (1967). 'Consistency, Derivability and Scientific Change.' *Journal of Philosophy*, 64: 231–40.
Grünbaum, A. (1976). 'Can a Theory Answer More Questions than One of its Rivals?' *British Journal for Philosophy of Science*, 27: 1–23.
Hooker, C. (1974). 'Systematic Realism.' *Synthese*, 26: 409–97.
Koertge, N. (1973). 'Theory Change in Science.' In G. Pearce and P. Maynard (eds.), *Conceptual Change*, 167–98. Dordrecht: Reidel.
Kordig, C. (1971). 'Scientific Transitions, Meaning Invariance, and Derivability.' *Southern Journal of Philosophy*, ••: 119–25.
Krajewski, W. (1977). *Correspondence Principle and Growth of Science*. Dordrecht: Reidel.
Laudan, L. (1976). 'Two Dogmas of Methodology.' *Philosophy of Science*, 43: 467–72.
——(1977). *Progress and its Problems*. Berkeley: University of California Press.
——(1978). 'Ex-Huming Hacking.' *Erkenntnis*, 13: 417–35.
Margenau, H. (1950). *The Nature of Physical Reality*. New York: McGraw-Hill.
McMullin, E. (1970). 'The History and Philosophy of Science: A Taxonomy.' In R. Stuewer (ed.), *Minnesota Studies in the Philosophy of Science*, v. 12–67,
Newton-Smith, W. (1978). 'The Underdetermination of Theories by Data.' *Aristotelian Society*, suppl. vol. 52: 71–91.
——(1981): 'In Defense of Truth.' In O. Jensen and R. Harré (eds.), *The Rationality of Science*, ch. 8. London: Routledge.
Niiniluoto, I. (1977). 'On the Truthlikeness of Generalizations.' In R. Butts and J. Hintikka (eds.), *Basic Problems in Methodology and Linguistics*, 121–47. Dordrecht: Reidel.

—— (1980). 'Scientific Progress.' *Synthese*, 45: 427–62.
Popper, K. (1959). *The Logic of Scientific Discovery*. New York: Basic Books.
—— (1963). *Conjectures and Refutations*. London: Routledge and Kegan Paul.
Post, H. R. (1971). 'Correspondence, Invariance and Heuristics: In Praise of Conservative Induction.' *Studies in the History and Philosophy of Science*, 2.
Putnam, H. (1975). *Mathematics, Matter, and Method*, 2 vols. Cambridge: Cambridge University Press.
—— (1976). 'What is Realism?' *Proceedings of the Aristotelian Society*, 76: 177–94.
—— (1978). *Meaning and the Moral Sciences*. London: Routledge and Kegan Paul.
Sellars, W. (1963). *Science, Perception and Reality*. New York: Humanities Press.
Sklar, L. (1967). 'Types of Inter-Theoretic Reductions.' *British Journal for Philosophy of Science*, 18: 190–224.
Szumilewicz, I. (1977). 'Incommensurability and the Rationality of the Development of Science.' *British Journal for Philosophy of Science*, 28: 345–50.
Watkins, J. (1978). 'Corroboration and the Problem of Content-Comparison.' In G. Radnitzky and G. Anderson (eds.), *Progress and Rationality in Science*, 339–78. Dordrecht: Reidel.

VII

STRUCTURAL REALISM: THE BEST OF BOTH WORLDS?

JOHN WORRALL

Presently accepted physical theories postulate a curved space-time structure, fundamental particles, and forces of various sorts. What we can know for sure on the basis of observation, at most, are only facts about the motions of macrosopic bodies, the tracks that appear in cloud chambers in certain circumstances, and so on. Most of the content of the basic theories in physics goes 'beyond' the 'directly observational'—no matter how liberal a conception of the 'directly observational' is adopted. What is the status of the genuinely theoretical, observation-transcendent content of our presently accepted theories? Most of us unreflectingly take it that the statements in this observation-transcendent part of the theory are attempted descriptions of a reality lying 'behind' the observable phenomena: that those theories really do straightforwardly assert that space-time is curved in the presence of matter, that electrons, neutrinos, and the rest exist and do various funny things. Furthermore, most of us unreflectingly take it that the enormous empirical success of these theories legitimizes the assumption that these descriptions of an underlying reality are accurate, or at any rate 'essentially' or 'approximately' accurate. The main problem of scientific realism, as I understand it, is that of whether or not there are, after reflection, good reasons for holding this view that most of us unreflectingly adopt.

There are, of course, several anti-realist alternatives on offer. The most widely canvassed is some version of the pragmatic or instrumentalist view that the observation-transcendent content of our theories is not in fact, and despite its apparent logical form, *descriptive* at all, but instead simply 'scaffolding' for the experimental laws. Theories are codification schemes; theoretical terms like 'electron' or 'weak force' or whatever should not be taken as even intended to refer to real entities, but instead as fictional names introduced simply to order our experimental laws into a

Reprinted from *Dialectica*, 43/1–2 (1989): 99–124, by permission of Societe Dialectica.

system.[1] A more recent anti-realist position—that of van Fraassen—holds that theortical terms *do*, at any rate purportedly, refer to real entities (and are not, e.g., simply shorthand for complex observational terms), but that there is no reason to assume that even our best theories are true nor even 'approximately' true, nor even that the *aim* of science is to produce true theories; instead, acceptance of a theory should be taken to involve *only* the claim that the theory is 'empirically adequate', that it 'saves the phenomena'.[2]

I can find no essentially new arguments in the recent discussions (see Worrall 1982). What seem to me the two most persuasive arguments are very old—both are certainly to be found in Poincaré and in Duhem. The main interest in the problem of scientific realism lies, I think, in the fact that these two persuasive arguments appear to pull in opposite directions; one seems to speak for realism and the other against it; yet a really satisfactory position would need to have both arguments on its side. The concern of the present paper is to investigate this tension between the two arguments and to *suggest* (no more) that an old and hitherto mostly neglected position may offer the best hope of reconciling the two.

The main argument (perhaps 'consideration' would be more accurate) likely to incline someone towards realism I shall call the 'no miracles' argument (although a version of it is nowadays sometimes called the 'ultimate argument' for realism—see Musgrave 1988). Very roughly, this argument goes as follows. It would be a miracle, a coincidence on a near-cosmic scale, if a theory made as many correct empirical predictions as, say, the general theory of relativity or the photon theory of light *without* what that theory says about the fundamental structure of the universe being correct or 'essentially' or 'basically' correct. But we shouldn't accept miracles, not at any rate if there is a non-miraculous alternative. If what these theories say is going on 'behind' the phenomena is indeed true or 'approximately true', then it is no wonder that they get the phenomena right. So it is plausible to conclude that presently accepted theories are indeed 'essentially' correct. After all, quantum theory gets certain phenomena, like the Lamb shift, correct to, whatever it is, 6 or 7 decimal places; in the view of some scientists, only a philosopher, overly impressed by merely logical possibilities, could believe that this is compatible with the

---

[1] According to a famous remark of Quine's, for instance, the theoretical entities involved in current science (like electrons) are epistemologically on a par with the Greek gods—both are convenient fictions introduced in the attempt to order (empirical) reality (Quine 1953: 44).

[2] Van Fraassen 1980. Van Fraassen calls his position 'constructive empiricism' (for criticisms see my 1983 review of his book).

quantum theory's failing to be a fundamentally correct description of reality.

Notice, by the way, that the argument requires the empirical success of a theory to be understood in a particular way. Not every empirical consequence that a theory has and which happens to be correct will give intuitive support for the idea that the theory must somehow or other have latched on to the 'universal blueprint'. Specifically, any empirical consequence which was *written into* the theory *post hoc* must be excluded. Clearly it is no miracle if a theory gets right a fact which was already known to hold and which the theory had been engineered to yield. If the fact concerned was used in the construction of the theory—for example, to fix the value of some initially free parameter—then the theory was *bound* to get that fact right. (On the other hand, if the experimental result concerned was *not* written into the theory, then the support it lends to the idea that the theory is 'essentially correct' is surely independent of whether or not the result was already known when the theory was formulated.[3])

This intuitive 'no miracles' argument can be made more precise in various ways—all of them problematic and some of them more problematic than others. It is, for instance, often run as a form of an 'inference to the best explanation' or Peircian 'abduction'.[4] But, as Laudan (1981) and Fine (ch. I, this volume) have both pointed out, since the anti-realist is precisely in the business of denying the validity of inference to the best explanation in science, he is hardly likely to allow it in philosophy as a means of arguing for realism. Perhaps more importantly, and despite the attempts of some philosophers to claim scientific status for realism itself on the basis of its explanatory power,[5] there is surely a crucial, pragmatic difference between a good scientific explanation and the 'explanation' afforded by the thesis of realism for the success of our present theories. A requirement for a convincing *scientific* explanation is *independent* testability—Newton's explanation of the planetary orbits is such a good one because the theory yields so much else that is testable besides the orbits: the oblateness of the earth, return of Halley's comet, and so on. Yet in the

---

[3] I have argued for this notion of empirical support and against the idea that temporal novelty is epistemically important in my 1985 paper and especially in my 1989a paper, which includes a detailed historical analysis of the famous 'white spot' episode involving Fresnel and Poisson, and often taken to provide support for the 'novel facts count more' thesis.

[4] This form of the argument is strongly criticized by Larry Laudan (1981). Strong and cogent reservations about the alleged explanation that realism supplies of science's success were also expressed in Howard Stein's paper delivered at the Neuchâtel conference.

[5] This position seems to have been held by Boyd, Niiniluoto, and others. It is disowned by Putnam (1978): 'I think that realism is like an empirical hypothesis in that it could be false, and that facts are relevant to its support (or to criticizing it); but that doesn't mean that realism is scientific (in any standard sense of "scientific"), or that realism is a hypothesis.'

case of realism's 'explanation' of the success of our current theories there can of course be no question of any independent tests. Scientific realism can surely not be *inferred* in any interesting sense from science's success. The 'no miracles' argument cannot *establish* scientific realism; the claim is only that, other things being equal, a theory's predictive success supplies a prima-facie plausibility argument in favour of its somehow or other having latched on to the truth.

Certainly the psychological force of the argument was sharply felt even by the philosophers who are usually (though, as we shall see, mistakenly) regarded as the great champions of anti-realism or instrumentalism: Pierre Duhem and Henri Poincaré. Here, for example, is Duhem:

The highest test, therefore of [a theory] is to ask it to indicate in advance things which the future alone will reveal. And when the experiment is made and confirms the predictions obtained from our theory, we feel strengthened in our conviction that the relations established by our reason among abstract notions only correspond to the relations among things. (1906: 28)

And here Poincaré:

Have we any right, for instance, to enunciate Newton's law? No doubt numerous observations are in agreement with it, but is not that a simple fact of chance? And how do we know besides, that this law which has been true for so many generations, will not be untrue in the next? To this objection the only answer you can give is: It is very improbable. (1905: 186)

So the 'no miracles' argument is likely, I think, to incline a common-sensical sort of person towards some sort of scientific realist view. But he is likely to feel those realist sentiments evaporating if he takes a close look at the *history* of science and particularly at the phenomenon of *scientific revolutions.*

Newton's theory of gravitation had a stunning range of predictive success: the perturbations of the planetary orbits away from strict Keplerian ellipses, the variation of gravity over the earth's surface, the return of Halley's comet, precession of the equinoxes, and so on. Newtonians even turned empirical difficulties (like the initially anomalous motion of Uranus) into major successes (in this case the prediction of a hitherto unknown trans-Uranian planet subsequently christened Neptune). Physicists were wont to bemoan their fate at having been born after Newton—there was only one truth to be discovered about the 'system of the world', and Newton had discovered it. Certainly an apparently hugely convincing 'no miracles' argument could be—and was—constructed on behalf of Newton's theory. It would be a miracle if Newton's theory got the planetary motions so precisely right, that it should be right about Neptune and about Halley's comet, that the motion of incredibly distant objects like some binary stars should be in accordance with the theory—it would be a miracle if this were true but the theory is not. However, as we all know, Newton's theory was rejected in favour of Einstein's in the early twentieth century.

This would pose no problem if Einstein's theory were simply an extension of Newton's; that is, if it simply incorporated Newton's theory as a special case, and then went on to say more. In general, if the development of science were cumulative, then scientific change would pose no problem either for the realist or for his 'no miracles' argument. The reason why Newton's theory got so many of the phenomena correct could still be that it was true, just not the whole truth.

Unfortunately Einstein's theory is not simply an extension of Newton's. The two theories are logically inconsistent: if Einstein's theory is true, then Newton's has to be false.[6] This is of course accepted by all present-day realists. The recognition that scientific progress, even in the 'successful', 'mature' sciences, is not strictly cumulative at the theoretical level, but instead involves at least an element of modification and revision is the reason why no present-day realist would claim that we have grounds for holding that presently accepted theories are *true*. Instead, the claim is only that we have grounds for holding that those theories are 'approximately' or 'essentially' true. This last claim might be called 'modified realism'. I shall, for convenience, drop the 'modified' in what follows, but it should be understood that my realists claim only that we have grounds for holding that our present theories in *mature* science are *approximately* true.

[6] Professor Agazzi in his paper at Neuchâtel took the view that Newtonian physics remains true of objects in its intended domain and that quantum and relativistic physics are true of objects in quite different domains. But this position is surely untenable. Newton's theory was not about (its 'intended referent' was not) macroscopic objects moving with velocities small compared with that of light. It was about *all* material objects moving with *any* velocity you like. And that theory is *wrong* (or so we now think), gloriously wrong, of course, but wrong. Moreover, it isn't even, strictly speaking, right about certain bodies and certain motions and 'only' wrong when we are dealing with microscopic objects or bodies moving at very high velocities. If relativity and quantum theory are correct, then Newton's theory's predictions about the motion of *any* body, even the most macroscopic and slowest-moving, are *strictly* false. It's just that their falsity lies well within experimental error. That is, what is true is that Newton's theory is an *empirically* faultless approximation for a whole range of cases. It's also true, as Agazzi claimed, that scientists and engineers still often see themselves as applying classical physics in a whole range of areas. But the only clear-sighted account of what they are doing is, I think, that they are *in fact* applying the best-supported theories available to them—viz. quantum mechanics and relativity theory. It's just that they know that these theories themselves entail the meta-result that, for their purposes (of sending rockets to the moon or whatever), it will make no practical difference to act *as if* they were applying classical physics, and indeed that it would be from the empirical point of view a waste of effort to apply the mathematically more demanding newer theories only for that sophistication to become entirely irrelevant when it comes to *empirical* application.

This realist claim involves two terms which are notoriously difficult to clarify. I shall propose my own rough characterization of the 'mature' sciences shortly. As for 'approximately true', well-known and major difficulties stand in the way of any attempt at precise analysis. Indeed, various attempted characterizations (such as Popper's in terms of 'increasing verisimilitude') have turned out to be formally deeply flawed.[7] Although we do often operate quite happily at the intuitive level with the notion of approximate truth, it is surely not the sort of notion which can happily be left as a primitive. For one thing: if the notion is going to do the work that realists need it to do, it is going to have to be transitive. Realists need to claim that although some presently accepted theory may subsequently be modified and replaced, it will still look 'approximately true' in the light not just of the next theory which supersedes it, but also in the light of the theory (if any) which supersedes the theory which supersedes it, etc. But is transitivity a property that the notion of approximate truth possesses even intuitively?

But there is anyway an important *prior* question here: that of whether or not, talking intuitively, in advance of formal analysis, the history of science (or some selected part of it) speaks in favour of successive scientific theories being increasingly good 'approximations to the truth'. This clearly depends on just *how radical* theory change has standardly been in science. Again, of course, we are dealing in unfortunately vague terms. But surely the realist claim—that we have grounds for holding that our present theories are approximately true—is plausible only to the extent that it seems reasonable to say that Newton's theory, for example, 'approximates' Einstein's, and that, in general, the development of science (at any rate the development of successful, 'mature' science) has been 'essentially' cumulative, that the deposed theories themselves, *and not just their successful empirical consequences*, have generally lived on, albeit in 'modified form', after the 'revolution'. If, on the contrary, theory change in science has often involved 'radical' shifts—something like the complete rejection of the genuinely theoretical assumptions (though combined of course with retention of the successful empirical content)— then realism is in dire straits. Before going further, let's be clear on the dependence of realism on the claim that theory change has been 'essentially cumulative'.

Assume, first, that the realist has convinced us that the development of theoretical science has indeed been 'essentially cumulative'. He could then argue for his realism roughly as follows. The development of the 'mature'

---

[7] See Tichy 1974 and Miller 1974.

sciences has so far been 'essentially' cumulative at all levels—theoretical as well as observational. It seems reasonable, therefore, to infer inductively that that development will continue to be 'essentially cumulative' in the future. This presumably means that, even should our present theories be replaced, they will continue to appear 'approximately' correct in the light of the successor theories. Such a development is, of course, logically compatible with the genuinely theoretical assumptions, both of presently accepted theories and of those destined to be accepted in the future, being entirely untrue. However, this is highly implausible, since it would make the empirical success of all these theories entirely mysterious; while, on the other hand, the assumption that our present theories are approximately true is enough to explain the empirical success as non-miraculous.

No one, I take it (reiterating the point made earlier), would claim that this argument is completely watertight. The inductive 'inference' from 'essential cumulativity' in the past to 'essential cumulativity' in the future could of course be questioned. Moreover, there is still the problem of what exactly is involved in approximate truth; and indeed the problem of whether or not the assumption of the approximate truth of our present theories really would explain their empirical success. It might seem plausible, intuitively speaking, to suppose that if a theory is 'approximately' or 'essentially' true, then it is likely that most of its consequences will themselves be 'essentially' correct. To take a straightforwardly empirical example, say that I make a slight arithmetical error in totting up my bank balance and come to the strictly mistaken view that my total worldly fortune is £100, when the truth is that it is £103. Will it seem 'miraculous' if this strictly false theory none the less supplies a quite reliable guide to life? After all, it might be claimed, most of the consequences that I am likely to be interested in—for example, that I can't afford a month's holiday in Switzerland—will in fact be consequences both of the false theory, that I hold, and of the truth. None the less, plausible or not, there are formidable formal difficulties here.[8] Every false theory, of course, has infinitely many false consequences (as well as infinitely many true ones), and there are things that my 'nearly true' theory gets totally wrong. For example, the truth is that my total fortune expressed in pounds sterling is a prime number, whereas the 'nearly true' theory I hold says—entirely incorrectly—that it's composite. Moreover, the argument seems committed to the claim that if theory T 'approximates' theory T', which in turn

---

[8] Two recent attempts to overcome these difficulties are Oddie 1986 and Niiniluoto 1987—though both attempts involve substantive, non-logical, and therefore challengeable assumptions.

'approximates' T'', then T 'approximates' T''. (The theories which eventually supersede our presently accepted ones, might themselves—presumably *will*—eventually be superseded by still further theories. The realist needs to be assured that any presently accepted theory will continue to look approximately correct, even in the light of the further theories in the sequence, not just in the light of its immediate successor.) But is this transitivity assumption correct? After all, if we took a series of photographs at one-second intervals, say, of a developing tadpole, each photograph in the sequence would presumably 'approximate' its predecessor, and yet we start with a tadpole and finish with a frog. Does a frog 'approximate' a tadpole? I propose, however, that, for present purposes, we put all these difficulties into abeyance. If he can sustain the claim that the development of the 'mature' sciences has been 'essentially cumulative', then the realist has at least some sort of argument for his claim.

If, on the contrary, the realist is forced to concede that there has been *radical* change at the theoretical level in the history of even the mature sciences, then he surely is in deep trouble. Suppose that there are cases of mature theories which were once accepted, were predictively successful, and whose underlying theoretical assumptions none the less now seem unequivocally entirely false. The realist would have encouraged the earlier theorist to regard his theory's empirical success as giving him grounds for regarding the theory itself as approximately true. He now encourages scientists to regard their newer theory's empirical success as giving them grounds for regarding that newer theory as approximately true. The older and newer theories are radically at odds with one another at the theoretical level. Presumably, if we have good grounds for thinking a theory T approximately true, we equally have good grounds for thinking that any theory T' radically at odds with T is false (plain false, not 'approximately true'). So the realist would be in the unenviable position of telling us that we now have good grounds to regard as false a theory which he earlier would have told us we had good grounds to believe approximately true. Why should not his proposed judgement about presently accepted theories turn out to be similarly mistaken?

Assuming, then, that the realist is not talking about 'good grounds' in some defeasible, conjectural sense,[9] realism is not compatible with the existence of radical theoretical changes in science (or at any rate in mature science). The chief argument against realism—the argument from scientific revolutions—is based precisely on the claim that revolutionary changes have occurred in accepted scientific theories, changes in which the

---

[9] See below, pp. 150–151.

old theory could be said to 'approximate' the new only by stretching the admittedly vague and therefore elastic notion of 'approximation' beyond breaking-point. At first glance, this claim appears to be correct. Consider, for example, the history of optics. Even if we restrict this history to the modern era, there have been fundamental shifts in our theory about the basic constitution of light. The theory that a beam of light consists of a shower of tiny material particles was widely held in the eighteenth century. Some of its empirical consequences—such as those about simple reflection, refraction, and prismatic dispersion—were correct. The theory was, however, rejected in favour of the idea that light consists, not of matter, but of certain vibratory motions set up by luminous bodies and carried by an all-pervading medium, the 'luminiferous aether'. It would clearly be difficult to argue that the theory that light is a wave in a mechanical medium is an 'extension', or even an 'extension with slight modifications', of the idea that light consists of material particles: waves in a mechanical medium and particles travelling through empty space seem more like chalk and cheese than do chalk and cheese themselves. Nor was that all: Fresnel's wave theory was itself soon replaced by Maxwell's electromagnetic theory. Maxwell, as is well known, strove manfully to give an account of the electromagnetic field in terms of some underlying mechanical medium; but his attempts and those of others failed, and it came to be accepted that the electromagnetic field is a primitive. So again, a fundamental change in the accepted account of the basic structure of light seems to have occurred—instead of vibrations carried through an elastic medium, it becomes a series of wave-like changes in a disembodied electromagnetic field. A mechanical vibration and an electric ('displacement') current are surely radically different sorts of thing. Finally, the acceptance of the photon theory had light consisting again of discrete entities, but ones which obey an entirely new mechanics.

In the meanwhile, as *theories* were changing light from chalk to cheese and then to superchalk, there was a steady, basically cumulative development in the captured and systematized empirical content of optics.[10] The material particle theory dealt satisfactorily with simple reflection and re-

---

[10] Genuine examples of 'Kuhn loss' of captured *empirical* content are remarkably thin on the ground—*provided*, that is, that *empirical* content is properly understood. Feyerabend and Kuhn both use examples of 'lost' content which are either clearly highly theoretical (Feyerabend even uses 'The Brownian particle is a perpetual motion machine of the second kind' as an example of an empirical statement!) or highly vague (Kuhn claims, e.g, that while phlogiston theory could explain why metalas are 'similar' to one another, the superseding oxygen theory could not). For a criticism of Feyerabend on facts see Worrall 1991; for a criticism of Kuhn see Worrall 1989b.

fraction and little else; the classical wave theory added interference and diffraction and eventually polarization effects too; the electromagnetic theory added various results connecting light with electrical and magnetic effects; the photon theory added the photoelectric effect and much else besides. The process at the empirical level (properly construed) was essentially cumulative. There were *temporary* problems (e.g. over whether or not the classical wave theory could deal with the phenomena which had previously been taken to support the ('essentially') rectilinear propagation of light), but these were invariably settled quickly and positively.[11]

Or take the Newton–Einstein case again. At the *empirical* level it does seem intuitively reasonable to say that Einstein's theory is a sort of 'extension with modifications' of Newton's. It is true that, even at this level, if we take the maximally precise consequences about the motion of a given body yielded by the two theories, they will always strictly speaking contradict one another. But for a whole range of cases (those cases, of course, in which the velocities involved are fairly small compared to the velocity of light), the predictions of the two theories will be strictly different but observationally indistinguishable. It is also true, of course, that Newton's equations are limiting cases of corresponding relativistic equations. However, there is much more to Newton's theory than the laws of motion and the principle of universal gravitation considered simply as mathematical equations. These equations were interpreted within a set of very general theoretical assumptions which involved amongst other things the assumption that space is infinite, that time is absolute, so that two events simultaneous for one observer are simultaneous for all, and that the inertial mass of a body is constant. Einstein's theory entails, on the contrary, that space is finite (though unbounded), that time is not absolute in the Newtonian sense, and that the mass of a body increases with its velocity. All these are surely out-and-out contradictions.

---

[11] The case of rectilinear propagation of light provides an illustrative example both of the essential empirical continuity of 'mature' science and of what it is about this process that leads Feyerabend and Kuhn to misrepresent it. Certain *theories* become so firmly entrenched at certain stages of the development of science, so much parts of 'background knowledge', that they, or at any rate particular experimental situations *interpreted in their light*, are readily talked of as 'facts'. This was certainly true of the 'fact' that light, if left to itself, is rectilinearly propagated. Here then is surely a 'fact' which was 'lost' in the wave revolution, since Fresnel's theory entails that light is *always* diffracted—it's just that in most circumstances the difference between the diffraction pattern and the predictions of geometrical optics is well below the observational level. But this last remark gives the game away. The idea that light is (*rigidly*) rectilinearly propagated was never an empirical result (not a 'crude fact' in Poincaré's terminology). The real empirical results—certain 'ray tracings', inability to see round corners or through bent opaque tubes, etc.—were not 'lost' but simply re-explained as a result of the shift to the wave theory.

The picture of the development of science certainly seems, then, to be one of essential cumulativity at the empirical level, accompanied by sharp changes of an entirely non-cumulative kind at the top theoretical levels.[12] This picture of theory change in the past would seem to supply good inductive grounds for holding that those theories presently accepted in science will, within a reasonably brief period, themselves be replaced by theories which retain (and extend) the empirical success of present theories, but do so on the basis of underlying theoretical assumptions entirely at odds with those presently accepted. This is, of course, the so-called *pessimistic induction*—usually regarded as a recent methodological discovery, but in fact already stated clearly by Poincaré.[13] How can there be good grounds for holding our present theories to be 'approximately' or 'essentially' true, and at the same time seemingly strong historical-inductive grounds for regarding those theories as (probably) ontologically false?

Unless this picture of theory change is shown to be inaccurate, then realism is surely untenable, and basically only two (very different) possibilities open. The first can be motivated as follows. Science is the field in which rationality reigns. There can be no rational acceptance of claims of a kind which history gives us grounds to think are likely later to be rejected. The successful empirical content of a once accepted theory *is* in general carried over to the new theory, but its basic theoretical claims are not. Theories, then, are best construed as making no real claims beyond their directly empirical consequences; or, if they *are* so construed, acceptance of these theoretical claims as true or approximately true is no part of the rational procedures of science. We are thus led into some sort of either pragmatic or 'constructive' anti-realism.

Such a position restores a pleasing, cumulative (or quasi-cumulative) development to science (i.e. to the 'real part' of science); but it does so at the expense of sacrificing the 'no miracles' argument entirely. After all, the theoretical science which the pragmatist alleges to be insubstantial and to play a purely codificatory role has, as a matter of fact, often proved *fruitful*. That is, interpreted literally and therefore treated as claims about the structure of the world, theories have yielded testable consequences over and above those they were introduced to codify, and those consequences have turned out to be correct when checked empirically. Why? The pragmatist asserts that there is no answer.

[12] That this is the intuitive picture was fully emphasized by Poincaré and Duhem, rather lost sight of by the logical positivists, and re-emphasized by Popper and those influenced by him (such as John Watkins and Paul Feyerabend).

[13] See Putnam 1978: 25; and Poincaré 1905: 160 (quoted below, p. 157).

The other alternative for someone who accepts the empirically cumulative, theoretically non-cumulative picture of scientific change, but who wishes to avoid pragmatism is pure, Popperian *conjectural realism*. This is Popper's view stripped of all the verisimilitude ideas, which always sat rather uncomfortably with the main theses. On this conjectural realist view, the genuinely theoretical, observation-transcendent parts of scientific theories are not just codificatory schemes, they are *attempted* descriptions of the reality hidden behind the phenomena. And our present best theories are our present best shots at the truth. We certainly have reason to think that our presently best theories are our present best shots at the truth (they stand up to the present evidence better than any known rival), but we have no real reason to think that those present theories are true or even closer to the truth than their rejected predecessors. Indeed, it can be accepted that the history of science makes it very unlikely that our present theories are even 'approximately' true. They do, of course, standardly capture more empirical results than any of their predecessors, but this is no indication at all that they are any closer to capturing 'God's blueprint of the universe'. The fully methodologically aware theoretical scientist nobly pursues his unended quest for the truth knowing that he will almost certainly fail and that, even if he succeeds, he will never know, nor even have any real indication, that he has succeeded.

Conjectural realism is certainly a modest, unassuming position. It can be formulated as a version of realism in the senses we have so far discussed—as saying in fact that we do have the best possible grounds for holding our present best theories to be true (they are best confirmed or best 'corroborated' by the present evidence); we should not even ask for better grounds than these; but since the best corroborated theory tomorrow may fundamentally contradict the best corroborated theory of today, the grounds that we have for thinking the theories true are inevitably conjectural and (practically, not just in principle) defeasible. I defended this conjectural realist view myself in an earlier paper: presentations of the view frequently (almost invariably) met with the response that there is little, if any, difference of substance between it and anti-realism.[14] The main problem, I take it, is again that conjectural realism makes no concessions to the 'no miracles' argument. On the conjectural realist view, Newton's theory does

[14] For my defence of conjectural realism see Worrall 1982. The response of 'no real difference' between conjectural and anti-realism was made many times in seminars and private discussions (by van Fraassen amongst others). See also, e.g., Newton-Smith 1981, where realism is *defined* as including an 'epistemological ingredient' foreign to this conjecturalist approach. I should add that I am of course giving up the conjectural realist position in the present paper only in the sense that I am now inclined to think that a *stronger* position can be defended.

assert that space and time are absolute, that there are action-at-a-distance forces of gravity, and that inertial mass is constant; all this was entirely wrong, and *yet* the theory based on these assumptions was highly empirically adequate. This just has to be recorded as a fact. And if you happen to find it a rather surprising fact, then that's your own business—perhaps due to failure to internalize the elementary logical fact that all false theories have true consequences (in fact, infinitely many of them).

Both the pragmatist and the conjectural realist can point out that we can't, on pain of infinite regress, account for everything, and one of the things we can't account for is why this stuff that allegedly does no more than streamline the machinery of scientific proof or that turns out to be radically false should have turned out to be fruitful. There obviously can be no question of any 'knockdown refutation' of either view. None the less, if a position could be developed which accommodated some of the intuitions underlying the 'no miracles' argument and yet which, at the same time, cohered with the historical facts about theory change in science, then it would arguably be more plausible than either pragmatism or conjectural realism.

Is it possible to have the best of both worlds, to account (no matter how tentatively) for the empirical success of theoretical science without running foul of the historical facts about theory change? Richard Boyd and occasionally Hilary Putnam have claimed that realism is itself already the best of both worlds. They have claimed, more or less explicitly, that the picture of scientific change that I have painted is inaccurate, and so the argument from scientific revolutions is based on a false premiss: the history of science is *not* in fact marked by radical theoretical revolutions (at any rate, not the history of 'mature' science). On the contrary, claims Boyd:

The historical progress of the mature sciences is largely a matter of successively more accurate approximations to the truth about both observable and unobservable phenomena. Later theories typically build upon the (observational and theoretical) knowledge embodied in previous theories. (1984: 41–2)[15]

Elsewhere he asserts that scientists generally adopt the (realist) principle that 'new theories should ... resemble current theories with respect to their accounts of causal relations among theoretical entities' (Boyd 1973:

[15] In discussion Richard Boyd acknowledged that he made no claim of approximate continuity for the 'metaphysical' components of accepted scientific theories. But I had thought that was what the debate is all about: does the empirical success of theories give us grounds to think that their basic ('metaphysical', observation-transcendent) description of the reality underlying the phenomena is at any rate approximately correct? Several of Richard Boyd's comments suggested to me, at least, that he defends not a full-blown realism, but something like the structural realism that I try to formulate below.

8). Similarly, Putnam once claimed (1978: 20) that many historical cases of theory change show that 'what scientists try to do' is to preserve 'as often as possible' the 'mechanisms of the earlier theory' or 'to show that they are "limiting cases" of new mechanisms'. I want first to explain why I think that these claims are wrong as they stand. I shall then argue that valid intuitions underlie the claims, but these intuitions are better captured in a rather different position which might be called *structural* or *syntactic realism*.

Larry Laudan has objected to Boyd and Putnam's claims by citing a whole list of theoretical entities, like phlogiston, caloric, and a range of ethers, which, he insists, once figured in successful theories but have now been *totally* rejected (1982: 231). How, Laudan wants to know, can newer theories resemble older theories 'with respect to their accounts of causal relations among theoretical entities' if the newer theories entirely reject the theoretical entities of the old? How can relativistic physics be said to preserve 'the mechanisms' of, say, Fresnel's account of the transmission of light, when, according to Fresnel's account, transmission occurs via periodic disturbances in an all-pervading elastic medium, while, according to relativity theory, no such medium exists at all? How can later scientists be said to have applied to Fresnel's theory the principle that 'new theories should . . . resemble current theories with respect to their accounts of causal relations among theoretical entities' when these later theories entirely deny the existence of the core theoretical entity in Fresnel's theory? Boyd alleges that the mechanisms of classical physics reappear as limiting cases of mechanisms in relativistic physics. Laudan replies that, although it is of course true that *some* classical laws are limiting cases of relativistic ones,

there are other laws and general assertions made by the classical theory (e.g., claims about the density and fine structure of the ether, general laws about the character of the interaction between ether and matter, models and mechanisms detailing the compressibility of the ether) which could not conceivably be limiting cases of modern mechanics. The reason is a simple one: a theory cannot assign values to a variable that does not occur in that theory's language . . . Classical ether physics contained a number of postulated mechanisms for dealing inter alia with the transmission of light through the ether. Such mechanisms could not possibly appear in a successor theory like the special theory of relativity which denies the very existence of an etherial medium and which accomplishes the explanatory tasks performed by the ether via very different mechanisms. (Laudan 1982: 237–8)

Does the realist have any legitimate come-back to Laudan's criticisms? Certainly *some* of Laudan's examples can be dealt with fairly straightforwardly. Boyd and Putnam have been careful to restrict their claim of 'essential' cumulativity to 'mature' science only. Pre-Lavoisierian chemistry is their chief example of an immature science, so they would be happy to concede that phlogiston has been entirely rejected by later science.[16] Presumably, some of the other items on Laudan's list of once scientifically accepted but now non-existent entities would receive similar treatment.

The cogency of this reply clearly depends to a large extent on whether or not some reasonably precise account can be given of what it takes for a science to achieve 'maturity'. Neither Boyd nor Putnam has anything very precise to say on this score, and this has naturally engendered the suspicion that the realist has supplied himself with a very useful *ad hoc* device: whenever it seems clear that the basic claims of some previously accepted theory have now been totally rejected, the science to which that theory belonged is automatically counted as 'immature' at the time that theory was accepted.

What is needed is a reasonably precise and *independent* criterion of maturity. And this can, it seems to me, in fact be 'read off' the chief sustaining argument for realism—the 'no miracles' argument. This argument, as I indicated before, applies only to theories which have enjoyed genuine predictive success. This must mean more than simply having correct empirical consequences—for these could have been forced into the framework of the theory concerned after the effects they describe had already been observed to occur. The undoubted fact that various chemical experimental results could be incorporated into the phlogiston theory does not on its own found any argument, even of the intuitive kind we are considering, to the likely truth of the phlogiston theory. Similarly, the fact that creationist biology can be made empirically adequate with respect to, say, the fossil record clearly founds no argument for the likely truth of the Genesis account of creation. Such empirical adequacy can of course easily be achieved—for example, by simply making Gosse's assumption that God created the rocks with the 'fossils' there already, just as they are found to be. (Perhaps God's purpose in doing this was to test our faith.) But the fact that this elaborated version of creationism is then bound to imply the empirical details of the fossil record is, of course, neither a miracle nor an indication that the theory 'is on the right track'. The explanation for this predictive 'success' is, of course, just that it is often easy to incorporate already known results *ad hoc* into a given framework. Nor is the success of a theory in predicting *particular* events of an *already known kind* enough on its own to sustain a 'no miracles' argument in favour of a theory. Even the most *ad hoc*, 'cobbled up' theory will standardly be predictive in the

---

[16] '[W]e do not carry [the principle of the benefit of the doubt] so far as to say that "phlogiston" referred' (Putnam 1978: 25).

sense that it will entail that the various results it has been made to absorb
will continue to hold in the future. (For example, the heavily epicyclic
corpuscular theory of light developed in the early nineteenth century by
Biot, having had various parameters fixed on the basis of certain results in
crystal optics, implied, of course, that the 'natural' generalizations of those
results would continue to hold in the future.) Theories will standardly
exhibit this *weak* predictiveness because, Popper or no, scientists do in-
stinctively inductively generalize on the results of well-controlled experiments which have so far always yielded the same results. But the success of
such inductive manœuvres, though no doubt miraculous enough in itself,
does not speak in favour of the likely truthlikeness of any particular
explanatory theory. The sort of predictive success which seems to elicit the
intuitions underlying the 'no miracles' argument is a much stronger, more
striking form of predictive success. In the stronger case, not just a new
instance of an old empirical generalization, but an entirely new empirical
generalization follows from some theory, and turns out to be experimentally confirmed. Instances of this are the prediction of the existence and
orbit of a hitherto unknown planet by Newton's theory and the prediction
of the white spot at the centre of the shadow of an opaque disc and of
the hitherto entirely unsuspected phenomenon of conical refraction by
Fresnel's wave theory of light. So my suggestion is that, instead of leaving
the notion of maturity as conveniently undefined, a realist should take it
that a science counts as mature once it has theories within it which are
predictive in this latter demanding sense—predictive of general *types* of
phenomena, without these phenomena having been 'written into' the
theory.

With this somewhat more precise characterization of maturity, Laudan's
list of difficult cases for the modified realist can indeed be pared down
considerably further. Laudan must be operating with some much weaker
notion of empirical success than the idea of *predictive* success just explained when he cites the gravitational ether theories of Hartley and
LeSage as examples of 'once successful' theories.[17] Presumably he means
simply that these theories were able successfully to accommodate various
already known observational results. But if we require *predictive* success of
the strong kind indicated above, then surely neither Hartley's nor LeSage's
speculative hypothesis scored any such success.

However there is no doubt that, no matter how hard-headed one is
about predictive success, some of Laudan's examples remain to challenge

---

[17] I have criticized Laudan on this point in Worrall 1988*b*.

the realist. Let's concentrate on what seems to me (and to others[18]) the sharpest such challenge: the ether of classical physics. Indeed, we can make the challenge still sharper by concentrating on the elastic solid ether involved in the classical wave theory of light proposed by Fresnel.

Fresnel's theory was based on the assumption that light consists in periodic disturbances originating in a source and transmitted by an all-pervading, mechanical medium. There can be no doubt that Fresnel himself believed in the 'real existence' of this medium—a highly attenuated and rare medium all right, but essentially an ordinary mechanical medium which generates elastic restoring forces on any of its 'parts' that are disturbed from their positions of equilibrium.[19] There is equally no doubt that Fresnel's theory enjoyed genuine predictive success—not least, of course, with the famous prediction of the white spot at the centre of the shadow of an opaque disc held in light diverging from a single slit. If Fresnel's theory does not count as 'mature' science, then it is difficult to see what does.[20]

Was Fresnel's elastic solid ether retained or 'approximately retained' in later physical theories? Of course, as I have repeatedly said and as realists would admit, the notion of one theoretical entity approximating another or of one causal mechanism being a limiting case of another is extremely vague, and therefore enormously elastic. But if the notion is stretched *too* far, then the realist position surely becomes empty. If black 'approximates' white, if a particle 'approximates' a wave, if a space-time curvature 'ap-

---

[18] See e.g. Hardin and Rosenberg 1982, which tackles this challenge on behalf of the realist (see below, pp. 156–7).

[19] This is not to deny, of course, that Fresnel was also guided by what was already known empirically about light. It is also true that at the time of Fresnel's work, much remained to be discovered about the dynamical properties of elastic solids. As a result, Fresnel's theory was dynamically deficient in certain respects (especially when viewed in hindsight). But the fact that he failed to construct a fully dynamically adequate theory of light as a disturbance in an elastic solid medium (or better: the fact that his theory ran into certain fundamental dynamical problems) does *not* mean that Fresnel did not even aim at such a theory, nor that he did not intend the theory he produced to be interpreted in this way. He clearly thought of light as a disturbance in an elastic medium, and dynamical and mechanical considerations (often of an abstract, mathematical sort) certainly guided his research, along with the empirical data on light.

There is no doubt that, as Whittaker pointed out (1951: 116), some aspects of Fresnel's theory—in particular the discontinuity of the normal component of the displacement across the interface between two media—cohere rather better with Maxwell's notion of a displacement current than they do with the idea of an ordinary dynamical displacement. But, *contra* Hardin and Rosenberg (who cite Whittaker), this doesn't mean that Fresnel was talking about displacement currents all along; instead, he was talking—in a flawed and problematic way—about elastic displacements.

[20] Cf. Laudan 1982: 225 (also p. 115, this volume): 'If that [Fresnel's prediction of the "white spot"] does not count as empirical success, nothing does!'

proximates' an action-at-a-distance force, then no doubt the realist is right that we can be confident that future theories will be approximately like the ones we presently hold. This won't, however, be telling us very much. It does seem to me that the only clear-sighted judgement is that Fresnel's elastic solid ether was entirely overthrown in the course of later science. Indeed, this occurred, long before the advent of relativity theory, when Maxwell's theory was accepted in its stead. It is true that Maxwell himself continued to hold out the hope that his electromagnetic field would one day be 'reduced' to an underlying mechanical substratum—essentially the ether as Fresnel had conceived it. But in view of the failure of a whole series of attempts at such a 'reduction', the field was eventually accepted as a primitive entity. Light became viewed as a periodic disturbance, *not* in an elastic medium, but in the 'disembodied' electromagnetic field. One would be hard pressed to cite two things more different than a displacement current, which is what this electromagnetic view makes light, and an elastic vibration through a medium, which is what Fresnel's theory had made it.

Hardin and Rosenberg (1982), replying to Laudan, suggest that, rather than trying to claim that Fresnel's elastic solid ether was 'approximately preserved' in Maxwell's theory, the realist can 'reasonably' regard Fresnel as having been talking about the electromagnetic field all along. This is certainly a striking suggestion! As someone influenced by Lakatos, I certainly would not want entirely to deny a role to rational reconstruction of history. Indeed, it does seem reasonable for a historian to reserve the option of holding that a scientist did not *fully* understand his own theory; but to allow that he may have totally *mis*understood it and, indeed, that it could not really be understood until some 50 years after his death, to hold that Fresnel was 'really' talking about something of which we know he had not the slightest inkling, all this is surely taking 'rational reconstruction' too far. Even 'charity' can be overdone.[21] Fresnel was *obviously* claiming that the light-carrying 'luminiferous aether' is an elastic solid, obeying, in essence, the ordinary laws of the mechanics of such bodies: the ether has 'parts'; restoring elastic forces are brought into play when a part is disturbed out of its equilibrium position. He was obviously claiming this, and it turned out that, if later science is right, Fresnel was wrong. Hardin and Rosenberg's claim has a definite air of desperation about it.

---

[21] Putnam has a well-known (and notoriously vague) 'principle of charity' (or 'benefit of the doubt') which says that 'when speakers specify a referent for a term they use by a *description* and, because of mistaken factual beliefs that those speakers have, that description fails to refer, we should assume that they would accept reasonable reformulations of their descriptions' (1978: 23–4).

None the less, there is *something* right about what they, and Boyd, say. There *was* an important element of continuity in the shift from Fresnel to Maxwell—and this was much more than a simple question of carrying over the successful *empirical* content into the new theory. At the same time, it was rather less than a carrying over of the full theoretical content or full theoretical mechanisms (even in 'approximate' form). And what was carried over can be captured without making the very far-fetched assumption of Hardin and Rosenberg that Fresnel's theory was 'really' about the electromagnetic field all along. There was continuity or accumulation in the shift, but the continuity is one of *form* or *structure*, not of content. In fact, this claim was already made and defended by Poincaré. And Poincaré used the example of the switch from Fresnel to Maxwell to argue for a general sort of *syntactic* or *structural realism* quite different from the anti-realist instrumentalism which is often attributed to him.[22] This largely forgotten thesis of Poincaré's seems to me to offer the only hopeful way of *both* underwriting the 'no miracles' argument *and* accepting an accurate account of the extent of theory change in science. Roughly speaking, it seems right to say that Fresnel completely misidentified the *nature* of light; but, none the less, it is no miracle that his theory enjoyed the empirical predictive success that it did; it is no miracle because Fresnel's theory, as science later saw it, attributed to light the right *structure*.

Poincaré's view is summarized in the following passage from *Science and Hypothesis,* which begins by clearly anticipating the currently fashionable 'pessimistic induction':

The ephemeral nature of scientific theories takes by surprise the man of the world. Their brief period of prosperity ended, he sees them abandoned one after the other; he sees ruins piled upon ruins; he predicts that the theories in fashion today will in a short time succumb in their turn, and he concludes that they are absolutely in vain. This is what he calls the *bankruptcy of science.* (1905: 160)

But this passage continues:

His scepticism is superficial; he does not take into account the object of scientific theories and the part they play, or he would understand that the ruins may still be good for something. No theory seemed established on firmer ground than Fresnel's, which attributed light to the movements of the ether. Then if Maxwell's theory is preferred today, does it mean that Fresnel's work was in vain? No; for Fresnel's object was not to know whether there really is an ether, if it is or is not formed of

---

[22] One critic who explicitly does not classify Poincaré as an instrumentalist is Zahar (see his 1983*b*). The term 'structural realism' was also used by Grover Maxwell for a position which he derived from Russell's later philosophy (see Maxwell 1970*a*,*b*). Maxwell's position grows out of different (more 'philosophical') concerns, though it is clearly related to that of Poincaré (one of the points for further research is to clarify this relationship).

atoms, if these atoms really move in this way or that; his object was to predict optical phenomena.[23]

This Fresnel's theory enables us to do today as well as it did before Maxwell's time. The differential equations are always true, they may be always integrated by the same methods, and the results of this integration still preserve their value.

So far, of course, this might seem a perfect statement of positivistic instrumentalism: Fresnel's theory is really just its empirical content, and this is preserved in later theories. However, Poincaré goes on to make it quite explicit that this is *not* his position.

It cannot be said that this is reducing physical theories to simple practical recipes; these equations express relations, and if the equations remain true, it is because the relations preserve their reality. They teach us now, as they did then, that there is such and such a relation between this thing and that; only the something which we then called *motion*, we now call *electric current*. But these are merely names of the images we substituted for the real objects which Nature will hide for ever from our eyes. The true relations between these real objects are the only reality we can attain. (1905: 162)

Poincaré is claiming that, although from the point of view of Maxwell's theory, Fresnel entirely misidentified the *nature* of light, his theory accurately described not just light's observable effects but its *structure*. There is no elastic solid ether. There is, however, from the later point of view, a (disembodied) electromagnetic field. The field in no clear sense approximates the ether, but disturbances in it do obey *formally* similar laws to those obeyed by elastic disturbances in a mechanical medium. Although Fresnel was quite wrong about *what* oscillates, he was, from this later point of view, right, not just about the optical phenomena, but right also that these phenomena depend on the oscillations of something or other at right angles to the light.

Thus, if we restrict ourselves to the level of mathematical equations— *not*, notice, the phenomenal level—there is in fact complete continuity between Fresnel's and Maxwell's theories. Fresnel developed a famous set of equations for the relative intensities of the reflected and refracted light beams in various circumstances. Ordinary unpolarized light can be analysed into two components: one polarized in the plane of incidence, the other polarized at right angles to it. Let I^2, R^2, and X^2 be the intensities of the components polarized in the plane of incidence of the incident, reflected, and refracted beams respectively; while I'^2, R'^2, and X'^2 are the components polarized at right angles to the plane of incidence. Finally, let

---

[23] Poincaré is quite wrong about Fresnel's 'object' (see above, n. 19). However, the normative philosophical question of how a theory *ought* to be interpreted is, of course, logically independent of the historical, psychological question of what its creator in fact believed.

i and r be the angles made by the incident and refracted beams with the normal to a plane reflecting surface. Fresnel's equations then state

$$R/I = \tan(i-r)/\tan(i+r)$$
$$R'/I' = \sin(i-r)/\sin(i+r)$$
$$X/I = (2\sin r . \cos i)/(\sin(i+r)\cos(i-r))$$
$$X'/I' = 2\sin r . \cos i/\sin(i+r)$$

Fresnel developed these equations on the basis of the following picture of light. Light consists of vibrations transmitted through a mechanical medium. These vibrations occur at right angles to the direction of the transmission of light through the medium. In an unpolarized beam, vibrations occur in all planes at right angles to the direction of transmission—but the overall beam can be described by regarding it as the composition of two vibrations: one occurring in the plane of incidence and one occurring in the plane at right angles to it. The bigger the vibrations, that is, the larger the maximum distance the particles are forced from their equilibrium positions by the vibration, the more intense the light. I, R, X, etc. in fact measure the amplitudes of these vibrations, and the intensities of the light are given by the squares of these amplitudes.

From the vantage-point of Maxwell's theory as eventually accepted, this account, to repeat, is entirely wrong. How could it be anything else when there is no elastic ether to do any vibrating? None the less, from this vantage-point, Fresnel's theory has exactly the right structure—it's 'just' that what vibrates according to Maxwell's theory are the electric and magnetic field strengths. And in fact, if we interpret I, R, X, etc. as the amplitudes of the 'vibration' of the relevant electric vectors, then Fresnel's equations are directly and fully entailed by Maxwell's theory. It wasn't, then, just that Fresnel's theory *happened* to make certain correct predictions; it made them because it had accurately identified certain relations between optical phenomena. From the standpoint of this superseding theory, Fresnel was quite wrong about the nature of light; the theoretical mechanisms he postulated are *not* approximations to, or limiting cases of, the theoretical mechanisms of the newer theory. None the less, Fresnel was quite right not just about a whole range of optical phenomena, but right that these phenomena depend on something or other that undergoes periodic change at right angles to the light.

But then, Poincaré argued, his contemporaries had no more justification for regarding Maxwell as having definitively discovered the nature of light, as having discovered that it *really* consists in vibrations of the electromagnetic field, than Fresnel's contemporaries had had for regarding Fresnel as having discovered the nature of light. At any rate, this attitude towards Maxwell would be mistaken if it meant any more than that Maxwell built on the relations revealed by Fresnel and showed that further relations existed between phenomena hitherto regarded as purely optical on the one hand and electric and magnetic phenomena on the other.

This example of an important theory change in science certainly appears, then, to exhibit cumulative growth at the structural level combined with radical replacement of the previous ontological ideas. It speaks, then, in favour of a *structural* realism. Is this simply a feature of this particular example, or is preservation of structure a general feature of theory change in mature (i.e. successfully predictive) science?

This particular example is in fact unrepresentative in at least one important respect: Fresnel's equations are taken over completely intact into the superseding theory—reappearing there newly interpreted but, as mathematical equations, entirely unchanged. The much more common pattern is that the old equations reappear as *limiting cases* of the new—that is, the old and new equations are strictly inconsistent, but the new tend to the old as some quantity tends to some limit.

The rule in the history of physics seems to be that, whenever a theory replaces a predecessor, which has however itself enjoyed genuine predictive success, the 'correspondence principle' applies. This requires the *mathematical equations* of the old theory to re-emerge as limiting cases of the mathematical equations of the new. As is being increasingly realized,[24] the principle operates, not just as an after-the-event requirement on a new theory if it is to count as better than the current theory, but often also as a heuristic tool in the actual development of the new theory. Boyd (1984) in fact cites the general applicability of the correspondence principle as evidence for his realism. But the principle applies *purely* at the mathematical level, and hence is quite compatible with the new theory's basic theoretical assumptions (which *interpret* the terms in the equations) being entirely at odds with those of the old. I can see no clear sense in which an action-at-a-distance force of gravity is a 'limiting case' of, or 'approximates', a space-time curvature. Or in which the 'theoretical mechanisms' of action-at-a-distance gravitational theory are 'carried over' into general relativity theory. Yet Einstein's equations undeniably go over to Newton's in certain limiting special cases. In this sense, there is 'approximate continuity' of *structure* in this case. As Boyd points out, a new theory could capture its predecessor's successful empirical content in ways other than yielding the

---

[24] See e.g. Zahar 1983*b* and Worrall 1985, as well as Boyd 1984.

equations of that predecessor as special cases of its own equations.[25] But the general applicability of the correspondence principle certainly is not evidence for full-blown realism—but, instead, only for structural realism.

Much clarificatory work needs to be done on this position, especially concerning the notion of one theory's structure approximating that of another. But I hope that what I have said is enough to show that Poincaré's is the only available account of the status of scientific theories which holds out realistic promise of delivering the best of both worlds: of underwriting the 'no miracles' argument, while accepting the full impact of the historical facts about theory change in science. It captures what is right about Boyd's realism (there is 'essential accumulation' in 'mature' science at levels higher than the purely empirical) and at the same time what is right about Laudan's criticism of realism (the accumulation does not extend to the fully interpreted top theoretical levels).

As one step towards clarifying the position further, let me end by suggesting that one criticism which, rightly or wrongly, has been levelled at scientific realism does not affect the structural version. Arthur Fine has strikingly claimed that

Realism is dead . . . Its death was hastened by the debates over the interpretation of quantum theory where Bohr's non-realist philosophy was seen to win out over Einstein's passionate realism. (p. 21, this volume)

But realism has been pronounced dead before. Some eighteenth-century scientists believed (implicitly, of course; they would not have expressed it in this way) that realism's death had been hastened by debates over the foundations of the theory of universal gravitation. But it is now surely clear that in this case realism was 'killed' by first saddling it with an extra claim which then proved a convenient target for the assassin's bullet. This extra claim was that a scientific theory could not invoke 'unintelligible' notions, such as that of action-at-a-distance, as primitives. A realist interpretation required intelligibility, and intelligibility required

---

[25] Putnam gives this account of Boyd's position in his 1978, adding that applying the correspondence principle 'is often the *hardest* way to get a theory that keeps the old observational predictions'. I find this last remark very difficult to understand. How exactly could it be done otherwise? (I am assuming that what comes out is required to be a theory in some recognizable sense rather than simply any old collection of empirical statements.) Zahar has shown (see n. 35) how the correspondence principle can be used as a definite heuristic principle supplying the scientist with real guidance. But suppose a scientist set out to obtain a theory which shares the successful empirical consequences of its predecessor in some other way than by yielding it predecessor's equations as limiting cases—surely he would be operating completely in the dark without any clear idea of how to go about the task. (I am assuming that various logical 'tricks' are excluded on the grounds that they would fail to produce anything that anyone (including the anti-realist) would regard as a theory.)

interpretation of the basic theoretical notions in terms of some antecedently accepted (and *allegedly* antecedently 'understood') metaphysical framework (in the Newtonian case of course this was the framework of Cartesian action-by-contact mechanics). Without claiming to be an expert in the foundations of quantum mechanics (and with all due respect for the peculiarities of that theory), it does seem to me that, by identifying the realist position on quantum mechanics with Einstein's position, Fine is similarly saddling realism with a claim it in fact has no need to make. The realist is forced to claim that quantum-mechanical states cannot be taken as primitive, but must somehow be understood or reduced to or defined in classical terms.

But the structural realist at least is committed to no such claim—indeed, he explicitly disowns it. He insists that it is a mistake to think that we can ever 'understand' the *nature* of the basic furniture of the universe. He applauds what eventually happened in the Newtonian case. There the theory proved so persistently successful empirically and so persistently resistant to 'mechanistic reduction' that gravity (understood as a genuine action-at-a-distance force) became accepted as a primitive irreducible notion. (And action-at-a-distance forces became perfectly acceptable, and realistically interpreted, components of other scientific theories, such as electrostatics.) On the structural realist view, what Newton really discovered are the relationships between phenomena expressed in the mathematical equations of his theory, the theoretical terms of which should be understood as genuine primitives.[26]

Is there any reason why a similar structural realist attitude cannot be adopted towards quantum mechanics? This view would be explicitly divorced from the 'classical' metaphysical prejudices of Einstein: that dynamical variables must always have sharp values and that all physical events are fully determined by antecedent conditions. Instead, the view would simply be that quantum mechanics does seem to have latched on to the real structure of the universe, that all sorts of phenomena exhibited by microsystems really do depend on the system's quantum state, which really does evolve and change in the way quantum mechanics describes. It is, of course, true that this state changes discontinuously in a way which the theory does not further explain when the system interacts with a 'macroscopic system'—but then Newton's theory does not *explain* gravitational interaction, but simply postulates that it occurs. (Indeed, no theory, of course, can explain everything on pain of infinite regress.) If such

---

[26] See, in particular, Poincaré's discussion of the notion of force (1905: 89–139).

discontinuous changes of state seem to cry out for explanation, this is because of the deeply ingrained nature of certain classical metaphysical assumptions (just as the idea that action-at-a-distance 'cried out' for explanation was a reflection of a deeply ingrained prejudice for Cartesian-style mechanics).

The structural realist simply asserts, in other words, that, in view of the theory's enormous empirical success, the structure of the universe is (probably) something like quantum-mechanical. It is a mistake to think that we need to understand the nature of the quantum state at all, and, *a fortiori*, a mistake to think that we need to understand it in classical terms. (Of course, this is not to assert that hidden variables programmes were obvious non-starters, that working on them was somehow obviously mistaken—no more than the structural realist needed to assert that the attempts at a Cartesian reduction of gravity were doomed from the start. The only claim is that ultimately evidence leads the way: if, despite all efforts, no scientific theory can be constructed which incorporates our favourite metaphysical assumptions, then no matter how firmly entrenched those principles might be, and no matter how fruitful they may have proved in the past, they must ultimately be given up.)

It seems to me, then, that, so long as we are talking about *structural* realism, the reports of realism's death at the hands of quantum mechanics are greatly exaggerated.[27]

REFERENCES

Boyd, R. (1973). 'Realism, Underdetermination and a Causal Theory of Evidence.' *Noûs*, 7: 1–12.
——(1984). 'The Current Status of Scientific Realism.' In Leplin (ed.) 1984: 41–82.
Duhem, P. (1906). *The Aim and Structure of Physical Theory.* (Page references to the translation by Philip Wiener (New York: Atheneum, 1962).)

[27] It is not in fact clear to me that Fine's NOA (the natural ontological attitude) is substantially different from structural realism. Structural realism perhaps supplies a banner under which *both* those who regard themselves as realists *and* those who regard themselves as antirealists of various sorts can unite.

Similar remarks about the 'anti-realist' consequences of quantum mechanics are made—though without reference to Fine—by McMullin (1984: 13). In allegedly defending realism, McMullin *also* seems to me in fact to defend structural realism, See my review of the Leplin volume (Worrall 1988a) in which McMullin's article appears.

These last remarks on quantum mechanics were modified and elaborated in an attempt to meet the interesting objections raised in discussion at Neuchâtel by Professor d'Espagnat.

I wish to thank John Watkins for some suggested improvements to an earlier draft; Elie Zahar for numerous enlightening discussions on the topic of this paper; and Howard Stein for his comments on the version delivered at Neuchâtel.

164   JOHN WORRALL

Fine, A. (1984). 'The Natural Ontological Attitude.' In Leplin (ed.) 1984: 83–107;
    repr. as Ch. I.
Hardin, C. and Rosenberg, A. (1982). 'In Defence of Convergent Realism.' *Philoso-
    phy of Science*, 49: 604–15.
Laudan, L. (1981). 'A Confutation of Convergent Realism.' *Philosophy of Science*,
    48; repr. in Leplin (ed.) 1984: 218–49, and as Ch. VI.
Leplin, J. (ed.) (1984). *Scientific Realism*. Berkeley: University of California Press.
Maxwell, G. (1970a). 'Structural Realism and the Meaning of Theoretical Terms.' In
    S. Winokur and M. Radner (eds.), *Minnesota Studies in the Philosophy of Science*,
    vol. 181–92. Minnesota: University of Minnesota Press.
—— (1970b). 'Theories, Perception and Structural Realism.' In R. G. Colodny (ed.),
    The Nature and Function of Scientific Theories, 3–34. Pittsburgh: University of
    Pittsburgh Press.
McMullin, E. (1984). 'A Case for Scientific Realism.' In Leplin (ed.) 1984: 8–40.
Miller, D. (1974). 'Popper's Qualitative Theory of Verisimilitude'. *British Journal
    for the Philosophy of Science*, 25: 166–77.
Musgrave, A. (1988). 'The Ultimate Argument for Scientific Realism.' In R. Nola
    (ed.), *Relativism and Realism in Sciences*, 229–52. Dordrecht: Kluwer.
Newton-Smith, W. (1981). *The Rationality of Science*. London: Routledge and
    Kegan Paul.
Niiniluoto, I. (1987). *Truthlikeness*. Dordrecht: Reidel.
Oddie, G. (1986). *Likeness to Truth*. Dordrecht: Reidel.
Poincaré, H. (1905). *Science and Hypothesis*. Repr. New York: Dover, 1952. (Page
    references are to the Dover edn.)
Putnam, H. (1978). *Meaning and the Moral Sciences*. London: Routledge and Kegan
    Paul.
Quine, W. V. O. (1953). 'Two Dogmas of Empiricism.' In *From a Logical Point of
    View*, 20–46. Cambridge, Mass.: Harvard University Press.
Tichy, P. (1974). 'On Popper's Definition of Verisimilitude.' *British Journal for the
    Philosophy of Science*, 25: 155–60.
Van Fraassen, B. (1980). *The Scientific Image*. Oxford: Clarendon Press.
Whittaker, E. T. (1951). *History of Theories of Aether and Electricity. The Classical
    Theories*. London: Nelson.
Worrall, J. (1982). 'Scientific Realism and Scientific Change.' *Philosophical Quar-
    terly*, 32: 201–31.
—— (1983). 'An Unreal Image.' *British Journal for the Philosophy of Science*, 35:
    65–80.
—— (1985). 'Scientific Discovery and Theory Confirmation.' In J. Pitt (ed.), *Change
    and Progress in Modern Science*, 311–14. Dordrecht: Reidel.
—— (1988a). Review of Leplin (ed.) 1984. *Philosophical Quarterly*, 38: 370–6.
—— (1988b). 'The Value of a Fixed Methodology.' *British Journal for the Philo-
    sophy of Science*, 39: 263–75.
—— (1989a). 'Fresnel, Poisson and the White Spot: The Role of Successful Predic-
    tion in Theory-Acceptance.' In D. Gooding, T. Pinch and S. Schaffer (eds.),
    The Uses of Experiment—Studies of Experimentation in Natural Science, 135–57.
    Cambridge: Cambridge University Press.
—— (1989b). 'Scientific Revolutions and Scientific Rationality: The Case of the
    "Elderly Hold-Out"'. In C. Wade Savage (ed.), *The Justification, Discovery and
    Evolution of Scientific Theories*, 319–54. Minneapolis: University of Minnesota
    Press, 1990.

——(1991). 'Feyerabend and the Facts.' In G. Munévar (ed.), *Beyond Reason*, 329–53. Dordrecht: Kluwer Academic Press.

Zahar, E. G. (1983*a*). 'Logic of Discovery or Psychology of Invention?' *British Journal for the Philosophy of Science*, 34: 243–61.

——(1983*b*). 'Poincaré's Independent Discovery of the Relativity Principle.' *Fundamenta Scientiae*, 4: 147–75.

VIII

WHAT SCIENCE AIMS TO DO

BRIAN ELLIS

In *The Scientific Image*, Bas van Fraassen has mounted an impressive challenge to scientific realism and argued strongly for a new empiricism which he calls 'constructive empiricism'. I agree with van Fraassen that the form of scientific realism which he challenges is untenable. However, I cannot accept his constructive empiricism as a satisfactory alternative. It is, undoubtedly, much more acceptable than earlier, cruder forms of empiricism, since it concedes much that has been found objectionable in other empiricist theories. But, in the end, it suffers from some of the same basic defects as they do, and must, therefore, be rejected along with them.

Van Fraassen sees empiricism and scientific realism as being opposed to each other. He locates the basic difference between them as a difference of view about the *aims* of science. Empiricists, he says, see the aim as being the anticipation of nature, of producing theories which yield successful predictions—predictions which are borne out by experience. Scientific realists, on the other hand, see it as being to describe things as they really are or have come to be—as Duhem says (of explanation), it is 'to strip reality of the appearances covering it like a veil, in order to see the bare reality itself' (1954: 7).

Traditionally, empiricism has been opposed, not to scientific realism, but to rationalism, and to revelationism, as a theory about the *sources* of our knowledge. And most scientific realists would agree with the empiricists on this issue. For example, Galileo was an empiricist about the sources of our knowledge, and Cardinal Bellarmino, a revelationist. But, by van Fraassen's criteria, it is Bellarmino who was the empiricist and Galileo who was not. So van Fraassen's concept of empiricism is not the traditional one. Nevertheless, the disagreement *within* the empiricist tradition about the aim of science is an important one, and I see no point in arguing about labels.

Reprinted from P. Churchland and C. Hooker (eds.), *Images of Science* (Chicago: University of Chicago Press, 1985), 48–74, by permission of the author and the University of Chicago Press.

The different views about the aim of science lead naturally to different views concerning its theoretical achievements. For scientific realists, well-established and accepted theories must be considered to be literally true—not just useful models for predicting what will occur, or, to use the old-fashioned term, 'saving the phenomena'. For empiricists, such theories are just models of reality which are, and which we believe will continue to be, empirically adequate. Whether they are true or not, van Fraassen maintains, is of no importance—although he does not deny that they are either true or false.

According to van Fraassen, the scientific realist's position is this:

(1) Science aims to give us, in its theories, a literally true story of what the world is like; and acceptance of a scientific theory involves the belief that it is true. (Van Fraassen 1980: 8)

Van Fraassen's concept of literal truth is a correspondence concept: a statement is literally true if, literally interpreted, it accurately describes or corresponds to reality. The rules for literal interpretation are not clearly specified, but he has in mind at least this: any apparent reference to a theoretical entity is to be construed as a genuine attempt to refer, unless there are good specific reasons for not so construing it.

Against the scientific realists' conception of the aim of science, van Fraassen argues that it is not required either to motivate science or to explain its practice. It is enough for these purposes to see the aim as being the provision of empirically adequate theories. Moreover, on the basis of a general thesis of empirical underdetermination of theories, he argues that we cannot rationally demand more of any theory than that it be empirically adequate. For, once we get down to those theories which are acceptable by his criterion of empirical adequacy, the choice between them cannot be made on grounds which have any bearing on whether or not they are true. The choice has now to be a pragmatic one. So, if we required belief in the *truth* of a theory as a condition for its acceptance, we could not rationally accept any theories at all, even though we had good empirical *and* pragmatic reasons for doing so. Conversely, if we did accept any theories as being literally true, we should have to allow that the pragmatic reasons (simplicity, elegance, and the like) which we have for preferring them to other empirically equivalent theories are actually grounds for believing them to be true.

Van Fraassen thus arrives at the following position:

(2) Science aims to give us theories which are empirically adequate; and acceptance of a theory involves as belief only that it is empirically adequate. (Van Fraassen 1980: 12)

I assume that van Fraassen intends this statement to be understood normatively, and that the same applies to the statement of position he attributes to scientific realists, for there is no sociological investigation of
what scientists themselves see as the aims of science or of what beliefs they
have concerning the theories they accept. So, presumably, it is a question
of what we ought to consider the aims of science to be and of what beliefs
ought to be involved in accepting a scientific theory. Nevertheless,
since the principal theses are not presented as normative judgements, and
there is a lot of discussion of modern physical theories, I cannot help seeing
van Fraassen's position as being, at least in part, a reflection of what he
thinks are the attitudes of scientists to the dominant physical theories of
today.

At the time when the dominant theories in science were mechanistic, it
was easy to see the aim of science as being to discover and describe the
underlying mechanisms of nature. Think of nineteenth-century chemistry.
Of course, the atomic-molecular theories of the time were usually seen as
describing basic chemical processes; and to accept them was to believe that
they truly described these processes. But the image of science has changed
a great deal since then, and the dominant theories are no longer mechanistic. Think now of quantum mechanics and geometrodynamics. Is scientific realism any longer the philosophy of science which we feel naturally
compelled to accept? I should think that many space-time and quantum
physicists would be quite puzzled by the suggestion that the theories they
accept, and work with, might literally be true, since they have no clear
conception at all of the reality with which these theories might correspond.
And I can well see many of them agreeing with van Fraassen that the aim
of science is only to give us theories which are empirically adequate. I think
that van Fraassen, rightly or wrongly, draws some comfort from this, and
sees scientific realism as a philosophy of science more appropriate to
another age. On this point I would agree with him, given his characterization of the position.

In considering van Fraassen's position and his arguments for it, I find
myself also in agreement with much of what he says in detail. I am convinced, at least, that any scientific realist who accepts a correspondence
theory of truth and van Fraassen's distinction between empirical and pragmatic considerations is in serious trouble defending his position. For, given
that the only considerations relevant to the truth or falsity of a claim are
empirical or logical, and that such considerations alone can never determine what theories we should accept as being true (because of
underdetermination), it follows that we can never have any good reason to
believe that a theory is true. So anyone who believes in the literal truth of

the theories they accept must do so irrationally. To counter this argument, it would be necessary to challenge either van Fraassen's empirical underdetermination thesis or the correspondence theory of truth upon which the whole argument is based.

However, most scientific realists are in a serious bind, for they see acceptance of a correspondence theory of truth as being essential to their position. Coherence and pragmatic theories, they think, go together not with realism but with idealism. So most scientific realists are likely to focus on either the empirical underdetermination thesis or the empirical/pragmatic distinction as the weak point of the argument. However, the empirical underdetermination thesis is very widely accepted and seldom challenged; and most scientific realists would be reluctant to give up the empirical/pragmatic distinction which gives the argument for underdetermination such force, since doing so would seem to undermine the correspondence theory of truth which they believe to be essential to their position. Nevertheless, I think that they *must* do one of these things if they wish to defend scientific realism. The correspondence theory of truth is *not* essential to their position; nor is van Fraassen's empirical/pragmatic distinction. For scientific realism can be combined with a pragmatic theory of truth; and, given such a theory of truth, all of the criteria we use for the evaluation of theories, including the so-called pragmatic ones, can be seen as being relevant to their truth or falsity.

In opposition to van Fraassen's constructive empiricism, I would propose the pragmatist thesis:

(3) Science aims to provide the best possible explanatory account of natural phenomena; and acceptance of a scientific theory involves the belief that it belongs to such an account.

Now, I think that (3) is nearly right, and empiricists and scientific realists can both accept it. Empiricists can accept it by allowing that the best possible account is just any that is empirically adequate. Scientific realists can accept it by agreeing that the best possible account, if it exists, is *necessarily* the true one. I suppose not many scientific realists would accept this pragmatist defence of their position, since most of them would accept a correspondence theory of truth—not a pragmatic theory. Nevertheless, I do not think they have much choice. They have either to reject the correspondence theory of truth, and accept the position known as internal realism, or to follow van Fraassen along the road to his constructive empiricism.

My aim in this paper is to defend scientific realism from the perspective of an internal realist. From this perspective I shall argue that van

Fraassen's case against scientific realism fails. Many of the arguments I shall present should be acceptable to scientific realists whatever their metaphysical persuasions. And these arguments should be of independent interest to scientific realists. In particular, I shall argue that the arguments for conventionalism and strong underdetermination are not as good as they are usually thought to be. However, in the end, I think that van Fraassen's arguments must defeat the metaphysical realist version of scientific realism, for there remains a sceptical challenge to metaphysical realism, to which there is no adequate reply. The final choice, therefore, is between constructive empiricism and internal scientific realism.

I. SCIENTIFIC REALISM

We can isolate a number of strands in the thought of scientific realists, apart from those already mentioned; and perhaps it will be useful to do so, for most scientific realists see them as going together as a package deal.

First, there is a commitment to a physicalist ontology. The precise nature of this ontology is rarely spelled out, and there are disagreements between scientific realists about what it includes. It would be almost universally agreed that it includes the fundamental particles, but it would be disputed whether properties, relationships, forces, sets, or qualia should also be included. Some would include universals (properties, relationships) in their ontology but deny the existence of sets or other abstract particulars.[1] Others are happy about sets but unhappy about universals.[2] Some would offer a Humean account of causation, and hence of forces (as I did once (Ellis 1965)), while others would accept some doctrine of natural necessitation (see Harré and Madden 1975). And there are some who would admit the existence of epiphenomenal qualia while yet claiming to be scientific realists (as Keith Campbell does (1976)).

A second strand of scientific realism concerns the status of scientific laws and theories. All would say that these are to be understood realistically rather than instrumentally. The laws and theories of science are genuine claims about reality, not mere instruments of prediction which are more or less useful for this purpose. This is the *central thesis of scientific realism*. Every scientific realist subscribes to some version of it. Yet this thesis, like the doctrine of physicalism, is also subject to various interpretations. Basically, the idea is that there are things in the world to which our laws and

---

[1] This, as I understand it, is the position taken in Armstrong 1978.
[2] The position is Quine's, although Smart and others have adopted it.

theories refer and of which they are true or false. They are, that is, to be understood as referring to real existents and ascribing genuine properties to them. It is not clear, though, how even this claim is to be interpreted. On a naïve interpretation, we must suppose that there are Hilbert spaces, perfect gases, inertial systems, and ideal incompressible fluids in steady flow in uniform gravitational fields, for it is to things like these that many of our laws seemingly refer, if they are not just vacuously true. But my impression is that most scientific realists do not really want to take the ontological commitments of laws or theories of science in such a *naïvely* realistic way, but only to those which can be so understood conformably with their ontology. The rest would have to be suitably reduced, and the apparent references parsed away, to avoid any unwanted commitments.[3]

The third strand of scientific realism is the *objectivity thesis* that the laws and theories of science are objectively true or false. Every scientific realist is committed to some version of this. The objectivity thesis is rarely distinguished from the central thesis of scientific realism, and is often confused with it. But, as I understand them, they are quite distinct. One could hold, for example, *contra* the conventionalists, that it is objectively true or false whether space is Euclidean without believing that there is such a thing as space or property of Euclideanness. One who thinks that truth is what occurs in the ultimately best theory, for example, might think this. Moreover, one could believe that this was so even though one did not believe that statements about the geometry of space were reducible to statements about, say, spatial relationships between physical entities.

The two theses are often confused, because most scientific realists accept some form of the correspondence theory of truth. They hold that a statement is objectively true iff it corresponds to reality. Of course, they do not hold that all objectively true statements correspond to reality in the same way. Some do so more directly than others. Some statements, for example, some of those ascribing physical properties to physical objects, have a kind of *direct* correspondence to reality; and for these *basic statements* a Tarski T-sentence provides a sufficient explication of the correspondence relationship. Other statements, however, are related to reality in more complex ways, and to understand how they may correspond, or fail to, we must have analyses of their truth conditions which will enable us to understand them in terms of direct correspondences. Now, assuming that a statement is objectively true or false iff there exists such an analysis, and that the existence of such an analysis is both a necessary and sufficient condition for the possibility of interpreting a statement realistically, the distinction be-

[3] e.g. some of those who admit universals in their ontologies have tried to construe laws as relationships between universals (e.g. Armstrong 1978).

tween the central thesis of scientific realism and the objectivity thesis collapses. A statement is capable of a realistic interpretation iff it is objectively true or false. However, anyone who would reject the correspondence theory of truth can continue to make the distinction.

The main difficulty with the objectivity thesis derives from the arguments of the conventionalists which appear to establish a general thesis of empirical underdetermination. The thesis in question is not the weak one that our theories are inductively underdetermined, but the strong one that, even if *all* the empirical consequences of a theory could be checked and it could be *known* that they have all been checked, it would still be possible to construct an alternative theory which has a different set of ontological commitments, or is otherwise incompatible with the given theory, but which has precisely the same observational consequences. If this strong thesis is true, then it threatens the realists' ontology, or at least the claim that they would all certainly make that the ontology is empirically well supported.

A fourth tenet of scientific realism is the correspondence theory of truth. Most scientific realists believe that we must have such a theory if we are to be realists about scientific entities—or, for that matter, about anything else non-mental. In championing the correspondence theory of truth, realists thus see themselves as being opposed to idealism—to the view that reality is a construct out of experience, rather than something existing independently of it. In believing that there is this connection, I think they are profoundly mistaken. We need not have a correspondence theory of truth to accept a physicalist ontology or to believe in an independently existing reality. On the contrary, I think that anyone who accepts a physicalist ontology should not also hold a correspondence theory of truth, although this is not a claim that I wish to defend here.[4]

Clearly, I cannot discuss all of these issues properly in one paper. So I shall focus on those questions which seem to be most at issue between scientific realists and constructive empiricists. The question of what ontology one should have if one is a scientific realist can safely be set aside. The main question concerns how scientific theories are to be understood—realistically or otherwise.

II. THEORIES AND EXPLANATIONS

One thing you learn about the United States when you visit that country is that it is dangerous to generalize about it. And one thing philosophers of

---

[4] This is a major theme of a book on truth and realism which I am currently writing.

science should have learned about scientific theories and explanations is that it is also dangerous to generalize about them. The danger is that what seems to be true of theories and explanations in one field, or of some of them at any rate, may not be true in others, so that different theories of theories and explanations may seem plausible or not depending on what examples are taken. I think that van Fraassen is well aware of this danger—probably more so than most of his realist opponents. For scientific realists do have a rather lamentable tendency to take nice homely examples of causal explanations as typifying scientific explanations generally, and simple mechanistic theories taken from nineteenth-century physics or chemistry, which were obviously intended by their authors to be understood realistically, as paradigms of scientific theories. Their philosophical position is thus given a strong flavour of initial plausibility. Van Fraassen's position, on the other hand, probably derives more from his earlier work on space-time theories (e.g. 1969, 1970), where it is not at all obvious that the theories were intended to be anything more than models of some kind which could be used with greater or less facility to 'save the phenomena'.

In fact, there are different sorts of theories and explanations which arise as answers to different sorts of questions. I follow van Fraassen in thinking that any request for explanation is a request for information. A *causal explanation* is information about the causal history of something or about the causal processes which result in something.[5] A *functional explanation* is information about the role of something in some ongoing system—about the contribution it makes to sustaining it. A *model-theoretic explanation* is information about how (if at all) the actual behaviour of some system differs from that which it should have ideally if it were not for some perturbing influences and, where necessary, includes some information about what perturbing influences may be causing the difference. A *systemic explanation* is information about how the fact to be explained is systematically related to other facts.

Theories, on the other hand, provide us with the general schemata for giving such explanations. *Causal process theories* attempt to describe the basic causal processes of nature. *Functionalist theories* are concerned with ongoing systems of various kinds and with the kinds of mechanisms, described in terms of their functional roles, necessary for their maintenance. *Model theories* define norms of behaviour against which actual behaviour may be compared and (causally) explained. *Systemic theories* set forth some general organizational principles adequate to deter-

---

[5] David Lewis has elaborated on this theme in a paper entitled 'Causal Explanation', in his *Philosophical Papers*, vol. 2 (Oxford: Oxford University Press, 1986).

mine the basic structure of some system of relationships between things. Euclidean geometry, for example, is such a theory for the system of spatial relationships.

Now, the argument for scientific realism, in so far as it concerns the reality of theoretical entities, derives whatever force it has from taking causal process theories to be typical of scientific theories generally. For to accept that A is the cause of B is to accept that both A and B are real existents (or events). But no such argument applies to the theoretical entities of model theories, for the hypothetical entities of model theories are not the postulated *causes* of anything. Consequently, there is no parallel argument that to accept a model theory involves the belief that the entities to which it apparently refers really exist. We do not have to believe in the reality of Newtonian point masses or of Einsteinian inertial frames or in the existence of the perfectly reversible heat engines of classical thermodynamics to accept these various theories, for none of these entities has any causal role in the explanations provided by the theories in question. Consequently, it is not necessary to think of these theories as being literally descriptive of the underlying causal processes of nature. And, if we do so, we become committed to the absurd view that many, if not all, of the basic principles of Newtonian dynamics, of special relativity, and of thermodynamics are just vacuously true.

Van Fraassen's view, on the other hand, seems to come from taking model and systemic theories as typical, for the value of such theories derives, not from any insights they may provide about the workings of nature, but from their capacity to systematize our knowledge of it. Consequently, it does not matter whether these theories are literally true or false. What matters is whether they are adequate to the task for which they were devised. Plausibly, therefore, to accept such a theory involves no more than the belief that it is so adequate—that it is, in van Fraassen's sense, an empirically adequate theory.

Scientific realists run into trouble when they try to generalize about scientific theories. To cope with laws which hold only for ideal systems of some kind (and most of the laws of nature are like this) and which would strictly be false if they were meant to apply to actual systems, many scientific realists think of them as being only good *approximations* to the truth (e.g. Boyd 1976, 1980). So, instead of van Fraassen's characterization of their position, they would accept something like this:

(1a) Science aims to give us, in its theories, a literally true story of what the world is like; and acceptance of a scientific theory involves the belief that it is at least approximately true.

Thus, they would consider such laws as conservation of energy and conservation of momentum, which strictly apply only to closed and isolated systems, as being only good approximations to the truth about actual systems. What is strictly true, they would say, is that the energy and momentum of any more or less closed and isolated system are more or less conserved. And, as science progresses, these essentially 'inaccurate' laws should be replaced by more accurate ones involving less idealization and hence greater faithfulness to the actual relationships among things. Think, for example, of the replacement of the ideal gas laws by van de Waal's equation. Many scientific realists thus envisage the eventual replacement of model theories by systemic ones in which all of the laws and principles are just true generalizations about how actual things behave.

However, if science is aiming to achieve such a result, it seems often to be pointing in the wrong direction, for a great deal of theoretical scientific research goes into devising increasingly abstract model theories, and relatively little into reducing the degree of idealization involved in our theories in order to make them more realistic. It is true that in economic forecasting, and in the applied sciences generally, researchers labour to develop increasingly elaborate computer simulations of real systems, taking into account as many as possible of the relevant variables so as to maximize the accuracy of their predictions. And I do not deny the importance of such research. But basic theoretical development in science tends, if anything, to proceed in the opposite direction—to greater abstraction and generality.

Take the development of Newtonian mechanics from 1700 to about 1900 (see Dugas 1955). In this period, no major scientist working in the field of mechanics developed theories which they thought were closer to the truth than Newton's. Most of them would have said that Newton's theory was true already. Nor, for that matter, did they think they were developing theories which were empirically more adequate than Newton's. So it seems wrong to suppose that they were aiming to increase either the realism or the empirical adequacy of the basic theory they were working with. Yet the great works of classical mechanics of Euler, the Bernoullis, d'Alembert, Fermat, Lazare Carnot, Lagrange, Laplace, Gauss, Coriolis, Hamilton, and Jacobi surely contributed *something* to fulfilling the aims of science, for they improved greatly our knowledge and understanding of mechanical processes. They solved many previously unsolved problems; they applied Newtonian theory in new ways; they discovered new principles and unsuspected symmetries; and they invented powerful new mathematical techniques for handling complex mechanical problems. I think we may conclude that scientists, *qua* pure scientists, are not always greatly concerned to make their model theories more realistic or more adequate empirically. They have other much more interesting things to do.

Van Fraassen's constructive empiricism is in trouble for these reasons. But he adds to his troubles by construing all theories on the model of model theories, for he is now committed to saying that the postulated entities of causal process theories have no more claim to be considered real existents than the theoretical constructs of model theories. Atoms, creatures of the Jurassic period, inertial systems, and possible worlds are all on a par, according to him, and to accept theories which apparently make reference to such entities involves only the belief that the theories are empirically adequate. Now, if the theories we are talking about are special relativity and 'possible worlds' semantics, van Fraassen's position is at least plausible. But it loses all plausibility if the theories in question are historical theories or theories of chemical combination. And the reason, I think, why this is so is that the postulated causes of the phenomena must be supposed to exist if the theory is to be accepted as doing what it purports to do; and normally we should expect to be able to find independent confirmation of their existence from various sources. The situation is quite different with the theoretical entities of abstract model theories. Since they are not postulated as causes, they are not supposed to have any effects. So we should not expect them to leave any traces or to manifest themselves in other ways, or indeed in *any* way at all. And that is why, apparently, we can play with them as we like or assign any properties to them we wish to produce a better theory. We know that no astronomers are going to discover inertial frames which don't have the properties we assign to them and that travellers are not going to stumble across other possible worlds where they shouldn't be and so spike our theories about conditionals.

The status of the theoretical entities of causal process theories is not like this, however. When the theory is accepted, we think we know only *some* of their properties. We know we might be wrong about them in some ways, and we might expect our picture of them to change somewhat. But, typically, we expect to discover more about them—to add to and refine our knowledge of them, and to explain why they have the properties they do. That is, we expect them to be like other physical things and to participate in various ways in causal processes, depending on what their properties are and what the surrounding circumstances are like. In short, we think of them, and expect them to behave, as real things do.[6] Moreover, our reasons

---

[6] According to James Clark Maxwell and P. W. Bridgman, an entity is physically real if it manifests itself in more than one way. G. Schlesinger (1963) calls this the Maxwell–Bridgman criterion for physical reality. It is surprising that this important criterion is not more widely known or discussed.

for believing in them are not basically different from our reasons for believing in more ordinary things. If the existence of atoms or of the moons of Jupiter were a legal issue, I think almost any jury would find the case proved by the ordinary rules of evidence.

Now, I think that van Fraassen is aware of all this. I don't think he embraces constructive empiricism out of naïveté or any failure to distinguish between causal process theories and model theories. His philosophical position probably arose out of a less ambitious theory about the status of the laws of special relativity and other abstract model theories (see van Fraassen 1969, 1970). Originally he argued that many of the principles involved in these theories were conventional. For example, he argued that it is ultimately a matter of convention whether we say that the one-way speed of light is the same in all directions. About this he would have said, following Reichenbach and the other positivists, there is no truth of the matter. However, he has now come to believe that his arguments for the conventionality of such principles apply *right across the board*. So, consistently with his earlier position, he ought to conclude that *in general* there is no truth of the matter concerning the fundamental laws and theories of science. But, as I understand him, van Fraassen does *not* accept this general conclusion. There *is* a truth of the matter, he now wants to say, *but we cannot really know what it is*. For the best that we can ever hope to do is construct a system of empirically adequate theories. Anything beyond this is necessarily beyond our grasp.

III. CONVENTIONALISM

Conventionalists argue that there are many questions in science which call, not for further empirical investigation, but for decision. Is the one-way velocity of light equal to its round-trip velocity? Is the geometry of space Euclidean? Does a body which is not acted upon by forces continue in its state of rest or uniform motion in a straight line? These questions, and many others, have been said by conventionalists not to be questions about what is true of reality but ones which can be resolved only by stipulation or definition. Concerning these questions, there is said to be no truth of the matter.

There are three main arguments for considering the statement of a law or theory to be conventional. First, there is *the circularity argument*, which was frequently used by Reichenbach. For example, in arguing for the conventionality of the principle that the one-way velocity of light is the same in all directions, he tries to show that we should have to presuppose the principle in order to prove it. He argues that to measure the one-way velocity of light, we should need clocks in synchrony at different places; that is, reading the same at the same time. Consequently, we must know what it is for two events (readings) occurring at different places to be simultaneous. But ultimately, he claims, we cannot determine whether two such events are simultaneous unless we make some prior assumptions about the one-way velocity of light. And he then goes on to say that 'the occurrence of this circularity proves that simultaneity is not a matter of knowledge, but of a coordinative definition, since the logical circle shows that a knowledge of simultaneity is impossible in principle' (Reichenbach 1958: 127). He concludes that the law in question is conventional.

It has now, I think, been conclusively established that Reichenbach was wrong in thinking that one has to make some assumption about the one-way velocity of light to determine the simultaneity of distant events. There are, in fact, several procedures which could be used to establish clocks in a relationship of distant synchrony which, logically, do not depend on this assumption (as argued in Ellis and Bowman 1967). Consequently, there are several logically independent criteria for distant simultaneity, and the standard signal synchrony criterion is just one of them.

However, conventionalists have a second argument to fall back on—*the argument from the need for definition*. It may be conceded that distant simultaneity is a multi-criterial concept, so that any of a number of different criteria for distant simultaneity might be chosen to define this relationship. Still, a choice has to be made,[7] and the law or principle which underlies the choice will have to be considered to be true by definition. This principle might be the one-way-light principle (that the one-way velocity of light is constant) or the principle of slow clock transport (that locally synchronized identical clocks transported sufficiently slowly remain in synchrony), or it might be some other principle. But surely at least one of these laws or principles must be regarded as conventional if the

---

[7] The use of this argument is hard to document. But the assumption that there is a need for definition to fix meaning *before* any questions of truth or falsity can arise is a common background assumption of logical positivists. This assumption was challenged, most notably by Quine and Putnam, in the course of their attacks on the analytic/synthetic distinction in the 1950s and early 1960s. Nevertheless, the implications of their work for the fact/convention distinction in the philosophy of science were not understood. In 1967, Bowman and I (pp. 116–36) took the challenge into the centre of the philosophy of science camp by arguing specifically that their most cherished convention, distant simultaneity, was not in any interesting sense conventional. Grünbaum, Salmon, and van Fraassen (all in *Philosophy of Science*, 36 (1969)) replied in terms which left no doubt that they still considered the fact/convention distinction to be absolutely indispensable for the analysis of science, and stoutly defended the conventionality of distant simultaneity. The arguments which follow are derived from my reply to them (Ellis 1971).

concept of distant simultaneity is to be well defined. If the one-way-light principle is the chosen one, then it is conventional, and all the others are empirical.

It is, however, arbitrary to pick *one* law out of a law cluster like this, and say that it is conventional while the rest are empirical, if no good reason can be given for choosing it. And, if the choice is arbitrary, then we might as well have chosen another one. Consequently, any given law in such a law cluster might arbitrarily be regarded as empirical or conventional, depending on how we choose to axiomatize our system. Assuming that the 'simultaneity' law cluster is like this and that the one-way velocity of light has been set equal to the round-trip velocity by definition, then the principle of slow clock transport is empirical. It is, however, a matter of convention that the one-way-light principle is conventional, and it is also conventional that the principle of slow clock transport is empirical. For we could equally well have reversed the roles of these two principles. Assuming that the special theory of relativity is correct, we know that the 'simultaneity' law cluster exists. In all of the vast literature on distant simultaneity, there is no argument for preferring the standard light signal to the slow-clock-transport definition of distant simultaneity, or conversely; and we may reasonably assume that there is none. Therefore, it is arbitrary which principle we choose to call conventional and which empirical.

This being the case, why should we choose at all? What difference would it make, either to our practice or to our beliefs, if we thought of all the laws in the cluster as empirical laws? Suppose that, contrary to the predictions of the special theory of relativity, we found that clocks synchronized by slow transport were not in standard signal synchrony? Would it make any difference to how we should proceed which of the two principles we had chosen to call true by definition or convention? It is my belief that it would make no difference at all. A radically new space-time theory would be needed, and I do not believe that, in constructing it, we should feel in any way constrained to accept what we had earlier said was true by definition or to reject what we had said was empirical.

In general, the empirical/conventional distinction is of no practical importance in science. Scientific practice is not affected by what we might choose to say is conventional or what we might think of as empirical. So we can just as well regard all of the laws and theories of science as empirical— at least in the sense that they are open to revision in the light of experience. The argument from the need for definition is, therefore, no good argument for conventionalism, for, in general, there *is* no need for definition. It may be useful or even necessary to offer a definition of a term when introducing it to the profession or to students for the first time, but the statement of that definition enjoys no privileged immunity to revision or even rejection in the light of further experience.

The third and perhaps most important argument for conventionalism is that from *empirical underdetermination*.[8] We can discover empirically, perhaps, that the 'simultaneity' law cluster exists, but this does not bind us to accepting any or even the conjunction of all of the laws in the cluster as defining the relationship of distant simultaneity. For we are free to adopt some non-standard signal or transport definition of distant simultaneity, say one which makes the one-way velocity of light a function of direction, or relative clock rates a function of their relative positions, and use this to determine what all of the other laws in the cluster are. Given such a definition, there will still be a 'simultaneity' law cluster, but it will be a different one. No doubt, such a co-ordination would seriously complicate our physics, but, so the conventionalist maintains, it cannot be ruled out on any a priori or empirical grounds. Consequently, there can be no truth of the matter concerning *any* of the laws in the cluster. They are *all* conventional.

This move is the most serious one for the scientific realist who wishes to retain the objectivity thesis along with the correspondence theory of truth, for it threatens his whole programme of interpreting the laws and theories of science realistically. It is the main reason why those positivists, who accepted the correspondence theory, rejected the objectivity thesis for some form of conventionalism; and it is one of the main reasons why some, who wish to retain the objectivity thesis, have rejected the correspondence theory of truth in favour of some form of coherence theory. The point is that the coherence theorist can argue that what is true is what occurs in the best (ultimately best) theory, and is not embarrassed by this conventionalist argument; for no one would pretend that a non-standard definition of distant simultaneity would produce a theory which was anything like *as good* as the special theory of relativity. It might be empirically equivalent, but we should have to allow that there are some strange spatial asymmetries in many of our laws of nature for which we have no adequate explanation.

Now, van Fraassen, as I understand him, accepts the main argument for conventionalism—that from empirical underdetermination—and continues to accept the correspondence theory of truth; but he rejects the conventionalist conclusion that there is no truth of the matter concerning those laws and theories which are empirically underdetermined. He agrees with the scientific realists that they are either true or false; but, unlike

---

[8] This is the argument most commonly used by empiricists from Mach and Poincaré onward.

them, he considers that whether they are true or false is ultimately not an empirical question but a metaphysical one. Truth, for van Fraassen, is a metaphysical concept.

IV. EMPIRICAL UNDERDETERMINATION

The implications of the conventionalist argument from empirical underdetermination are serious for the scientific realist, for empirical underdetermination would appear to be a feature of *all* of our theories. Consequently, it threatens to render even the realists' belief in a physicalist ontology either false or metaphysical. If it is always possible to construct another theory which is incompatible with a given theory, but is empirically equivalent to it, then all theories are empirically underdetermined. Therefore, if one accepts both the correspondence theory of truth and the objectivity thesis, one may be forced to consider all theories to be either false or metaphysical—a conclusion which would be unavoidable if the general principle of empirical underdetermination were accepted.

This thesis of empirical underdetermination is not, of course, the obvious point that our theories are *inductively* underdetermined by the evidence we have for them. Things might turn out to be grue or bleen rather than green or blue. The point is that there are incompatible theories which are empirically equivalent in the sense that *no* evidence could distinguish between them; and the claim is that our theoretical understanding of the world could be underdetermined even by the supposed totality of empirical evidence. This thesis, if it is true, is the real threat to scientific realism; for what reason could we have to believe in the existence of any theoretical entity if we could be assured that there is another theory, empirically equivalent to the one in which it is postulated, which may assume the existence of other, quite different kinds of entities?

Although the thesis of empirical underdetermination is commonly asserted, the arguments for it are not as compelling as they are usually thought to be. First, there is the failure of Carnap's programme of reductive analysis (summarized in Carnap 1956). All attempts to define the theoretical terms of science in an observational language have met with failure. Even the simplest theoretical predicates have stoutly resisted such an analysis. Consequently, the conviction has grown that theories say more than anything that can be said in an observational language. It is not just that they go beyond the evidence. All inductive generalizations do that. They go beyond anything for which there could be inductive support. For they cannot even be expressed in a language the terms of which are purely observational. Consequently, it is held that all theories must be strongly underdetermined by evidence. Even if all possible observations had been made, and were known to have been made, and our theories were compatible with this supposed totality of evidence, our theories would still say more than this, and so be underdetermined by it. This argument is plausible, I think, but not entirely convincing, for it does seem to rely on a rather naïve inductivism—the crucial premiss being that if A cannot be expressed in an observation language L, then A cannot be supported by any evidence which is expressible in L. And I know of no good arguments for this premiss.

Second, and I think more importantly for van Fraassen, there are many persuasive examples of empirical underdetermination—ones which make it reasonable to believe that all theories are similarly underdetermined. At any rate, van Fraassen does argue for empirical underdetermination primarily by cases, and is not content to rely on sweeping general arguments such as the one I have briefly presented.

Without going into detail, I would admit that there are persuasive arguments for the existence of empirically equivalent theories, but, even if they are sound, they do not establish the general thesis. For one cannot establish that all theories are ultimately underdetermined by showing that some are. One might, for example, try to isolate our dynamical theories and our theories of space and time, for which the most detailed arguments for conventionality exist, and admit that at least some of the entities which occur in these theories, such as forces, may not be real, but preserve a realist interpretation for the entities of atomic and subatomic physics by arguing that the theories which postulate the existence of these entities are not conventional in the same way. This is the strategy I had in mind when I spoke of 'scientific entity' realism (Ellis 1979: 45 n. 15).

Now, I do not particularly wish to defend 'scientific entity' realism in this form, because I would now prefer the stronger internal realist position. However, it is worth pointing out that the main arguments for the conventionality of our space-time and other abstract model theories do not carry over as arguments for the conventionality of our causal process theories. For, if the entities postulated in these theories exist, we should expect them to manifest themselves in various ways and to participate in causal processes *other* than those described in the theories we have so far developed. Consequently, we should expect confirmation to come from unexpected sources as our causal process theories are developed. Hence, theories which cannot be distinguished between empirically in one theoretical context may be distinguished in another. New theoretical developments may show some empirical consideration to be differentially relevant to the truth of the two theories, which previously would have been considered to be empirically equivalent.

It is possible, for example, to construct two theories of chemical combination to explain why one volume of hydrogen combines with one of chlorine to form two volumes of hydrogen chloride (described fully in Ellis 1957). One is the classical theory of Avogadro; the other postulates the existence of certain 'gas numbers' characteristic of elemental and compound gases and certain laws relating these gas numbers to combining volumes. In terms of empirically testable consequences, the theories are equivalent. Nevertheless, they are very different theories. One postulates the existence of atoms and molecules and para-mechanical processes of chemical combination; the other does not. We may add the further hypothesis to the second theory that there are no such things or processes—just to make sure that the two theories are incompatible. Now, in spite of the fact that these two theories have the same empirically testable consequences, they clearly are not, and I want to say were never, empirically equivalent. For they offered quite different prospects for development, and, as new theories were developed, new facts became relevant to the acceptance of one of them (Avogadro's theory). The facts of electrolysis, for example, are not empirical consequences of Avogadro's theory, but they certainly supported it, because they could be readily explained on the atomic-molecular model which it used.

Many philosophers think of theories as sets of sentences implying observational conditionals, that is, sentences of the form

$$C_1 \wedge C_2 \wedge \ldots C_n \supset O$$

where $C_1, C_2, \ldots C_n$ are empirically determinable boundary conditions and O is an observation statement (e.g. an event of kind K is occurring in the space-time region S). This model of a theory is so widely accepted and used that it may reasonably be called 'the standard model'. On this model, it is natural enough to say that two theories are empirically equivalent iff they imply the same set of observational conditionals. For only if this is so will they have the same empirical consequences. Yet, according to this criterion, my 'gas number' theory is empirically equivalent to Avogadro's—which in my view is absurd. It is absurd because the evidence in favour of Avogadro's theory and against the gas number theory, is now overwhelming. This is so because the original theory of Avogadro has become embedded in a very general and powerful theory of chemical combination and has well-established links with (in the sense that it has hypotheses in common with) a wide range of other physical and chemical theories.

Avogadro's theory has gained support from these other theories because it has become an integral part of them. The evidence in favour of Avogadro's theory cannot be identified with the confirmation that its observational consequences have received—at least, not if 'observational consequences' is understood as narrowly as the standard model would suggest.

The point is a Duhemian one. Theories do not normally occur in isolation, and evidence for or against a theory can come from unexpected quarters. This evidence may be unexpected, not because we have failed to carry out the relevant deductions from the axioms, but *because it is not a consequence of these axioms at all*, at any rate not of these axioms *alone*. It may be evidence which can be seen to be relevant to the theory in question *only* because some new, linked theoretical development has occurred. Therefore, unless we can know in advance what theoretical developments might occur, we cannot say in advance what evidence, if any, might distinguish between theories which on present indications would appear to be empirically equivalent. Therefore, assuming that we cannot know what theoretical developments will occur, we cannot ever be fully justified in claiming that two incompatible theories are empirically equivalent in the sense that no evidence could ever distinguish between them. That they cannot be distinguished, given our present theoretical understanding of the world, does not imply that they cannot be distinguished. A scientific realist may therefore be able to defend his position against the empiricist/conventionalist arguments for the empirical underdetermination of theories, for he can argue that the underdetermination thesis *cannot be demonstrated*. It cannot be shown, except by fiat, that there is *any* genuine case of empirical underdetermination.

The reason for this perhaps needs emphasizing. It derives from the *openness of the field of evidence*. By the field of evidence of a theory I mean the set of possible empirical discoveries relevant to its truth or falsity. This set is to be clearly distinguished from the set of *empirical consequences* of the theory. The empirical consequences of a theory are the observational conditionals entailed by it. Hence, the set of such consequences is a subset of the set of logical consequences of a theory. But not so the field of evidence. There can be evidence for a theory which is in no way entailed by it, and evidence against it which does not contradict it. Faraday's laws of electrolysis, for example, certainly supported the atomic-molecular theory of Avogadro, since the atomic-molecular model proposed by Avogadro to explain certain facts about chemical combinations could readily be adapted to explain the facts of electrolysis. But Avogadro could not, without making additional assumptions, have deduced Faraday's laws from his original theory. The failure of the Michelson–Morley experiment, on the other hand, surely proved to be evidence against Newtonian mechanics, even though this null result was compatible both with Newton's laws of motion and with his law of gravity. Newton's laws of mechanics simply had *nothing to say* about the behaviour of electromagnetic radiation.

This point is of considerable importance in the present context, for it implies that claims of empirical equivalence should be treated with caution or be relativized to a given stage of theoretical development. From the fact that two logically distinct theories have the same empirical consequences (in the sense that they imply the same set of observational conditionals), it does not follow that there is no evidence which could distinguish between them. Nor does this follow from our inability to think of any way in which they might be so distinguished. For, so long as the theories we are concerned with are logically distinct, it is always logically possible that unforeseen theoretical developments will occur which will enable us to devise tests which would decide between them. To justify accepting any empirical equivalence claim, therefore, we should at least need a good theoretical argument that no such developments could occur.

Defenders of the thesis of the conventionality of distant simultaneity claim to have just such an argument. They claim that the facts relating to signal speeds and slow clock transport make it impossible for us ever to distinguish empirically between standard and non-standard relativity theories—for example, theories which make the one-way speed of light in an inertial system some kind of function of direction.[9] For reasons I have given at length elsewhere, I am not convinced by their arguments (Ellis 1971). However, even if it were established that there is some significant conventionality in our theories of space and time, this would not establish the *general* thesis of empirical underdetermination which van Fraassen needs for his argument to apply to all theories. For the theoretical entities of causal process theories are different from those of the model theories we have been considering. They are the sorts of thing which, if they exist, we can expect to discover a great deal more about as we improve our understanding of nature. And, because of the openness of the field of evidence concerning them, we can never be sure that logically distinct but empirically equivalent causal process theories cannot ultimately be a distinguished empirically. So the general empirical underdetermination thesis van Fraassen needs to establish his position is, as yet, not proved.

[9] See the panel discussion on distant simultaneity with papers by A. Grünbaum, W. C. Salmon, B. C. van Fraassen, and A. Janis (*Philosophy of Science*, 36 (1969)) for the detailed argument.

As far as I know, the only way of demonstrating the empirical underdetermination thesis for causal process theories is to build empirical equivalence into the specification of the supposed, logically non-equivalent, alternative theories. For example, instead of the theory T, we might have the theory T', where

T' = Although the world is not as theory T says it is, the world behaves, so far as we can ever tell, as if T were true.

Of course, there are all sorts of variants of this, like

T'' = Actually, we are all brains in a vat, but . . .

or

T''' = Actually, the world began five minutes ago (local time), but . . .

But the variations just illustrate the general strategy. So the fundamental issue which needs to be considered is whether the metaphysical realism presupposed by such specifications of alternative theories is acceptable to a scientific realist. If it is, then I think van Fraassen wins, for the following theses cannot be consistently maintained:

1. All theories are empirically underdetermined in the sense that logically non-equivalent but empirically equivalent alternative theories always exist.
2. The choice between empirically equivalent theories cannot be made on any grounds except pragmatic ones (simplicity, symmetry, elegance, explanatory power, etc.).
3. Pragmatic considerations have no relevance to truth or falsity.
4. We have good reason to believe that the world is more or less how our best theories say it is.

In discussing the issue of metaphysical realism, let me contrast it with my own position of internal realism. For I want to show how scientific realism can be defended by rejecting thesis 3 and accepting a pragmatic theory of truth. Also, in retrospect, I think that this has always been the basic issue on which van Fraassen and I have disagreed.

V. INTERNAL AND METAPHYSICAL REALISM

To know that something is the case, we must somehow be justified in believing it, and, moreover, it must be true. I say 'somehow' because apparently not every case of justified true belief is a case of knowledge. But the refinements here need not concern us, since the point I want to make is independent of how the concept of justification may be spelled out. So let us just say that knowledge is true and justified belief and ignore the refinements. Then, apparently, there are two kinds of conditions for knowledge: one concerns the way the world is, and the other the justification of our beliefs about it. Let us call them the *ontological* and the *epistemic* conditions, respectively. Now, I think that nearly everyone would agree that satisfaction of the epistemic conditions does not entail satisfaction of the ontological ones, or conversely. We may be justified in believing what is false and not justified in believing what is true. This is common ground between internal and metaphysical realists. Where they differ is in the extrapolation of this point to the limit of what is ultimately justifiable. For the internal realist, what is true, if anything, *is* just what is ultimately justifiable. For the metaphysical realist, however, truth remains independent of rational evaluation, even in the projected limit of this process.

It is important to be clear that the internal realist does *not* equate truth with *warranted assertability* or *reasonable belief*. For these notions are tied to existing or available evidence, and perhaps also to certain background assumptions and theories which, in the given context, are not in question. No, for the internal realist, truth is a kind of limit notion of reasonable belief. We all believe that we are limited in what we know and that probably some of the things we believe are false, but we do not, for these reasons, think that it is unreasonable for us to believe what we do. Circumstances can change what we may reasonably believe. If new evidence is discovered, if a better theory is proposed, if results previously accepted are brought into question, then what we may reasonably believe is affected by such changes. Moreover, we should normally consider these changes to be ones for the better—to be changes which *improve* our knowledge and understanding of the world. Truth, for the internal realist, is the supposed perfection of this process. It is what we should believe, if our knowledge were perfected, if it were based on total evidence, was internally coherent and was theoretically integrated in the best possible way. There are, of course, very great problems in trying to explicate this concept of perfection. But for the internal realist, this is what truth, if it exists, must be.

The metaphysical realist, on the other hand, has a different concept of truth. For him, truth is a relationship between what is true and what it is true of, which holds independently of our epistemic values. Our epistemic values are those which come obviously into play whenever we are involved in evaluating (for truth or falsity) what other people say or believe or in re-evaluating our own beliefs. We judge *how well* something has been attested or supported or explained or *how well* it coheres with other things that we think we know. And to do this, we must have some system of preferences. The values which give rise to these preferences are our epistemic values. The metaphysical realist normally does, and certainly should, admit that we have such values. But truth, he would say, is not dependent on what these values may be. We may, as a matter of fact, have evolved criteria for epistemic evaluation which are well adapted to the goal of discovering what is true. But, then again, we may not. Even the perfect theory (supposing it existed) might not be true, though the world and our systems of epistemic values might be such that we could never turn up (or have turned up) any evidence which would give us good reasons to doubt it.

One's concept of truth need not greatly affect one's beliefs about reality. Consequently, internal realists and metaphysical realists may have quite similar ontologies. There is, for example, no reason why an internal realist should not be a physicalist, for there is no reason, which is not a reason for anyone else, why an internal realist should not consider the physicalist ontology to be the theory that it is most rational to believe. On the contrary, there would appear to be very good reason why he/she should do so. Moreover, an internal realist can, and indeed should, be a realist in the quite full-blooded sense of believing that there is a reality which exists independently of anyone's knowledge of it or beliefs about it. For surely, on the best available theories that we have concerning the nature of the physical world, reality would be much the same even if the human race had never existed. Any sensible person must believe that, and internal realists can be eminently sensible. Moreover, the independent reality which the internal realist may believe in is not a featureless noumenon, but a world of physical objects with physical properties interacting with each other in the usual ways. If it is right to believe that these things exist, and if truth is what it is right to believe, then, of course, it is true that they do.

Internal and metaphysical realists differ from each other essentially only in that they have different concepts of truth. But this difference gives rise to others. For example, the metaphysical realist is open to a kind of sceptical challenge: if truth is independent of our epistemic values, then what reason do we have to believe that those theories which we should judge to be the best are nearer to the truth? How do we know that our epistemic values are well adapted to discovering what is true? We can investigate nature and develop a theoretical understanding of the world, but we cannot compare what we think we know with the truth to see how well we are doing. We cannot even be assured that science has made progress toward its goal of discovering the true nature of reality. The internal realist, on the other hand, is not plagued by these sceptical doubts. Our epistemic values *must* be adapted to the end of discovering what is true, because *truth* is just the culmination of the process of investigating and reasoning about nature in accordance with these values.

Given these characterizations of the two positions, I think most scientific realists would consider themselves to be metaphysical realists, for this is the position which must be associated with adherence to the correspondence theory of truth. The idea that one could be a scientific realist and yet have a concept of truth so clearly reminiscent of Bradley's would strike most scientific realists as absurd. Nevertheless, I think they are wrong about this. I think they go wrong because they systematically misunderstand the position of the internal realist. They take internal realists to be denying that there is a reality which exists independently of anyone's knowledge or understanding of it. But, of course, they do not, or at least they *need* not, do this. I certainly believe in the existence of such a reality, because that is what I think any sane person *ought* rationally to believe. The position I should defend, therefore, is not an idealist one in the usual sense, even though my concept of truth is similar in some ways to the coherence concept defended by some idealists.

The point that an internal realist does not have to deny the existence of an independent physical reality needs to be stressed. But I fear that the name that Putnam (1979) has given to the kind of position I want to defend will ensure that it is constantly misunderstood. There is, however, nothing 'internal' about my ontology, and it is not my view that reality is a construction out of ideas or sense impressions or whatever. Reality exists and has virtually all of the properties that it has, independently of anyone's knowledge or belief about it. That is, the following counterfactual conditional is one that I think is true, that is, right to believe:

If human beings had never existed, the world would not be so very different from the way it is.

Of course, there would be no cities, ploughed fields, books, or theories about the world, and maybe some species of animals would not now be extinct, if we had never existed. But, overall, the difference would not be very great. The difference would be even less great if we just had different beliefs about the world. It would still be different, because our beliefs are parts of reality and beliefs have effects, but the basic ontology to which we should subscribe would be unaffected by our having or coming to have different beliefs about it.

An internal realist can thus believe in the existence of an independent reality. That is, he can believe that the ontology to which he subscribes and the basic laws of nature would be the same, whatever he might believe. Indeed, it is part of his understanding of the world that this should be so. The independence here is a *causal* independence. One cannot change the basic structure of the world by changing one's beliefs about it.

But what about changes in our epistemic values? Wouldn't a change in our epistemic values change the way the world is by making it right for us to believe things about the world which it would not now be right for us to believe? Isn't an internal realist not, after all, a kind of idealist?

First, let me distinguish between two sorts of changes. If I change the position of my little finger, then in one sense I change Sirius because I change its relationship to my little finger. But in another sense I do not change Sirius, because there is no causal influence—at least nothing which will be felt on Sirius for a long time. Changes of the former kind are called 'Cambridge' changes. What is characteristic of such changes is that they are brought about *without causal influence* on the object that is said to be changed. The causal explanation of the change in the position of Sirius relative to my little finger will refer only to events occurring in me. Now, in this sense a change in our epistemic values effects only a Cambridge change on the world. The explanation of what has occurred will refer only to events occurring in us. So anyone with our present epistemic values would not see the world as having changed. He would see only a set of changes in the way people think about it.

But suppose we all changed, so that we all became different sorts of beings equipped with a different system of epistemic values. Then, presumably, there would be a different way the world is for the beings we then became. Yes, I think that is right. The way the world is, is relative to the sorts of beings we are. That is one of the consequences of internal realism. It does not make it wrong for us to believe what we do about the world. Nor would it necessarily be wrong for us to believe something different if we were different sorts of beings. Nor is there any third standpoint from which the belief systems of two different sorts of beings could be compared and evaluated, or, if there is, then it enjoys no privileged position. So, according to the internal realist, there is no way that the world is absolutely, only ways in which it is relative to various kinds of beings, none of which can claim absolute priority.

Does this then mean that an internal realist is after all a kind of idealist? Yes, in so far as he believes that the way the world is for us is dependent, although not causally dependent, on the sorts of being we are, an internal realist might fairly be called an idealist of sorts. But he is not, or certainly need not be, an idealist in the sense that he denies the existence of an independent reality. For he may well think, as I do, that it is right to believe in the existence of a reality which is not affected by our beliefs about it and is not dependent on our experience of it. There is, therefore, no strong incompatibility between internal realism and scientific realism. An internal realist can accept most of the principal theses of scientific realism.[10] And this paper has been a defence of scientific realism from an internalist perspective.

But what of the alternative? What is wrong with metaphysical realism? Is there any reason why a scientific realist should not be a metaphysical realist? Well, yes, I think there is. First, metaphysical realism is essentially a sceptical position. If even the perfection of human knowledge by human standards does not necessarily lead to truth, then the truth is essentially unknowable, as van Fraassen has shown. We can have no reason to think that improving our knowledge and understanding of the world brings us any nearer to the truth. Second, there cannot be a good argument for metaphysical realism. If the assumption that there is a way the world is, independent of our *epistemic* values, had any explanatory power, then it would be right for us to believe in it. But in that case it would be part of our world and not an absolutely independent reality. We do, I think, have good arguments that there are things which exist independently of our *knowledge and understanding* of the world. The internal realist accordingly embraces them. Likewise, if we had good arguments for the existence of a transcendental noumenon, then it would be right for us to believe in it, too. But let us not pretend that we can frame an argument which would be persuasive whatever our epistemic values might be and, so, binding on all thinking creatures. Third, the assumptions of metaphysical realism are unnecessary. Human truth is all we can ever aspire to. If there is another kind of truth for beings of another kind, then that's their problem. There is not, and cannot be, any absolute truth, and therefore there cannot be any way that the world is, independently of how we or some other kind of creature would evaluate its beliefs about it. So I conclude that, if you want to be a scientific realist, then you had better be an internal realist.

But isn't there, finally, a crucial objection to this position? I say that there is a way the world is for us, and that perhaps for other sorts of beings there is a way, a different way, it is for them, and that neither can claim priority. But surely there has to *be* a world, a *common* world, to which the members of both species are reacting. Well, yes, I think that is right. I am not denying the existence of a world or asserting the existence of a multiplicity of worlds. What I am denying is that there is any *way* that the world

---

[10] But not, of course, the correspondence theory of truth.

192   BRIAN ELLIS

is, independently of how it is for various kinds of beings. Different beings may have different perspectives on the world, but there is no reason to think that there is any perspective which can claim priority. But perhaps if we met some aliens who had a different perspective, we could explain why they see the world as they do, differently from us. So perhaps in this way we could achieve a better understanding of how the world really is. Maybe so, but we still could not achieve any truly objective stance, for the theory we constructed to explain the belief systems of creatures different from us would still be a theory *of ours* which we should have to evaluate as we would evaluate any other theory. And, if the aliens constructed a theory about how *we* thought about the world and evaluated it by *their* standards, then there is no guarantee that it would be the same. So no objective stance which would define the way the world is independently of our epistemic values seems to be possible. Yet the metaphysical realist believes that there is a way the world is independently of our epistemic values. But I confess that I cannot make much sense of this view. Nor can I see any reason at all for holding it. Why do things behave as if the theory T were true? The perfectly adequate answer is that it is because the world is, for us, a T world.

The concept of absolute or metaphysical realism is like the concept of absolute space. It does no useful work. We can explain anything that can be explained about how bodies move without supposing that there is any absolute motion. Likewise, we can explain anything that can be explained about our knowledge of reality—for example, the phenomena of epistemic convergence—without supposing that there is any way that the world is absolutely. It is enough if there is a way (for us) that the world is *for us*.

REFERENCES

Armstrong, D. M. (1978). *Universals and Scientific Realism*. 2 vols., Cambridge: Cambridge University Press.
Boyd, R. (1976). 'Approximate Truth and Natural Necessity.' *Journal of Philosophy*, 73: 633–5.
——(1980). 'Scientific Realism and Naturalistic Epistemology.' *PSA vol. 2*, 613–62. East Lansing: Philosophy of Science Association.
Bradley, F. H. (1914). *Essays on Truth and Reality*, ch. 7. Oxford: Clarendon Press.
Campbell, K. K. (1976). *Metaphysics: An Introduction*. Encino, Calif.: Dickenson Publishing Co.
Carnap, R. (1956). 'The Methodological Character of Theoretical Concepts.' In H. Feigl and M. Scriven (eds.), *Minnesota Studies in the Philosophy of Science*, i. 38–76. Minneapolis: University of Minnesota Press.
Dugas, R. (1955). *A History of Mechanics*, tr. J. R. Maddox. Neuchâtel, Switzerland: Éditions du Griffon.

Duhem, P. (1954). *The Aim and Structure of Physical Theory*, tr. P. P. Wiener. Princeton: Princeton University Press.

Ellis, B. D. (1957). 'A Comparison of Process and Non-Process Theories in the Physical Sciences.' *British Journal for the Philosophy of Science*, 8: 45–56.

—— (1965). 'The Origin and Nature of Newton's Laws of Motion.' In R. G. Colodny (ed.), *Beyond the Edge of Certainty*, 29–68. Englewood Cliffs, NJ: Prentice-Hall.

—— (1971). 'On Conventionality and Simultaneity—A Reply.' *Australasian Journal of Philosophy*, 49: 177–203.

—— (1979). *Rational Belief Systems*. Oxford: Basil Blackwell.

—— and Bowman, P. (1967). 'Conventionality in Distant Simultaneity.' *Philosophy of Science*, 34: 116–36.

Grünbaum, A. (1969). 'Simultaneity by Slow Clock Transport in the Special Theory of Relativity.' *Philosophy of Science*, 36: 5–43.

Harré, R., and Madden, E. H. (1975). *Causal Powers*. Oxford: Basil Blackwell.

Janis, E. I. (1969). 'Synchronism by Slow Transport of Clock in Noninertial Frames of Reference.' *Philosophy of Science*, 36: 74–81.

Putnam, H. (1979). 'How to Be an Internal Realist and a Transcendental Idealist (at the same time).' In *Language, Logic and Philosophy: Proceedings of the Fourth International Wittgenstein Symposium*, 100–8.

Reichenbach, H. (1958). *The Philosophy of Space and Time*, tr. M. Reichenbach and J. Freund. New York: Dover.

Salmon, W. (1969). 'The Conventionality of Simultaneity.' *Philosophy of Science*, 36: 44–63.

Schlesinger, G. (1963). *Method in the Physical Sciences*. London: Routledge and Kegan Paul.

Smart, J. J. C. (1969). 'Quine's Philosophy of Science.' In D. Davidson and J. Hintikka (eds.), *Words and Objections*, 3–14. Dordrecht: Reidel. Also see Quine's reply in the same volume, 292–4.

Van Fraassen, B. C. (1969). 'Conventionality in the Axiomatic Foundations of the Special Theory of Relativity.' *Philosophy of Science*, 36: 64–73.

—— (1970). *An Introduction to the Philosophy of Time and Space*. New York: Random House.

—— (1980). *The Scientific Image*. Oxford: Clarendon Press.

IX

PROGRESS OR RATIONALITY? THE PROSPECTS FOR NORMATIVE NATURALISM

LARRY LAUDAN

I. INTRODUCTION

The theory of scientific methodology ('methodology' for short) appears to have fallen on hard times. Where methodology once enjoyed pride of place among philosophers of science, many are now sceptical about its prospects. Feyerabend claims to have shown that every method is as good (and thus as bad) as every other:[1] Kuhn insists that methodological standards are too vague ever to determine choice between rival theories.[2] Popper generally treats methodological rules as conventions, between which no rational choice can be made.[3] Lakatos goes so far as to assert that the methodologist is in no position to give warranted advice to contemporary scientists about which theories to reject or accept, thereby robbing methodology of any prescriptive force.[4] Quine, Putnam,[5] Hacking, and Rorty, for different reasons, hold that the best we can do is to *describe* the methods used by natural scientists, since there is no room for a normative methodology which is prescriptive in character. To cap things off, everyone in the field is mindful of the fact that the two most influential programmes in twentieth-century epistemology, associated with the inductivists and the

Reprinted from *American Philosophical Quarterly*, 24/1 (1987): 19–31, by permission.

[1] See esp. Feyerabend 1975.

[2] Kuhn: 'I am denying . . . neither the existence of good reasons nor that these reasons are of the sort usually described. I am, however, insisting that such reasons constitute values to be used in making choices rather than rules of choice. Scientists who share them may nevertheless make different choices. . . . Simplicity, scope, fruitfulness and even accuracy can be judged differently . . . by different people' (1970: 262). Kuhn devotes an entire chapter of his 1977 to developing the theme that all the rules of science radically underdetermine choice between rival theories.

[3] For Popper's line on methodological rules as conventions, see his 1959. Lakatos develops this theme in his (1971).

[4] Specifically, Lakatos holds that, although the methodologist can 'appraise' the relative merits of rival theories, he is never in a position—except long after the fact—to give any trustworthy 'advice' about which theories to accept and which to reject.

[5] See esp. Putnam 1981: 8.

Popperians respectively, have run into technical difficulties which seem beyond their resources to surmount.

Not everyone has given up on the methodological enterprise; but those who still see a prescriptive role for scientific methodology disagree about how to warrant that methodology. In part, this is a practical problem; there are several rival methodologies on the market, and they cannot all be right. But it is also a conceptual difficulty, for admitting that one has no non-arbitrary way of choosing between rival methods is to say that the epistemology of methodology (i.e. meta-methodology) is deeply suspect. Notoriously, practising methodologists cannot agree about the conditions under which a methodology of science would be warranted; small wonder indeed that they cannot agree about which methodology to accept. All of which invites the perception that there is no good reason for anyone to pay the slightest attention to whatever advice the methodologist might be moved to offer.

It is thus scarcely surprising that many methodologists are suffering from a loss of nerve. Yet, despite these gloomy facts, I believe that the methodological enterprise has been written off prematurely. I think it has been abandoned largely on the strength of a series of initially plausible, but ultimately specious, arguments. Still worse, its withdrawal from the field has conceded the high ground to various unsavoury forms of epistemic relativism, or to assorted, and almost equally suspect, forms of epistemic intuitionism.

But before I attempt to show that it is possible to breathe new life into the beast, I need to linger awhile on what has probably been the most influential argument in recent years against the methodological enterprise. That argument is rooted in what has been called the '*historical turn*'. That phrase refers to the views of writers like Kuhn, Feyerabend, Lakatos, Laudan, and Toulmin, who, during the 1960s and 1970s, argued that our philosophical notion of scientific rationality, as embodied in the various familiar methodologies of science, fails utterly to capture the rationality of the great historical achievements in science. These 'historicists', as I shall be calling them, claim to have shown that scientific giants like Galileo, Newton, Darwin, and Einstein violated all the familiar methodological canons of theory appraisal once advocated by philosophers. The historicists insist that the failure of great scientists to have made choices in conformity with the recommendations issuing from any methodology of science stands as a dramatic *reductio* of that methodology. They say that the failure of existing methodologies to allow a reconstruction of the action of former scientists *as rational* decisively shows the inadequacy of those methodologies. I believe that the requirement that a methodology or epistemology must exhibit past science as rational is thoroughly wrong-headed. Moreover, I hold that we can make reasonable and warranted choices between rival theories of method or knowledge without insisting on this requirement. The aim of this paper is to show that the requirement of rational reconstructibility is neither wanted nor needed.

II. THE REJECTION OF RATIONAL RECONSTRUCTIBILITY AS A META-METHODOLOGICAL CRITERION

The historicists believe that science has been a rational activity, and that a methodology of science is *eo ipso* a theory of rationality, whose adequacy is to be appraised by determining whether it exhibits the history of science as a rational activity. Both of these assumptions—that a methodology is an omnibus theory of rationality, and that we have independent and veridical means of judging the rationality of certain paradigmatic actions—beg several important questions. The idea that we can recognize certain actions or episodes as cognitively rational, independent of any theory about rational action, and thus that we can use those actions as litmus tests of the adequacy of a theory of methodology rests on a form of epistemic intuitionism which I have criticized at length elsewhere (Laudan 1986). Here, however, I shall focus not on the blatant intuitionism of this position, but rather on its over-hasty identification of methodological soundness with rationality.

Obviously, the historicists' strategy for evaluating methodologies requires a prior commitment to the thesis that science has been, at least in its key and formative episodes, a quintessentially rational activity. In their view, if a methodology gives advice at odds with the choices made by the Newtons, Einsteins, and Darwins of the world, then that is evidence—not that these giants were irrational[6]—but that the methodology under scrutiny has failed to explicate (our pre-analytic convictions) about that rationality. Virtually all the philosophers in this camp would subscribe to the spirit of Lakatos's assertion that:

A rationality theory [by which Lakatos specifically means a methodology] . . . is to be rejected if it is inconsistent with an accepted 'basic value judgment' of the scientific elite. (1971: 110)[7]

---

[6] For that would be to deny our 'intuitions' that great scientists have generally been rational cognitive actors.

[7] In the same vein, Lakatos says that 'better rational reconstructions . . . can always reconstruct more of actual great science as rational' (1971: 117).

Although some would quibble about whether a single such conflict is sufficient to reject a methodology, all historicists agree that a sound methodology must coincide with the bulk of the theory choices of the scientific élite. The meta-methodology associated with Kuhn, Feyerabend, and Lakatos proceeds in this way: one applies the methodology under review to prominent historical cases, for example, the choice between Cartesian or Newtonian physics, or between Newtonian and relativistic physics. In each case, one asks which rival the methodology would have picked out, assuming that it had available all and only the evidence which was accessible to the contending parties. One then compares the theory choices mandated by the methodology with the theory choices made by the consensus of great scientists who brought the episode to a close. If the methodology leads to choices congruent with those actually made by the scientific élite, then—according to the historicists—it has exhibited the rationality of those choices and the scientists who made them, and has established its own credentials. If the methodology under appraisal leads to preferences different from those actually made by the scientific élite, then that methodology should be rejected because of its unacceptable corollary that many great scientists have been irrational. In short, they take such non-congruence as a refutation of the methodology under appraisal.[8] Unfortunately, as I shall show shortly, *both* the assumption that a methodology has implications concerning the rationality of great scientists and the thesis that methodologies are to be evaluated in light of the degree to which they capture past science *as rational* are thoroughly wrong-headed.

As we have seen, the historicist subscribes to the following doctrines:

The rationality thesis (RT): most great scientists have made their theory choices rationally.

The meta-methodology thesis (MMT): a methodology of science is to be evaluated in terms of its ability to replicate the choices of past scientists as rational.

The historicists are surely right in thinking that existing methodologies often fail to pick out the theories which the scientific élite have chosen. Thus, Newton's physics was accepted long before it was known to have made any successful surprising predictions, thereby violating the rules of Popperian methodology. Galilean physics was accepted in preference to Aristotle's, despite the fact that Aristotle's physics was much more general than Galileo's, thereby violating Popper's and Lakatos's injunction that

8 See e.g. Kuhn 1970: 236.

successor theories should always be more general than their predecessors. Einstein accepted the special theory of relativity long before anyone had been able to show that all of classical mechanics was a limiting case of relativity, thereby running afoul of the methodologies of Popper, Lakatos, Putnam,[9] Reichenbach, Sellars, and most of the logical empiricists—all of whom require that a rationally accepted successor theory must explain everything explained by its predecessors.[10] For such reasons, Kuhn and Feyerabend reject all existing methodologies as inadequate. There is much that is fishy going on here, but, for purposes of this essay, I want to focus on just one of the howlers committed by most of the historicists. Its examination will show why methodology and rationality need to be sharply distinguished, and why MMT—the core meta-thesis of the historical school—is wholly unacceptable.

What in particular has gone wrong is that the historicists' meta-methodology has failed to reckon with the fact that *both* the aims *and* the background beliefs of scientists vary from agent to agent, and that this is particularly so when one is talking about scientific epochs very different from our own. If the aims of scientists have changed through time in significant respects, we cannot reasonably expect *our* methods—geared as they are to the realization of *our* ends—to entail anything whatever about the rationality or irrationality of agents with quite different aims.[11] Whatever else rationality is, it is agent- and context-specific. When we say that an agent acted rationally, we are asserting minimally that he acted in ways which he believed would promote his ends. Determining that an agent acted in a manner that he believed would promote his ends may or may not be sufficient to show the rationality of his actions; philosophers will quarrel about that matter. But few would deny that it is a *necessary* condition for ascribing rationality to an agent's action that he believed it would promote his ends. It follows that, whenever we judge an agent's rationality (or the rationality of an axiologically homogeneous community of agents), we must consider:

what actions were taken;
what the agent's ends or aims were;
the background beliefs which informed his judgements about the likely consequences of his possible actions.

[9] In referring to Putnam's methodology, I have in mind his methodological strictures about theory succession in his 1978, esp. lecture 2. He may himself have repudiated these doctrines in later writing.
[10] For a lengthy discussion of this example, see Laudan 1984: ch. 5.
[11] It is more than a little ironic that the 'historical' school in philosophy of science, which has insisted that methodology must capture the rationality of past science, has itself been chiefly responsible for teaching us that the aims of science have shifted significantly through time.

There is no viable conception of rationality which does not make these ingredients essential to, even if not exhaustive of, the assessment of an agent's rationality.

I shall now show that these ingredients are sufficient to expose the wrong-headedness of conflating methodology and rationality in the manner of the historicists. The argument is straightforward: to the extent that scientists of the past had aims and background beliefs different from ours, then the rationality of their actions cannot be appropriately determined by asking whether they adopted strategies intended to realize *our* aims. Yet our methodologies are precisely sets of tactical and strategic rules designed to promote our aims. It would be appropriate to use our methods to assess the rationality of past scientists only if their cognitive utilities were identical to ours, *and* only if their background beliefs were substantially the same as ours.

The key question here is whether past scientists had background beliefs and cognitive goals (or utility structures) substantially the same as ours. That the relevant background beliefs of historical agents differed significantly from ours seems too obvious to require elaborate argument. Scientists of the past clearly had theories about the world and causal beliefs different both from one another's and ours. Even if their cognitive aims were identical to ours, such differences in the relevant background beliefs between them and us would presumably lead them to assign very different utilities to various courses of action from those we would make.

The situation is made worse by the fact that the cognitive aims of past scientists differed significantly from ours. Indeed, historical scholarship of the last two decades—much of it coming from the pens of historicists themselves—has pretty definitively established that scientists of the past have had constellations of cognitive utilities very different from those which we currently entertain.[12] Newton, for instance, saw it as one of the central aims of natural philosophy to show the hand of the Creator in the details of his creation. It was, after all, he who insisted in the General Scholium to *Principia* that 'to discourse of [God] from the appearances of things, does certainly belong to Natural Philosophy'. (Boyle too saw the construction of a natural theology as a central task of science.[13]) Newton held that the scientist should aim at producing theories which were either certain or highly probable. Whenever Newton was confronted with a specific choice between rival theories, he would—for he probably was a rational man—tend to make the choice between them in light of his beliefs

---

[12] To say this is to assume, which is itself highly doubtful, that contemporary scientists and philosophers would all accept the same cognitive utility assignments.

[13] See esp. Boyle (1738).

about what would promote his cognitive ends. Does the mechanical philosophy tend to undermine any role for an active creator? If so, that constitutes grounds for rejecting it, on the grounds that it contradicts Newton's axiology. Does the Cartesian vortex theory lack secure 'proof' comparable to that for the inverse-square law? Then it should be jettisoned. The point is that judgements about the rationality or irrationality of Newton's theory choices have to be made in the light of Newton's cognitive values and against the background of Newton's prior beliefs. Ignore those ingredients, and one is no longer in a position even to address—let alone to settle—the question of Newton's rationality.

A critic might be prepared to grant that some of Newton's cognitive aims differed from ours, while insisting that certain Newtonian aims are identical to some of ours (the most familiar candidate: 'finding true theories about the world'). It is surely correct that past scientists have shared some of the aims we entertain as the appropriate aims for scientific enquiry, particularly if one couches them in such general terms that one glosses over relevant differences.[14] However, such commonality as there is proves to be insufficient to establish the appropriateness of assessing the rationality of the actions of former scientists against our aims. Such *partial* overlap of ends is of little avail, since one does not assess the rationality of an agent's action by determining whether he acted so as to promote only one among his cluster of aims. An agent's actions can be judged as rational only with respect to the weighted set of his cognitive utilities, not with respect to a proper subset of those utilities.[15] Hence, even if a Newton or a Darwin did share some of the aims which undergird our methodological musings, it does *not* follow that we can assess their rationality by determining whether their actions promoted *some* of their aims (specifically the ones they share in common with us). On the contrary, their actions can be determined to be rational only with respect to the suitably weighted product of their cognitive utilities. If the latter differs to any significant degree from ours, which a Newton's or a Darwin's or an Einstein's palpably does, then it is to no avail, in assessing the rationality of their actions, to ask whether they acted so as to promote our ends.

The problem we have our finger on is simply this: to the extent that our judging an agent's having acted rationally involves taking his aims seriously, then methodologies designed to promote aims different from those of the agent appear incompetent to pronounce on the agent's rationality. One might try to defend the general approach I have criticized here by

[14] Such as whether Newton, who surely aspired to the aim of 'true theories', operated with the same concept of true which we post-Tarskians do.
[15] Unless the utilities in the subset invariably co-vary with the larger utility structure.

saying that, in claiming a certain scientist to be rational, we are not alluding to his ends, but rather to the ends constitutive of scientific enquiry. On this way of approaching the issue, an agent is rational only in so far as his actions tend to promote these general 'aims of science', even if his intentions (i.e. the aims driving his actions) were quite different from those of science. This analysis makes it possible for rationality to involve a good deal of sleep-walking, in that agents may promote 'the genuine ends of science' (and supposedly thereby be rational) without intending to do so. I have no trouble with the suggestion that agents often end up furthering ends quite different from those which motivated their actions. Indeed, this seems to me to be quite a salient point (to which I shall return later). But I cannot accept the violence it does to our usual notion of rationality, entailing among other things that agents who acted effectively so as to promote their ends may turn out to be irrational (viz. if their actions failed to promote 'the' ends of science), and that agents who dismally failed to act so as to promote their ends can turn out to be rational (specifically, when their actions inadvertently further the aims of science). But there is an even more serious difficulty with this approach, for it assumes that there is a set of identifiable ends which are constitutive of science for all time. We have already seen that the aims of individual 'scientists' in one epoch are very different from those in another; it would be no more difficult to document the claim that the aims of the 'scientific' community change through time. (Contrast the aims of chemists during the period of Paracelsian dominance with the aims of chemists now.) But perhaps the proposal that certain aims are constitutive of science is meant to be stipulative rather than descriptive. But how could such a stipulative characterization of the axiology of science do any work for us whatever in determining the rationality or otherwise of the actions of Archimedes, Descartes, Newton, or Lavoisier? Do we decide that Lavoisier was irrational if his work failed to further ends which we now regard as constitutive of science? Surely not. Lavoisier's rationality could be assessed only by determining whether his actions furthered his own ends, but the status of Lavoisier's rationality in studying nature cannot coherently be made parasitic on whether his actions promoted ends which were quite different from his.

This much surely is clear. But its corollary is that it is equally inappropriate to judge the soundness of our methodologies by seeing whether they render rational the actions of great scientists of the past. *Because our aims and background beliefs differ from those of past scientists, determinations of the rationality of their actions and of the soundness of our methodological proposals cannot be collapsed into one and the same process. Rationality is one thing: methodological soundness is quite another.* Since that is so, the historicist's rejection of the methodological enterprise, like his rejection of specific methodologies, on the grounds that they render the history of science irrational, is a massive *non sequitur.*

But accepting the point that methodologies cannot be authenticated by asking whether they exhibit past science as rational seems to leave the philosopher of science more than a little vulnerable. The great attraction of MMT was that it promised a 'neutral' basis for choosing between rival methodologies. If we abandon MMT, then we have to ask ourselves afresh how one might go about warranting methodological proposals.

III. A NATURALISTIC META-METHODOLOGY

Let us start afresh by looking carefully at the epistemology of method-ology. A methodology consists of a set of rules or maxims, ranging from the highly general to the very specific. Typical methodological rules include the following:

Propound only falsifiable theories.

Avoid *ad hoc* modifications.

Prefer theories which make successful surprising predictions over the-ories which explain only what is already known.

When experimenting on human subjects, use blinded experimental techniques.

Reject theories which fail to exhibit an analogy with successful theories in other domains.

Avoid theories which postulate unobservable entities.

Use controlled experiments for testing causal hypotheses.

Reject inconsistent theories.

Prefer simple theories to complex ones.

Accept a new theory only if it can explain all the successes of its predecessors.

All these rules have been advocated by contemporary methodologists, and several remain hotly contested. The key question of meta-methodology is how one provides a warrant for accepting or rejecting such methodological rules.

The first point to note is that these rules or maxims have the form of commands. Their grammar is that of the injunction rather than that of the declarative statement. As such, they appear decidedly not to be the sort of utterance which could be true or false, but at best only useful. Their grammar and semantics have been the source of much grief to philosophers of science and epistemologists, for there is no received way of characterizing the warrant for such commands. To enquire concerning their truth conditions seems a mistake, for they appear to be quite unlike ordinary statements. Yet, if they have no truth conditions, what would it even mean to ask about their *warrant*? Warrant for what purposes? Perhaps a warrant for their use? But it is not even clear what that would mean.

I suggest that syntax, here as in so many other areas, is not only unhelpful, but misleading. Methodological rules do not emerge in a vacuum, and without context. Methodological rules or maxims are propounded for a particular reason, specifically because it is believed that following the rule in question will promote certain cognitive ends which one holds dear. By formulating methodological rules without reference to the axiological context which gives them their bite (as I just have in the list of rules above), one is systematically disguising the route to their warrant.

I submit that all methodological rules should be construed not (in the form illustrated above) as if they were categorical imperatives, but rather as *hypothetical* imperatives. Specifically, I believe that methodological rules, when freed from the elliptical form in which they are often formulated, take the form of hypothetical imperatives whose antecedent is a statement about aims or goals, and whose consequent is the elliptical expression of the mandated action. Put schematically, methodological rules of the form

(0) 'One ought to do x',

should be understood as having the form

(1) 'If one's goal is y, then one ought to do x'.

Thus, Popper's familiar rule, 'Avoid *ad hoc* hypotheses', is more properly formulated as the rule: 'If one wants to develop theories which are very risky, then one ought to avoid *ad hoc* hypotheses.' Two points need to be emphasized: (a) every methodological rule can be recast as a hypothetical imperative (once we know something about the aim-theoretic context in which it is embedded) and (b) the relevant hypothetical imperative will link a recommended action to a goal or aim.

Imperatives of the sort schematized by (1) above always assert a relation between means and ends. Specifically, every such rule presupposes that 'doing x' will, as a matter of fact, promote y or tend to promote y, or bring one closer to the realization of y. Methodological rules are thus statements about instrumentalities, about effective means for realizing cherished ends. It is clear that such rules, even if they do not yet appear to be truth-value-bearing statements themselves, none the less depend for their warrant on the truth of such statements. If I assert a rule of type (1), I am committed to believing that doing x has some prospect of promoting y. If it should turn out that we have strong reason to believe that no amount of an agent's doing x will move him closer to the realization of y, then we will have strong grounds for rejecting the methodological rule (1). If, on the other hand, we find evidence that doing x does promote y, and that it does so more effectively than any other actions we have yet devised, then we would regard (1) as warranted advice. I am suggesting that we conceive rules or maxims as resting on claims about the empirical world, claims to be assayed in precisely the same ways in which we test other empirical theories. Methodological rules, on this view, are a part of empirical knowledge, not something wholly different from it. Provided that we are reasonably clear about how low-level empirical claims (e.g. these alleged ends–means connections) are tested, we will know how to test rival methodologies. We thus have no need of a special meta-methodology of science; rather, we can choose between rival methodologies in precisely the same way we choose between rival empirical theories of other sorts. That is not to say that the task of choosing between rival methods will be any easier than it sometimes is to choose between rival theories. But it is to say that we have no need of a *sui generis* epistemology for methodology.

If this sounds vaguely familiar, let me stress that what I have in mind is very different from the naturalism of, say, a Quine. He and many of his fellow naturalists evidently believe that epistemology should be an entirely non-normative affair, in effect a branch of descriptive psychology, merely recording how we have come to construct the bodies of 'knowledge' which we call the 'sciences'. Quine apparently believes that, with foundationalism dead and the theory of knowledge 'scientized', there is no legitimate scope for the sorts of normative and prescriptive concerns which have traditionally preoccupied epistemologists and methodologists. As I shall try to show in detail below, such a denormativization of methodology is not entailed by its naturalization. Quite the contrary, *one can show that a thoroughly 'scientific' and robustly 'descriptive' methodology will have normative consequences.*

IV. METHODOLOGICAL 'RULES' AND METHODOLOGICAL 'FACTS'

To show the appropriateness of testing methodological rules in the same way as any other descriptive or theoretical assertion, I have to show that it is appropriate to regard methodological rules as parasitic on counterpart descriptive or theoretical statements. We are already half way to that result, for we have just seen that categorical methodological rules[16] of the form

(0) 'You (or one) ought to do y',

are merely elliptical versions of statements of the form (1) 'If your (one's) central cognitive goal is x, then you (one) ought to do y'.

Such imperatives as (1) are true (or warranted) just in case

(2) 'Doing y is more likely than its alternatives to produce x'

is true (or warranted).[17] When (2) is false, so is (1). But (2) turns out to be a conditional declarative statement, asserting a contingent relationship between two presumably 'observable' properties, namely, 'doing y' and 'realizing x'.[18] Specifically, (2) has the familiar form of a statistical law. I submit that *all* methodological rules (at least all of those rules and constraints of the sort usually debated among methodologists) can be recast as contingent statements of this sort about connections between ends and means.[19]

[16] I am not advocating the general thesis that all categorical imperatives can be reduced to hypothetical imperatives. But I do maintain that all methodological rules, even those which appear to be unconditional in form, are best understood as relativized to a particular cognitive aim.

[17] All the obvious riders and qualifications are needed to establish the warrant-conditional substitutivity of (2) for (1). We need to make sure, e.g., that doing y will not undermine other central cognitive or non-cognitive goals of the agent in question. Equally, we should not regard it as appropriate to do y if doing so would be prohibitive of time or expense. But, as a little reflection will show, all those and other similar considerations can be built into the characterization of the goal 'x', thus guaranteeing (1)'s epistemic reducibility to (2).

One might want to argue that, although (1) is true only when (2) is true and that (1) is false whenever (2) is false, statements (1) and (2) are not identical since—it might be thought—(1) might be false even when (2) is true. I do not believe that such a thesis could be successfully defended. However, it is sufficient for the analysis I offer in this section if one establishes the first two dependencies, even if (1) and (2) retain a degree of semantic autonomy.

[18] Of course, if one has adopted a transcendental aim, or one which otherwise has the character that one can never tell when the aim has been realized and when it has not, then we would no longer be able to say that a methodological rule asserts connections between detectable or observable properties. I believe that such aims are entirely inappropriate for science, since there can never be evidence that such aims are being realized, and thus we can never be warrantedly in a position to certify that science is making progress with respect to them. In what follows, I shall assume that we are dealing with aims which are such that we can ascertain when they have and when they have not been realized.

[19] The claim that all methodological rules are contingent may be too strong. One can imagine some ends–means connections which are, in effect, analytic, and whose truth or falsity can be established by conceptual analysis. But that does not undermine the strong analogy I am drawing between science and methodology, for there are plenty of conditional claims in the natural sciences which can be proved by analysis rather than experience (e.g. 'If this system is Newtonian, then all transfers of motion in it will be momentum-conserving').

Of course, this does not yet tell us specifically *how* such methodological rules are to be tested, any more than knowing that physics is an empirical discipline tells us how to test its theories. But what this insistence on the empirical and instrumental (i.e. ends–means character) of methodological rules suggests is that those rules do not have a peculiar or uniquely problematic status. Choosing between rival theories of methodology—conceived now as families of methodological rules—is no more (and, I hasten to add, no less) problematic than choosing between the rival empirical theories of any other branch of learning. The fact that methodological rules appear to have an imperative character, indeed that they *do* have an imperative character, does nothing to put them on an evidential or epistemic footing different in kind from that of more transparently descriptive claims.

But my philosophical critic may be quick to point out that, even if this manœuvre gets around the is/ought problem, it thus far ignores the fact that we could 'test' a methodological rule only by taking for granted the prior establishment of some other methodological rule, which will tell us how to test the former. And that latter rule, in its turn, will presumably require for its justification some previously established methodological rule, etc. We seem to be confronted by either a vicious circularity or an infinite regress, neither of which looks like a promising therapy for our meta-methodological anxieties. How do we either break the circle or block the regress?

The quick answer to that question is that we can avoid the regress provided that we can find some warranting or evidencing principle which all the disputing theories of methodology share in common. *If* such a principle—accepted by all of the contending parties—exists, then it can be invoked as a neutral and impartial vehicle for choosing between rival methodologies.[20] The worry is whether we have any grounds to believe that all of the major theories of scientific methodology, despite their many well-known differences, none the less share certain principles of empirical support, which can in turn be treated as 'uncontroversial' for purposes of choosing between them.

I believe that we have such a criterion of choice in our normal inductive convictions about the appraisal of policies and strategies. In brief, and for these purposes, those convictions can be formulated in the following rule:

(R₁) If actions of a particular sort, *m*, have consistently promoted certain cognitive ends, *e*, in the past, and rival actions, *n*, have failed to do so,

20 In so far as those methodologies do differ.

then assume that future actions following the rule 'If your aim is e, you ought to do m' are more likely to promote those ends than actions based on the rule 'If your aim is e, you ought to do n'.[21]

I hasten to say that R_1 is neither a very sophisticated, nor a very interesting, rule for choosing between rival strategies of research. But then, we would be well advised to keep what we are taking for granted to be as rudimentary as possible. After all, the object of a formal theory of methodology is to develop and warrant more complex and more subtle criteria of evidential support. So let us start for now with R_1. Before we ask how much justificatory work R_1 might do for us, we need to ask whether we can treat R_1 as uncontroversial for purposes of choosing among extant, rival methodologies. Two points are central: (1) R_1 is arguably assumed universally among philosophers of science, and thus has promise as a quasi-Archimedean standpoint, and (2), quite independently of the sociology of philosophical consensus, it appears to be a sound rule of learning from experience. Indeed, if R_1 is not sound, no general rule is.

How safe is it to assume that R_1 is universally taken for granted? Few would doubt that the various schools of inductivists would give it the nod. So too would most members of the so-called historical school in philosophy of science, since their entire programme rests on the assumption that we can learn something from the past about how scientific rationality works. Take away R_1 and its near-equivalents, and the historical camp would be unable to mount any of its critiques of classical positivism. Among philosophers of science, that leaves the Popperians as the only group which might be inclined to reject R_1. And Popper, after all, is on record as opposing any form of inductivist inference. But, within the last decade, Popper has come to see that his own epistemic programme (particularly with its emphasis on corroboration and verisimilitude) is committed to a 'whiff of inductivism'. Indeed, as Grünbaum and others have observed, if Popper were to repudiate a principle like R_1, he would be without licence for his belief—central to his position—that theories which have previously stood up to severe testing should be preferred over theories which have not stood up to such tests (Grünbaum 1976). For reasons such as these, it seems plausible to hold that a broad consensus could be struck among philosophers of science about the appropriateness of presupposing something like R_1.

[21] Like all other general empirical claims, this one needs an appropriate *ceteris paribus* rider added to it. But such qualifications as would be called for would not be different in kind from those we usually associate with the laws of nature.

V. CHOOSING BETWEEN METHODOLOGIES

But, even if one grants that R_1 should be uncontroversial among contemporary methodologists, how does one get from there to a solution to our meta-methodological conundrums? The steps are quite simple. Assume, first, that R_1 is given. Assume, second, that we conceive of all methodological rules as asserting ends–means co-variances of the sort indicated in (2) above. Our task is then this: to ascertain whether, for any proposed methodological rule, there is evidence (of the sort countenanced by R_1) for the assertion of the co-variance postulated by the rule. If the answer to that question is affirmative, then we have shown the warrant for accepting that rule rather than any of its (extant) rivals. If the answer is negative, then we have grounds for rejecting the rule. In sum, I am proposing that the only important meta-methodological question is this:

Given any proposed methodological rule (couched in appropriate conditional declarative form), do we have—or can we find—evidence that the means proposed in the rule promotes its associated cognitive end better than its extant rivals?[22]

If we can get evidence that following a certain rule promotes our basic ends better than any of its known rivals does, then we have grounds for endorsing the rule. If we have evidence that acting in accordance with the rule has thwarted the realization of our cognitive ends, we have grounds for rejecting the rule. Otherwise, its status is indeterminate.[23] In this way, it should be possible in principle for us to build up a body of complex methodological rules and procedures by utilizing at first only principles of simple evidential support. I say 'at first' because one would expect that simple inductive rules (like R_1) will quickly give way to more complex rules of evidential support, so soon as we have a body of methodological rules which has been picked out by these simple test procedures.

Lest it be thought that R_1 is too weak to have any bite to it, one can show that many familiar methodological and epistemological rules of investigation are discredited by it. As I have shown elsewhere (Laudan 1984: ch. 5), such familiar rules as 'If you want true theories, then reject proposed theories unless they can explain everything explained by their predecessors', 'If you want theories with high predictive reliability, reject ad hoc hypotheses', 'If you want theories likely to stand up successfully to subsequent testing, then accept only theories which have successfully made

---

[22] A more precise formulation would be couched in comparative language concerning the evidence that the means proposed in the rule promotes its associated ends better than its known rivals.

[23] Although not necessarily indeterminable.

surprising predictions' all fail to pass muster, even utilizing a selection device as crude as R_1.[24]

Notice that within this account of meta-methodology, we need not concern ourselves with questions about the rationality or irrationality of particular episodes or actors in the history of science. Nor need we invoke shared intuitions about cases, whether real or imaginary. We need no presumptions about the rationality of past scientists and no shared intuitions about concrete cases in order to decide, on this approach, whether one methodology is better than another. We simply enquire about which methods have promoted, or failed to promote, which sorts of cognitive ends in the past. Sometimes it will be easy to answer such questions; other times, it will be very difficult. But here, too, we simply replicate a distinction familiar in every other area of empirical inquiry.

VI. A KEY ROLE FOR HISTORY

It should already be clear that, although this approach severs a link between methodology and the rationality of historical agents, it none the less brings the history of science back to centre stage in evaluation of proposed methodological rules. That centrality can perhaps be made clearer by an example. Suppose that we are appraising the familiar methodological rule to the effect that 'if one is seeking reliable theories, then one should avoid *ad hoc* modifications of the theories under consideration'. Assume, for the sake of argument, that we have reasonably clear conceptions of the meanings of relevant terms in this rule; indeed, without them, the rule could never be tested by anyone's meta-methodology.[25] Now, as I have proposed construing such rules, this one asserts that a certain strategy of research (i.e. avoiding *ad hoc* modifications) more often produces theories which stand up to subsequent testing than does any plausible contrary strategy (e.g. making frequent *ad hoc* modifications).

Formerly, philosophers debating this issue have resorted to trading methodological intuitions. That dialectic, as you might expect, has been largely inconclusive; some philosophers have maintained that *ad hoc* modifications should be avoided at all costs; others have argued that they are essential to scientific progress. But we are now in a position to see that such

---

[24] To be more precise, I have shown that these rules (which I state here for convenience in their categorical form), when cashed out in terms of the aims which they are designed to promote, persistently fail to promote those aims.

[25] Moreover, Lakatos and his followers have given several quite clear formulations of what *ad-hoc*-ness consists in.

armchair bickering is largely beside the point; for the question can in principle be settled by invoking the relevant facts of the matter. Is it true or false that theories which have been *ad hoc* have, up to now, generally proved less reliable than theories which are non-*ad hoc*? Of course, we do not yet know the answer to that question because philosophers of science have not bothered to look carefully at the historical record. But if we were to look, is there not reason to expect that we would be able to determine the status of this widely cited methodological dictum? Similarly with virtually all other methodological rules. Once they are cast as conditional declaratives of the appropriate sort, it becomes possible to test them against the historical record in the same way that any other hypothesis about the past can be tested against the record. And once we know the results of such a test, we will have no need for our 'pre-analytic intuitions' about concrete cases, or for value profiles of 'the scientific élite', or for any other form of intuitionism about concrete cases.

The so-called problem of meta-methodology is hereby seen to be largely bogus, at least in so far as it was prompted by a conviction that methodological rules have a character and status different from ordinary empirical claims, and thus call for a special sort of warranting process.

This analysis should explain why I want to resist the insistence of Lakatos and Kuhn that methodology has, as its primary subject-matter, our intuitions about the rationality of great scientists. To those who take for granted that, as Janet Kourany once put it (1982: 541), 'the aim of a methodology is to articulate' the 'criteria of evaluation actually employed in the greatest or most successful science', I reply that such is not the most important aim of methodology; indeed, if I am right, it is no part of the aim of methodology to articulate such criteria. In my view, the chief aim of the methodological enterprise is to discover the most effective strategies for investigating the natural world. That search may or may not involve us in articulating the criteria of evaluation used by past scientists. But the latter task is, at best, a means to an end rather than an end in itself.

The naturalistic meta-methodologist, as I have described him, needs no pre-analytic intuitions about cases, no information about the choices of the scientific élite, no detailed knowledge (as Carnap required) of the nuances of usage of methodological terminology among scientists, and no prior assumptions about which disciplines are 'scientific' and which are not. What it does need, and in abundance, is data concerning which strategies of enquiry tend to promote which cognitive ends.

VII. PROGRESS, NOT RATIONALITY

I have said that the key role for history *vis-à-vis* methodology is that of providing evidence about ends–means connections. But methodology also has an important role to play in explaining some striking features about the history of science. However, it has nothing to do with exhibiting or explicating the rationality of past scientists. What does require explanation is the fact that science has been so surprisingly successful at producing the epistemic goods. We take science seriously precisely because it has promoted ends which we find cognitively important. More than that, it has become *progressively* more successful as time goes by. If you ask, 'Successful according to whom?' or 'Progressive according to what standards?', the answer, of course, is: successful by *our* lights, progressive according to *our* standards. Science in our time is better (by our lights, of course) than it was 100 years ago, and the science of that time represented progress (again, by our lights) compared with its state a century earlier.

We can readily make these claims about the progress of science, even if we know nothing whatever about the aims or the rationality of earlier scientists, and we can make them *even though* the course of science has dismally failed to realize many of the aims of science as these earlier actors construed them. We can do this because, unlike rationality, progress need not be an agent-specific notion. We can, and often do, talk without contradiction about a certain sequence of events representing progress, even though the final products of that sequence are far from what the actors intended. It is a cliché that actions have unintended consequences. Because they do, it can happen that those unintended consequences eventually come to be regarded as more worthwhile than the goals which the actors were originally striving for. Just so long as the actions of those agents brought things closer to states of affairs which we hold to be desirable, we will view those actions as progressive. Thus, a Social Democrat can view the signing of the Magna Carta as a progressive step towards the more equitable distribution of political power, even though nothing could have been further from the minds of its aristocratic framers. Hard-line empiricists can view Cartesian optics as a progressive improvement on the optics of Descartes's predecessors, even though Descartes would share few of the cognitive goals of the empiricists. Instrumentalists can, and do, regard Newtonian physics as better than Cartesian counterparts, even though the instrumentalists share few of Newton's views about the aims of science.

Hence, the history of science has to be reckoned with, not because scientists are always or more often rational than anyone else (I rather doubt that they are), but rather because the history of science—unlike that of many other disciplines—offers an impressive record of actions and decisions moving closer through time to a realization of ends that most of us hold to be important and worthwhile.[26] The record that is the history of science shows us what sorts of cognitive ambitions have been realized and what sorts have not. If we were today espousing cognitive aims which had not been progressively realized in the development of science, then that history would put few constraints on our methodological musings.[27] Under those circumstances, we should be forced to do methodology largely as Aristotle or Bacon were forced to do it: a priori. But as soon as there is a record of people whose behaviour has been largely successful at realizing many of the cognitive aims which we hold dear, then a proposed methodology of science cannot afford to ignore that record. What the sciences offer us is a set of implicit strategies which have already shown themselves to be successful at promoting our ends.

Under such circumstances, it is entirely appropriate to ask of any proposed methodology whether, had it been explicitly utilized at various important junctures in the history of science, it would have led to theory choices which would have contributed to progress. If a certain methodology of science would have led us to prefer all the discredited theories in the history of science and to reject all the accepted ones, then we have prima-facie grounds for rejecting that methodology. But if this were to come to pass (and it does with respect to many of the best-known methodologies of our time), we would then have grounds for rejecting that methodology—*not* because it led to the conclusion that past scientists were irrational (a conclusion it lacks the generality to sanction)—but because it would have led to choices which were arguably less progressive at promoting our ends than other strategies have been.[28]

I have said that a methodology is one key part of a theory of scientific

---

[26] This should not be taken as suggesting that all of 'us' have the same cognitive ends. I do not believe that we do. But I *do* believe that science manages to promote cognitive ends which most of us hold dear. (Consider the fact that both instrumentalists and realists can point with pride to many achievements in science.)

[27] Although even then it would not be irrelevant to the appraisal of methodologies, since it would provide ample evidence about what sorts of strategies of appraisal and acceptance fail to further our *cognitive* aims.

[28] Of course, one must be careful about applying this criterion too strictly. After all, it is possible that some of the theories which were rejected in the history of science might have proved eventually to be more successful or more progressive than their rivals which survived; our belief otherwise might be a result of the fact that no one bothered to develop them further.

progress. But there is another equally central part of a theory of cognitive progress. We have so far been assuming that all aims were on a par, and that a methodology's task was simply to investigate, in an axiologically neutral fashion, which means promote those aims. On this analysis, the construction of a methodology of science is the development of a set of methodological rules, conceived as hypothetical imperatives, parasitic on a given set of cognitive or epistemic ends. Yet, although this is an attractive conception of methodology, it scarcely addresses the full range of epistemic concerns germane to science. I suspect that we all believe that some cognitive ends are preferable to others. Methodology, narrowly conceived, is in no position to make those judgements, since it is restricted to the study of means and ends. We thus need to supplement methodology with an investigation into the legitimate or permissible ends of enquiry. That is, a theory of scientific progress needs an axiology of enquiry, whose function is to certify or de-certify certain proposed aims as legitimate. Limitations of space preclude a serious treatment of this question here, although I have elsewhere (Laudan 1984) attempted to describe what such an axiology of enquiry might look like. For our purposes here, it is sufficient to note the fact that the axiology of enquiry is a grossly underdeveloped part of epistemology and philosophy of science, whose centrality is belied by its crude state of development. Methodology gets nowhere without axiology.

VIII. CONCLUSION

I began by examining the claim that methodological rules are to be assayed by determining whether they exhibit as rational the theory choices of major scientists. I showed that this was no legitimate test of a methodological rule because those choices were typically made by scientists with axiologies and background beliefs different from ours. Their choices may have been rational, but it is inappropriate to expect our methodologies to reveal that. Having rejected rationality reconstructibility as a suitable meta-criterion, I proposed to fill the void by conceiving of methodological doctrines as statements asserting contingent linkages between ends and means. I showed that the strategies of research incorporated in methodological rules can be tested by ascertaining whether we have plausible arguments and evidence that following the rule in question will enable us to make progress towards the realization of our cognitive ends. The appearance of question-begging involved in using empirical methods to 'test' empirical methods is avoided by invoking principles of evidential support

214   LARRY LAUDAN

which are universally accepted by methodologists. What we thus have
before us is the sketch of a *naturalistic* theory of methodology which
preserves an important critical and prescriptive role for the philosopher of
science, and which promises to enable us to choose between rival
methodologies and epistemologies of science. What it does *not* promise is
any a priori or incorrigible demonstrations of methodology; to the contrary, it makes methodology every bit as precarious epistemically as science itself. But that is just to say that our knowledge about how to conduct
enquiry hangs on the same thread from which dangle our best guesses
about how the world is. There are those who would like to make methodology more secure than physics; the challenge is rather to show that it is as
secure as physics.[29]

REFERENCES

Boyle, W. (1738). *A Disquisition about the Final Causes of Natural Things*. In
    Works, 2nd edn. London: Innys and Monby.
Feyerabend, P. (1975). *Against Method*. London: NLB.
Grünbaum, A. (1976). 'Is Falsificationism the Touchstone of Scientific Rationality?
    Karl Popper versus Inductivism.' In R. Cohen, R. K. Feyerabend, and M.
    Wartofsky (eds.), *Essays in Memory of Imre Lakatos*, 213–52. Dordrecht: Reidel.
Kourany, J. (1982). 'Towards an Empirically Adequate Theory of Science.' *Philoso-
    phy of Science*, 49: 526–48.
Kuhn, T. (1970). 'Reflections on my Critics.' In Lakatos and Musgrave (eds.): 231–
    78.
——(1977). *The Essential Tension*. Chicago: University of Chicago Press.
Lakatos, I. (1971). 'History of Science and its Rational Reconstructions.' In R. Buck
    and R. Cohen (eds.), *Boston Studies in the Philosophy of Science*, viii. 110–47.
    Dordrecht: Reidel.
——and Musgrave, A. (eds.) (1970). *Criticism and the Growth of Knowledge*. Cam-
    bridge: Cambridge University Press.
Laudan, L. (1984). *Science and Values*. Berkeley: University of California Press.
——(1986). 'Intuitionist Meta-Methodologies.' *Synthese*, 67: 115–29.
Popper, K. (1959). *Logic of Scientific Discovery*. London: Hutchinson & Co. Ltd.
Putnam, H. (1978). *Meaning and the Moral Sciences*. London: Routledge and Kegan
    Paul.
——(1981). *Reason, Truth and History*. Cambridge: Cambridge University Press.

[29] An earlier version of this paper was read at the 1985 Chapel Hill Philosophy Colloquium.
I am very grateful for critical comments on it to Richard Burian, Jarrett Leplin, Alan Musgrave,
Phil Quinn, Alex Rosenberg, and Dudley Shapere.

X

REALISM, APPROXIMATE TRUTH, AND PHILOSOPHICAL METHOD

RICHARD BOYD

1. INTRODUCTION

1.1. Realism and Approximate Truth

Scientific realists hold that the characteristic product of successful scientific research is knowledge of largely theory-independent phenomena, and that such knowledge is possible (indeed actual) even in those cases in which the relevant phenomena are not, in any non-question-begging sense, observable (Boyd 1982). The characteristic philosophical arguments for scientific realism embody the claim that certain central principles of scientific methodology require a realist explication. In its most completely developed form, this sort of abductive argument embodies the claim that a realist conception of scientific enquiry is required in order to justify, or to explain the reliability with respect to instrumental knowledge of, all of the basic methodological principles of mature scientific enquiry (Boyd 1973, 1979, 1982, 1983, 1985*a,b,c*; Byerly and Lazara 1973; Putnam 1972, 1975*a,b*).

The realist who offers such arguments is not committed to the view that rationally applied scientific method will always lead to progress towards the truth, still less to the view that such progress would have the exact truth as an asymptotic limit (Boyd 1982, 1988). Nevertheless, it would be difficult to defend scientific realism without portraying the central developments of twentieth-century physical science, for example, as involving a dialectical and progressive interaction of theoretical and methodological commitments (Boyd 1982, 1983).

A defence of realism along these lines requires two things. In the first

Reprinted from C. Wade Savage, *Scientific Theories*, Minnesota Studies in the Philosophy of Science, vol. 14 (Minneapolis: University of Minnesota Press, 1990), 355–91, by permission of the University of Minnesota Press.

place, the realist must be able to defend a historical thesis regarding the recent history of relevant sciences according to which their intellectual achievements involve *approximate* theoretical knowledge, and according to which theoretical progress within them has been (to a large extent) a process of (not necessarily converging) *approximation*. No realist conception that does not treat theoretical knowledge and theoretical progress as involving approximations to the truth is even prima facie compatible with the actual history of science. The realist must, therefore, employ a conception of approximate theoretical knowledge and of theoretical progress through approximation that makes historical sense of the recent development of scientific theories.

Secondly, the realist must be able to establish that her historical appeal to approximate theoretical knowledge and to theoretical progress by successive approximation is appropriate by philosophical as well as by historical standards. Neither the realist's historical account nor her appeal to it in the defence of scientific realism as a philosophical thesis should be undermined by any of the distinctly philosophical considerations characteristic of anti-realist positions in the philosophy of science. Important challenges to scientific realism arise from doubts that a realist conception of approximate truth and of the growth of approximate knowledge is available that satisfies both of these constraints. The appropriate realist responses to these challenges and the philosophical implications of those responses are the subject of the present essay.

1.2. Challenges to a Realist Treatment of Approximation

A number of philosophers (realists included) have had serious concerns about the realist's ability to provide an adequate account of the development of scientific theories as involving the growth of approximate theoretical knowledge. The *locus classicus* of objections to realism reflecting these concerns is surely Laudan 1981 (Ch. VI, this volume) (see also Fine 1984, Ch. I, this volume). That there should be such concerns is, in significant measure, a reflection of the striking difference between the depth of our understanding of the notion of (exact) truth and that of our understanding of approximate truth.

Since the work of Tarski in the 1930s we have had a systematic, general, and topic-and-context-independent mathematical and philosophical theory of (exact) truth. By contrast, there is no generally accepted general and systematic theory of approximate truth. We have available from the various special sciences a very large number of well-worked-out examples of particular instances of approximation, but the details in these cases depend not only on contingent and often esoteric facts about the relevant natural phenomena, but also upon the particular context of application within which the approximate theories and models are to be applied. In part because of the complexities created by such topic and context dependence, we do not have as clear a general understanding of what the epistemological relevance of appeals to approximate truth should be. Moreover, as we shall see, the dependence of the relevant details upon a posteriori theoretical claims raises special problems of philosophical method when an appeal to conception of approximate truth is to be made in the course of a defence of scientific realism.

I have argued elsewhere (Boyd 1982, 1983, 1985a,b,c, 1988) that the scientific realist must adopt distinctly naturalistic conceptions of philosophical methodology and of central issues in epistemology and metaphysics. My aim in the present paper will be to show how the distinctly naturalistic arguments for realism that I have developed in the papers cited can be extended to provide an adequate realist treatment of approximate truth.

Instead of replying to particular anti-realist arguments in the literature, I shall respond to four objections that capture, I believe, the deep philosophical concerns that the realist's conception of approximate theoretical knowledge properly occasion. My expectation is that the responses to those objections will provide an adequate basis for a realist's response to other objections regarding her conception of approximate truth and approximate knowledge. The objections I shall consider are these:

1 (*The historical objection*). Realists are simply mistaken as a matter of historical fact: many important scientific advances seem to have been grounded in what (by realist standards) were deep errors in background theories. Approximately true background theoretical knowledge is thus not required to explain reliability of scientific practices.

2 (*The triviality objection*). The realist might reply (following Hardin and Rosenberg 1982, for example) about many of the advances in question that the relevant background theories were *to some extent* or *in some respects* approximately true.

Here the realist's philosophical project is in danger of being reduced to *triviality*. The problem is that we lack altogether a general theory of approximation: we have no general characterization of what it is for a sentence to be approximately true, to be approximately true to a specified degree or in a specified respect, or to be more nearly true (in specific respects or in general) than some other sentence. If we had such a general theory, then the realist could appeal to it in refining the thesis that the relevant historical episodes reflect some respects of approximation to the truth. As it is, we are faced with the fact that *any* consistent theory is approximately true in some respect or other, and *any* sequence of such theories will reflect progress towards the truth in some respect or other.

3 (*The contrivance objection*). The realist might next reply by distinguishing between relevant and irrelevant respects of approximation to the truth regarding matters theoretical, and by claiming that the growth of scientific knowledge characteristically involves the former. Here the realist avoids triviality at the expense of a contrived or *ad hoc* conception of approximate truth—indeed, at the expense of both contrivance (objection 3) and circularity (see objection 4).

The contrivance in question arises from the important difference just mentioned between extant theories of truth and of approximate truth respectively. In the case of truth *simpliciter*, Tarski's strategy for defining truth (Tarski 1951) provides a uniform treatment that is largely independent of the particular subject-matter or of the particular historical episodes or context of application under consideration. By contrast, our conception of relevant approximation reflects considerations specific to the particular theory or theories, historical settings, and contexts of application under consideration.

Thus, for example, if the realist sees relativistic mechanics as growing out of previously acquired approximate theoretical knowledge, her conception of the relevant respects of approximation reflected in Newtonian mechanics will emphasize numerical accuracy for systems of particles with relative velocities low with respect to that of light, the identification of, and the development of reliable measurement procedures for, various physical magnitudes, and the central role assigned to certain fundamental laws. It will de-emphasize, for example, numerical accuracy for high relative velocities, or of soundness of the Newtonian theoretical conception of space and time.

Here the distinctions between relevant and irrelevant respects of approximation reflect judgements, based on current theoretical conceptions, about the respects in which Newtonian mechanics happened to be approximately true, and similarly theory-dependent judgements about the role that such approximations played in the successful development of relativistic mechanics. Since we lack a general theory of approximation, the realist's appeal to relevant respects of approximation in response to the triviality objection will always have to be grounded in just this sort of topic-and-episode-sensitive conception.

We can now see why the realist's treatment of respects of approximation will involve an *ad hoc* or contrived element. For each of the episodes of scientific enquiry typically considered by philosophers of science, there is a standard realist picture (or, at any rate, a narrow range of such pictures) of how the relevant approximations to the truth have gone and what contributions, if any, they have made to the subsequent growth of scientific knowledge. The realist, in defining the relevant sense(s) of approximation, will rely on such a picture. But such a picture merely reflects the realist research tradition in the history and philosophy of science. Since there is no topic-and-episode-neutral conception of relevant approximation with respect to which her proposed definitions may be assessed, the realist will simply be presupposing the soundness of the 'findings' of her own tradition when she defines the difference(s) between relevant and irrelevant respects of approximation. It is no surprise—and certainly no basis for an abductive argument for realism—that the realist can construct a realist account of approximate truth when she is permitted to beg questions in so thoroughgoing a way.

4 (*The circularity objection*). There is some precedent in scientific enquiry, especially historical enquiry, for explanatory concepts that lack topic-and-episode-neutral general specifications of the sort alluded to above: sometimes theoretical considerations that resist incorporation into a fully general definition can justify the (topic-and-episode-non-neutral) ways in which such concepts are applied in particular cases. Let us suppose for the sake of argument that this is the case with respect to the employment of the concept of approximate truth in the various historical explanations of scientific progress (or its absence) that are offered in the realist tradition. Even if the realist's accounts of the relevant episodes are thus methodologically acceptable *as explanations in the history of science*, they will involve an unacceptable circularity if they are understood to address the *philosophical* issue between scientific realists and anti-realists.

Here's why. Any realist explanation of the growth of knowledge and of reliable methodology in a particular field must involve an account of the kinds of epistemically relevant causal interactions that exist(ed) between members of the relevant scientific community and the features of the world that were (or are) the alleged objects of their study. Thus, for example, a realist account of such developments in atomic theory will incorporate a causal account of how scientists gain(ed) epistemic access to various subatomic particles, and the realist's claim that atomic theory is about such unobservable theory-independent particles will depend on that account (see sects. 2.1.3 and 2.1.4). The realist's account of epistemic access to subatomic particles will be grounded in the best available theory of such particles together with related contemporary physical theories.

Suppose, now, that the realist's explanation of the development of some field, including the relevant account of epistemic access, is advanced in defence of realism as a philosophical thesis. Plainly, the resulting defence of realism is cogent only if the realist's explanation, and her account of epistemic access in particular, are understood *realistically*. For example, only if the account of epistemic access to subatomic particles is understood realistically is the realist's case that atomic theory has an unobservable and theory-independent subject-matter advanced. But, on the realist's own account, her explanation and the account of epistemic access it incorporates are ordinary scientific theories themselves grounded in the very research tradition regarding which a defence of realism is sought. To insist on a realistic interpretation of the realist's explanation would thus *presuppose* realism regarding the tradition in question. Thus the realist's appeal to her explanation of the development of instrumentally reliable methodology in an abductive argument for realism as a philosophical thesis is question-beggingly circular.

1.3. An Argumentative Strategy

The challenges we are considering seem to fall into two classes. The first three represent an essentially pre-philosophical critique of the realist's historical explanations: they deny that the realist's conception of the role of approximate truth regarding theoretical matters in the growth of scientific knowledge represents the best explanation for the relevant episodes in the history of science. The fourth offers a distinctly philosophical challenge: it argues that even if the *realist*'s account of the growth of scientific knowledge does provide the best explanation, inductive inference to *realism* begs the philosophical question at issue.

After some philosophical preliminaries, I propose to respond to the challenges in two distinct stages, corresponding to these two classes. In the first stage of my response, I treat the characteristic realist explanatory appeal to approximate truth as an ordinary piece of historical explanation. I identify a general methodological problem of *parametric specification* in explanatory contexts, of which the deeper problems raised by the first three challenges are special cases, and I identify the generally appropriate solution to that problem. I then indicate why it is plausible that the realist's explanatory appeal to approximate truth satisfies the methodological demands dictated by the solution in question.

With respect to the fourth challenge, I assume for the sake of argument that the realist's historical explanations have been confirmed, and I enquire whether they are to be understood realistically or whether, instead, such an understanding—which is essential to the realist's case—begs the question against the anti-realist. Here too I argue that the methodological question regarding the realist's appeal to approximate truth—in this case, a question about *philosophical* method—is a special case of a more general methodological question about the appropriate interaction between philosophical considerations and empirical findings in the philosophy of science. I define the notion of a large-scale *philosophical package*, and I indicate why the incorporation of realistically understood scientific theories into the realist philosophical package is compatible with (and indeed required by) an adequate *and non-circular* defence of the realist package against rival philosophical conceptions.

On now to the philosophical preliminaries.

2. PHILOSOPHICAL PRELIMINARIES

2.1. The Abductive Argument for Scientific Realism

The challenges we are considering arise in the context of a class of abductive arguments for realism according to which we must recognize approximate knowledge of unobservable (and approximately mind-independent) 'theoretical entities' in order to adequately explain the growth of even instrumental knowledge in recent science. To assess the realist's arguments and the appeals to the notion of approximate truth embodied in them, we need an understanding of just what those arguments are. In what follows of this section I'll indicate, in broad outline, how the abductive arguments for realism go.

2.1.1. *Objective Knowledge from Theory-Dependent Method* By the 'instrumental reliability' of a scientific theory, I mean the extent of its capacity to make approximately true observational predictions about observable phenomena—the extent of its approximate empirical adequacy. By the 'instrumental reliability' of some body of methods, I mean the extent to which their practice is conducive to the acceptance of instrumentally reliable theories. The abductive arguments for scientific realism take place in a dialectical situation in which scientific realists and their philosophical opponents largely agree that the methods of actual recent scientific practice are significantly instrumentally reliable.

The abductive arguments for realism are in the first instance directed against the empiricist, who denies the possibility of 'theoretical' knowledge—knowledge of 'unobservables'. Against the empiricist, the realist argues that only by accepting the reality of approximate theoretical knowledge can we adequately explain the (uncontested) instrumental reliability of apparently theory-dependent scientific methods. In the present paper I shall focus my attention primarily on the dispute between realists and empiricists, reserving attention to the corresponding dispute between realists and constructivists largely to a later paper. I discuss the realism–constructivism dispute briefly in section 2.4 and briefly discuss the distinctly constructivist version of the circularity objection in section 4.3.

The case for realism lies largely in the recognition of the extraordinary role that theoretical considerations play in actual (and patently successful) scientific practice. To take the most striking example, scientists routinely modify or extend operational 'measurement' or 'detection' procedures for 'theoretical' magnitudes or entities on the basis of new theoretical developments. The reliability and justifiability of this sort of methodology is perfectly explicable on the realist's conception of measurement and of theoretical progress. Accounts of the revisability of operational procedures that are compatible with an empiricist position appear inadequate to explain the way in which theory-dependent revisions of 'measurement' and 'detection' procedures make a positive methodological contribution to the progress of science.

There are two important consequences of the realist explanation of the reliability of the methodology in question. First, scientific research, when it is successful, is *cumulative by successive (but not necessarily convergent) approximations to the truth.* Second, this cumulative development is possible because *there is a dialectical relationship between current theory and the methodology for its improvement.* The approximate truth of current theories explains why our existing measurement procedures are (approximately) reliable. That reliability, in turn, helps to explain why our experimental or observational investigations are successful in uncovering new theoretical knowledge, which, in turn, may produce improvements in measurement techniques, etc.

Theory dependence of methods and the consequent dialectical interaction of theory and method are entirely general features of all aspects of scientific methodology—principles of experimental design, choices of research problems, standards for the assessment of experimental evidence and for assessing the quality and methodological import of explanations, principles governing theory choice, and rules for the use of theoretical language. In all cases, there is a pattern of dialectical interaction between accepted theories and associated methods of just the sort exemplified in the case of the theory dependence of measurement and detection procedures. Moreover, this pattern of theory dependence contributes to the reliability of scientific methodology rather than detracting from it (Boyd 1972, 1973, 1979, 1980, 1982, 1983, 1985*a,b,c*; Kuhn 1970; Putnam 1972, 1975*a,b*; van Fraassen 1980).

According to the realist, the only scientifically plausible explanation for the reliability of a scientific methodology that is so theory-dependent is a thoroughgoingly realistic explanation: scientific methodology, dictated by currently accepted theories, is reliable at producing further knowledge *precisely because, and to the extent that, currently accepted theories are relevantly approximately true.* Scientific method provides a paradigm-dependent paradigm-modification strategy: a strategy for modifying or amending our existing theories and methods in the light of further research that is such that its methodological principles at any given time will themselves depend upon the theoretical picture provided by the currently accepted theories. If the body of accepted theories is itself relevantly sufficiently approximately true, then this methodology operates to produce a subsequent dialectical improvement both in our knowledge of the world and in our methodology itself. It is not possible, according to the realist, to explain even the instrumental reliability of actual recent scientific practice without invoking this explanation and without adopting the realistic conception of scientific knowledge that it entails (Boyd 1972, 1973, 1979, 1982, 1983, 1985*a,b,c*).

2.1.2. *Projectability, Evidence, Theoretical Plausibility, and the Evidential Indistinguishability Thesis*   If the realist's abductive argument is correct, a dramatic rethinking of our notion of scientific evidence is required. Consider the question of the 'degree of confirmation' of a theory given a body of observational evidence. To a very good first approximation, a theory receives significant evidential support from a body of successful predictions (or other evidentially favourable observations) just in case (a) the theory is itself 'projectable' (see Goodman 1973), (b) the observations in question pit the theory's predictions (or, in other contexts, its explanations) against those of its projectable rivals; and (c) in the relevant experiments or observational settings, there have been suitable controls for those possible artefactual influences that are themselves suggested by projectable theories of those settings (Boyd 1982, 1983, and especially 1985*a*).

Central to the realist's argument is the observation that projectability judgements are, in fact, judgements of theoretical plausibility: we treat as projectable those proposals that relevantly resemble our existing theories (where the determination of the relevant respects of resemblance is itself a theoretical issue). The reliability of this conservative preference is explained by the approximate truth of existing theories, and one consequence of this explanation is that *judgements of theoretical plausibility are evidential*. The fact that a proposed theory is plausible in the light of previously confirmed theories is some evidence for its (approximate) truth. Judgements of theoretical plausibility are matters of inductive inference from (partly) theoretical premisses to theoretical conclusions; precisely these inferences justify, and explain the reliability of, 'inductive inference to the best explanation' (Boyd 1972, 1973, 1979, 1982, 1983, 1985*a,b,c*).

The claim that judgements of theoretical plausibility are evidential affords the realist a reply to the deepest empiricist argument against realism. The empiricist appeals (tacitly or explicitly) to a principle that I have called the *evidential indistinguishability thesis*. In its most plausible form, it holds that for any two empirically equivalent total sciences, the empirical support or disconfirmation that one receives, given a given body of observational data, will be just the same as that received by the other. The empiricist's conclusion that knowledge of unobservables is impossible is a straightforward application of this thesis, which can be thought of as an empiricist analysis of the claim that all scientific knowledge is empirical knowledge. The realist accepts the latter claim, but rejects the empiricist analysis. Instead, the realist holds, evidential considerations regarding theoretical plausibility are indirectly experimental, and can serve to distinguish total sciences that embody or naturally extend the current total science (that are favoured by those considerations) from empirically equivalent total sciences which significantly depart from the prevailing total science (which such considerations reject as unprojectable). (See Boyd 1982, 1983, and sect. 2.2.)

2.1.3. *Natural Definitions*   Locke speculates at several places in book IV of the *Essay* (see e.g. IV. iii. 25) that when kinds of substances are defined, as empiricism requires, by purely conventional 'nominal essences', it will be impossible to have a general science of, say, chemistry. There is no reason to believe that kinds defined by nominal essences will reflect actual causal structure, and thus be apt for the formulation or confirmation of general knowledge of substances. Only if we are able to sort substances according to their hidden real essences will systematic general knowledge of substances be possible.

Locke was right (at any rate, so the realist thinks). Only when kinds (properties, relations, magnitudes, etc.) are defined by natural rather than conventional definitions is it possible to obtain the theory-dependent solutions to the problem of projectability just described (Putnam 1975*b*; Quine 1969*b*; Boyd 1979, 1982, 1983). It is thus central to the realist's abductive argument that most scientific terms be seen as possessing natural, rather than conventional, definitions. Such terms are defined in terms of properties, relations, etc. that render the kinds (etc.) to which they refer appropriate to particular sorts of scientific or practical reasoning. In the case of such terms, proposed definitions are always in principle revisable in the light of new evidence or new theoretical developments, and it is possible for people to refer to the same kind (property, magnitude, etc.) by a term, while disagreeing about what its correct a posteriori natural definition is. This last consequence of the naturalistic conception of definitions is essential to the realist's dialectical conception of the development of scientific knowledge and methods. The realist will (at least typically) need to portray developments in which mature scientific communities change their conception of the definitions of kinds, relations, magnitudes, etc. as dialectical advances (or, if things go badly, set-backs) rather than as changes of subject-matter (Putnam 1972, 1975*a*,*b*; Boyd 1979, 1980, 1988). (For more on naturalistic definitions see sect. 2.5.)

2.1.4. *Reference and Epistemic Access*   If the traditional empiricist account of definition is to be abandoned for scientific terms in favour of a naturalistic account, then a naturalistic conception of reference is required for such terms. An account of the appropriate sort is provided by recent causal theories of reference (see e.g. Feigl 1956, Kripke 1972, Putnam 1975*b*). The reference of a term is established by causal connections of the right sort between the use of the term and (instances of) its referent.

The connection between naturalistic theories of reference and of knowledge (see sect. 2.2) is quite intimate: reference is itself an epistemic notion, and the sorts of causal connections that are relevant to reference are just those that are involved in the reliable regulation of belief (Boyd 1979, 1982). *Roughly*, and for non-degenerate cases, a term t refers to a kind (property, relation, etc.) k, just in case there exist causal mechanisms whose tendency is to bring it about, over time, that what is predicated of the term t will be approximately true of k. In such a case, we may think of the properties of k as regulating the use of t, and we may think of what is said using t as providing us with socially co-ordinated *epistemic access* to k. t refers to k (in non-degenerate cases), just in case the socially co-ordinated use of t provides significant epistemic access to k, and not to other kinds (properties, etc.) (Boyd 1979, 1982). The mechanisms of reference *just are* the mechanisms of reliable belief regulation.

RICHARD BOYD

Thus, just as the realist conception requires, two different terms, or the same term in two historically different settings, may afford epistemic access to, and thus may refer to, the same kind (property, etc.) even though the definitions associated with them by the relevant linguistic communities are quite different or even inconsistent.

One further feature of the naturalistic conception of reference is important to an understanding of the realist's conception of the growth of approximate knowledge. In many scientifically important cases the use of a term may afford epistemic access to more than one kind (property, relation, . . .), but our knowledge may be insufficient for us to recognize that this is so, and we may consequently have a conception of, as it seems to us, one kind (etc.) that conflates information regarding several distinct kinds.

Field (1973, 1974) calls the relation thus established between a term and several kinds (etc.) *partial denotation*, and he calls the revision of language usage to eliminate such cases of ambiguity *denotational refinement*. On the realist's conception of the growth of approximate knowledge, one sort of approximate knowledge is that represented by a body of sentences involving a partially denoting term when what is predicated of that term in these sentences represents methodologically important approximations to the truth regarding one or more of the relevant *partial denotata* considered individually. In such cases, one characteristic form of subsequent improvement in approximation is the discovery of the ambiguity and the consequent denotational refinement (see Boyd 1979).

2.2. Naturalism and Radical Contingency in Epistemology

Modern epistemology has been largely dominated by 'foundationalist' conceptions: all knowledge is seen as grounded in certain foundational beliefs that have an epistemically privileged position. Other true beliefs are instances of knowledge only if they can be justified by appeals to foundational knowledge. It is an a priori question which beliefs fall in the privileged class. Similarly, the basic inferential principles that are legitimate for justifying non-foundational knowledge claims can be justified a priori; it is, moreover, an a priori question about a given inference whether it meets the standards set by those principles or not. We may fruitfully think of foundationalism as consisting of two parts: *premiss foundationalism*, which holds that all knowledge is justifiable from an a priori specifiable core of foundational beliefs, and *inference foundationalism*, which holds that principles of justifiable inference are reducible to inferential principles that are *a priori justifiable* and whose application is *a priori checkable*.

Recent work in 'naturalistic epistemology' (see e.g. Armstrong 1973; Goldman 1967, 1976; Quine 1969*a*) strongly suggests that foundationalism is fundamentally mistaken. For the typical case of perceptual knowledge, there seem to be neither premisses nor inferences; instead, perceptual knowledge obtains when perceptual beliefs are produced by epistemically reliable mechanisms. Even where premisses and inferences occur, it seems to be the reliable production of belief that distinguishes cases of knowledge from other cases of true belief. A variety of naturalistic considerations suggests that there are no beliefs that are epistemically privileged in the way foundationalism seems to require.

I have argued (see Boyd 1982, 1983, 1985*a,b,c*) that the abductive defence of scientific realism requires an even more thoroughgoing naturalism in epistemology and, consequently, an even more thoroughgoing rejection of foundationalism. In particular, *all* of the significant methodological principles of inductive inference in science are profoundly theory-dependent. They are a reliable guide to the truth only because, and to the extent that, the relevant background theories are relevantly approximately true. They are not reducible to some more basic rules whose reliability as a guide to the truth is independent of the truth of background theories. Since it is a contingent empirical matter which background theories are approximately true, the justifiability of scientific principles of inference rests ultimately on a contingent matter of empirical fact, just as the epistemic role of the senses rests upon the contingent empirical fact that the senses are reliable detectors of external phenomena. Thus, inference foundationalism is radically false; there are no a priori justifiable rules of non-deductive inference, and it is an a posteriori question about any such inference whether or not it is justifiable. The epistemology of empirical science is an empirical science (Boyd 1982, 1983, 1985*a,c*).

One consequence of this radical contingency of scientific methods is important to the realist's conception of the growth of approximate knowledge. The emergence of successful modern scientific methodology as we know it depended upon the logically, epistemically, and historically contingent emergence of a relevantly approximately true theoretical tradition. It is not possible to understand the initial emergence of such a tradition as the consequence of some more abstractly conceived scientific or rational methodology that itself is theory-independent. There is no such methodology. The theoretical innovations that established the first successful paradigm within a particular scientific discipline must be thought of as the beginnings of successful methodology within the field, not as consequences of it (for a further discussion see Boyd 1982).

Note that radical contingency in epistemology is central to the realist's case against empiricism. Against the evidential indistinguishability thesis, the realist argues that plausibility judgements grounded in the current total science afford evidential distinctions between empirically equivalent total sciences. But, according to the realist's account, it is not the *currency* of the current total science that makes plausibility judgements with respect to it epistemically reliable, but its approximate truth. That a time should have arisen in which total sciences embodied relevant approximations to the truth is of course radically contingent. Thus, central to the realist's rebuttal to empiricism are the epistemological principles that reflect that contingency.

2.3. Metaphysics and 'Metaphysics'

Logical positivists employed the term 'metaphysics' for the sort of enquiry about 'unobservables' that verificationism led them to reject. Most of what has traditionally fallen under that term was 'metaphysics' in the positivists' sense, but so was enquiry about, for example, the atomic structure of matter. If scientific realism is right, then it follows that scientists routinely do successful 'metaphysics'. With respect to metaphysics (as philosophers and others ordinarily use the term) the situation is more complex.

If scientific realism is true for any of the standard reasons, then scientists have discovered the real essences of chemical kinds (Kripke 1971, 1972), and have thus done some real metaphysics. Moreover, the fact that scientific knowledge of unobservables is possible makes it a serious question whether or not scientific findings have (or will have) resolved some traditional metaphysical questions. Certainly the recent near-consensus in favour of a materialist conception of mind reflects a realist understanding of the possibility of experimental metaphysics. Nevertheless, it does not follow from scientific realism that scientists routinely tend to get the right answers to the distinctly metaphysical questions that are the special concern of philosophers, even when their methods lead them to adopt theories that reflect answers to such questions.

In particular, when a scientific realist proposes to explain the reliability of the scientific methods employed at a particular historical moment by appealing to the approximate truth of the background theories accepted at that time, she need not hold that the metaphysical conceptions embodied in those theories represent a good approximation by philosophical standards. Two examples will illustrate the point.

Consider the way in which the reliability of the methods by which Darwin's account in the *Origin* was assessed is to be explained by reference to the approximate truth of much of the prevailing background biological theory. A great deal was known, for example, about species—

not just facts about particular species, but about anatomical, behavioural, genetic, and biogeographical generalizations that can only be formulated in terms of the notion of a species. The realist will hold that the approximations to the truth embodied in this lore of species is part of what explains the reliability of the research methods in biology employed by Darwin and his contemporaries.

Prior to Darwin's work, the prevailing conception made species membership in the first instance a property of individuals; after Darwin, we have correctly seen a species as in the first instance a family of populations. The background biological theories of Darwin's era got it profoundly wrong about the metaphysics of species. Nevertheless, the classificatory practices of pre-Darwinian biologists were reliable enough to serve to establish the rich and significantly accurate lore about species upon which the reliability of methodology in early evolutionary theory crucially depended—or, at any rate, so the realist may reasonably maintain.

Similarly, the realist will want to explain the reliability of the methods by which physicists assessed early developments in quantum theory by appealing to respects in which the pre-quantum theory of, say atoms and subatomic particles was approximately true. She will appeal to the correct identification of various subatomic particles and of (many of) the fundamental physical magnitudes, to the availability of reliable procedures for the detection of those particles and for the measurement of various of their physical properties, and to the classical insights reflected both in the formulation of the equation for the time-evolution of quantum-mechanical systems and in the techniques employed in practice in picking the appropriate Hamiltonian for quantum-mechanical systems.

Indeed, she will want to portray much of the early development of quantum theory as the gradual extension of the range of phenomena for which an adequate quantum-mechanical treatment had been provided. On such an account, at any given stage in the early development of quantum theory, the proposed models for physical systems were always a mixture of distinctly quantum-mechanical components together with essentially classical (or relativistic) components awaiting later quantum-mechanical reformulation. The realist will want to explain the reliability and justifiability of this sort of development by appealing to the respects of approximation to the truth of classical mechanics itself and of the successive stages in the development of the quantum theory.

Consider now the classical conception of atomic phenomena understood as a contribution to philosophical metaphysics. Arguably, the metaphysical component of that conception is some sort of mechanistic atomism: a picture of discrete and unproblematically individuated particles and their associated fields interacting in a deterministic fashion without action at a distance. Our current quantum-mechanical conception of matter rejects each component of this picture: for the atomist's discrete particles, we substitute entities with wave-like features for which particle-like individuation is sometimes impossible; we reject determinism; and we acknowledge that there are non-local effects that would surely be precluded by the classical philosophical rejection of action at a distance. Classical conceptions of the atomic world were, let us agree, poor approximations to the truth in metaphysics. Does this preclude their having been good enough approximations in other respects to sustain the realist's account of the development of quantum theory?

Plainly not. Whatever other objections there may be to the realist's account, it is not a cogent objection that the classical conception that her account treats as relevantly approximately true is not good metaphysics. All she need do is to explain how the metaphysical errors in the classical conception failed to vitiate the methodological contribution of its genuine insights. To this end she might, for example, appeal to the respects in which subatomic particles are (classical) particle-like, to the determinism of the time-evolution of quantum-mechanical systems prior to measurement, and to the wide variety of phenomena that do not significantly exhibit the effects of non-local 'action at a distance'. Perhaps in the case of the development of evolutionary theory, and certainly in the case of quantum mechanics, the realist's account will have scientists doing 'metaphysics' with some significant success; in neither case must she portray them as doing good metaphysics.

The cases just discussed illustrate an additional point. In each case, if the metaphysical criticism of the earlier theoretical tradition is sound, then it embodied, in addition to metaphysical errors, errors about the logical form of certain key propositions. Conspecificity is a relation between populations, not individuals; so pre-Darwinian biology embodied a mistake about the logical type of propositions regarding species membership. Similarly, quantum mechanics requires that we think of the classically acknowledged physical magnitudes as corresponding to Hermitian operators rather than to vector- or scalar-valued functions; in consequence, classical mechanics is mistaken about the logical form of, for example, attributions of position or momentum to particles. Neither error undermines the contribution that the approximate truth of the earlier theory is said to have made to the methodology by which the later theory was developed and confirmed. The realist need attribute to successful background theories neither metaphysical success nor logical exactitude. Approximation need not be philosophically clean. (Note that the distinctly realist naturalistic semantic conceptions are operative in this discussion. What evolutionary theory and quantum mechanics have taught us is that, as we might say, 'there are no classical species' and 'there are no classical particles'. Only naturalistic alternatives to the empiricist conceptions of definitions and reference permit the realist to say—as the account just given requires—that, nevertheless, Darwinian species and the particle-like phenomena acknowledged by quantum mechanics were the subjects of the relevant classical investigations.)

2.4. Realism Causation and Mind Independence

The realist conception of science contrasts with various neo-Kantian constructivist conceptions according to which, when scientific theories address fundamental questions, there is a deep element of social construction of reality reflected in what they say. It is sometimes said that realists and constructivists differ over the extent to which the reality studied by scientists is 'mind-independent' or is 'theory-independent'. In order to understand the demands placed on the realist by what we have called the 'circularity objection', we require some understanding of what is distinctly realist about the realist's explanatory appeal to approximate truth of theoretical presuppositions, given that the constructivist shares with the realist the conviction that scientific progress involves theoretical as well as instrumental knowledge and that scientific methods are deeply theory-dependent. In the present essay I'll touch on this issue only briefly.

2.4.1. *Defining Mind Independence* The realist and the constructivist each reject the Humean and verificationist claim that reference to hidden mechanisms, essences, and causal powers is, on 'rational reconstruction', eliminable from the findings of science. They agree that scientists' methods and conceptions are determined by ineliminably metaphysical conceptions about the basic sorts of mechanisms, processes, and forces that operate to produce the phenomena under study, and that this dependence is not merely a psychological quirk of the 'context of invention' to be rationally reconstructed away in the 'context of confirmation'. They agree too in rejecting the eliminative Humean or regularity account of the causal powers and relations discovered by scientific enquiry. So where does the difference lie? What is the import of the question of the mind or theory independence of reality, given that both parties reject empiricism?

The answer, subject to an important qualification, is that the realist denies, while the constructivist affirms, that the adoption of theories, paradigms, conceptual frameworks, perspectives—or the having of associated

interests, intentions, purposes, etc.—in some way constitutes, or contributes to the constitution of, the causal powers of, and the causal relations
between, the objects scientists study in the context of those theories,
interests, etc. Of course (here is the qualification), the realist does not deny
that the adoption of theories, etc., and the having of projects or interests,
are themselves causal phenomena, and thus contribute *causally* to the
establishment of, for example, those causal factors that are explanatory in,
for example, the history, philosophy, and sociology of science and that, in
consequence, the adoption of a theory in such a discipline could contribute
causally to the causal powers and relations that are the subject-matter of
the theory itself. What the realist denies is that there is some further sort
of contribution (logical, conceptual, socially constructive, or the like)
which the adoption of theories or the having of interests makes to the
establishment of causal powers and relations.

Thus the realist denies the *non-causal* contribution of minds and (the
adoption of) theories to the establishment of causal powers and relations,
whereas the constructivist insists that such a contribution is fundamental.
While the present paper focuses primarily on the realist's abductive argument against empiricism, it is important to note two constraints that a
suitably developed realist explanation of the reliability of scientific methods must meet if there is to be any prospect of its serving as the basis for
a rebuttal to constructivism. In the first place, the definitions of natural
kinds, categories, etc. to which the realist's explanation makes essential
reference are, in a certain sense, interest-dependent. The properties
and causal powers that are relevant to explanation or prediction depend
on the practical or theoretical projects being undertaken. Thus, appropriateness of definitions and conceptual frameworks depends upon the interests with respect to which they are to be employed. The realist must
acknowledge this fact in a way which is compatible with denying that the
interest dependence in question involves any non-causal contribution
of the adoption of interests or projects to the causal powers of the objects
of scientific study.

Similarly, as Quine and others have reminded us, even when an agenda
of interests and projects is fixed, there may be several ways of defining
kinds and categories—of 'cutting the world at its joints'—that are equally
adequate to the task of reflecting explanatorily significant causal relations
(even as the realist understands those relations). It may sometimes happen
that the theoretical commitments of two such frameworks will appear to
involve conflicting metaphysical conceptions. The choice between such
frameworks will be, for the realist, arbitrary. Thus, the realist's account of
approximation must not treat one such framework as more nearly approxi-

mately true than the others, despite apparent metaphysical conflicts; certainly, it must not treat the adoption of one rather than another as contributing non-causally to the establishment of causal relations or to similar settling of matters metaphysical.

It is by no means uncontroversial that arbitrariness and the interest dependence of kinds can be treated in the way the realist requires. For the purposes of the present essay I'll assume that an appropriate realist treatment is possible, while acknowledging that, in an essay in which constructivism rather than empiricism was the primary target, the question would require more extensive treatment. Two other issues regarding mind independence deserve our brief attention.

2.4.2. *Mind Independence and the Causal Role of Minds*   We have seen that the realist acknowledges the causal role of mental phenomena (since, e.g., she explains the reliability of scientific method by reference to the causal powers of approximately true beliefs) and differs from constructivists only in that she denies such phenomena a non-causal role in constituting causal structure. Nevertheless, there are cases in which the attribution of a plainly causal role to mental phenomena has been seen as supporting constructivism. Two such cases deserve attention. First, scholars who are impressed by the social role of ideology often claim that 'human nature' and the 'natures' of various socially defined groups are 'social constructions', and often they appear to mean by this, at least in the first instance, that the actual psychological capacities and tendencies exhibited by people generally or by members of socially defined groups are significantly determined by the ideologically established beliefs about psychological tendencies and capacities that are accepted in their own culture—determined in such a way as to tend to make their psychologies conform to the ideology.

Interestingly, many who make such claims seem to take this mode of social construction to be appropriate to a constructivist conception of reality and of knowledge. Plainly this is not so. Whatever the independent evidence for constructivism, the fact that culturally transmitted stereotypes causally influence the actual psychological make-up of those stereotyped provides no evidence of the sort of *non*-causal determination of causal structure by minds or theories that the constructivist requires.

The second case concerns solutions to the problem of defining the notion of measurement in quantum mechanics. According to one important conception, part of what characterizes measurements is that they are epistemically relevant interactions, so that measurement is defined in terms of knowledge—that is, in terms of something (one component of which is) mental—and it is a special sort of interaction with a knowing system that produces discontinuous changes in physical state and results in sharp values for measured quantities. It is sometimes added that the explanation for the fact that measurements are not governed by Schrödinger's equation is that they involve interactions between a physical system (whose isolated time-evolution is governed by that equation) and a nonphysical mind. Whether or not the second suggestion is adopted, it is sometimes suggested that the special role of knowing systems thus identified refutes realism because it shows that the phenomena studied by scientists—in particular, the results of their experimental measurements—are mind-dependent. Reflection shows that this interpretation (even in its dualist version) simply assigns a distinctive causal role to certain mental phenomena. No non-causal social construction of causal structure is suggested. Indeed, the development of quantum mechanics might well be cited as the most dramatic recent demonstration of our *inability* to define causal reality in accordance with our conceptual schemes (for an excellent discussion see McMullin 1984).

2.5. *Homeostatic Property-Cluster Definitions, Realism, and Bivalence*

There is an established practice of identifying realism regarding a body of enquiry with the view that all of the sentences in the vocabulary employed within it have determinate mind-independent truth-values, and such a conception of realism places a significant constraint on any realist account of the growth of approximate knowledge. We have just seen that the requirement of mind independence must be carefully qualified. Moreover, the role in approximation that the realist assigns to partial denotation and to denotational refinement (see sect. 2.1.4) precludes any understanding according to which scientific statements must have determinate truth-value: statements involving partially denoting expressions might be true on one denotational refinement and false on another.

There is a quite different way in which a realist conception of scientific language predicts failures of bivalence, and it is important to our understanding of the realist's explanatory project, both because it reflects another dimension of dialectical complexity in the realist's account of approximation and because it provides the philosophical machinery for a deeper analysis of the underlying notion of scientific rationality.

The sorts of essential definition of substances anticipated by Locke and reflected in the currently accepted natural definitions of chemical kinds by molecular formulas (e.g. 'water = H_2O') appear to specify necessary and sufficient conditions for membership in the kind in question. Recent *non*-naturalistic property-cluster or criterial attribute theories in the 'ordinary language' tradition suggest the possibility of definitions that do not provide necessary and sufficient conditions. Instead, some terms are said to be defined by a collection of properties such that the possession of an adequate number of those properties is sufficient for falling within the extension of the term. It is supposed to be a conceptual (and thus an a priori) matter what properties belong in the cluster and which combinations of them are sufficient for falling under the term. However, it is usually insisted that the kinds corresponding to such terms are 'open textured', so that there is some indeterminacy in extension legitimately associated with property-cluster or criterial attribute definitions. The 'imprecision' or 'vagueness' of such definitions is seen as a perfectly appropriate feature of ordinary linguistic usage, in contrast to the artificial precision suggested by rigidly formalistic positivist conceptions of proper language use.

I doubt that there are any terms whose definitions actually fit the ordinary-language model, because I doubt that there are any significant 'conceptual truths' at all. I believe, however, that terms with somewhat similar definitions are commonplace in the special sciences that study complex phenomena. Here's what I think often happens (I formulate the account for monadic property terms; the account is intended to apply in the obvious way to the cases of terms for polyadic relations, magnitudes, etc.):

(i) There is a family F of properties that are contingently clustered in nature in the sense that they co-occur in an important number of cases.

(ii) Their co-occurrence is, at least typically, the result of what may be metaphorically (sometimes literally) described as a sort of *homeostasis*. Either the presence of some of the properties in F tends (under appropriate conditions) to favour the presence of the others, or there are underlying mechanisms or processes that tend to maintain the presence of the properties in F, or both.

(iii) The homeostatic clustering of the properties in F is causally important: there are (theoretically or practically) important effects that are produced by a conjoint occurrence of (many of) the properties in F together with (some or all of) the underlying mechanisms in question.

(iv) There is a kind term *t* that is applied to things in which the homeostatic clustering of most of the properties in F occurs.

(v) *t* has no analytic definition; rather, all or part of the homeostatic cluster F, together with some or all of the mechanisms that underlie it, provide the natural definition of *t*. The question of just which properties and mechanisms belong in the definition of t is an a posteriori question—often a difficult theoretical one.

(vi) Imperfect homeostasis is nomologically possible or actual: some thing may display some, but not all, of the properties in F; some, but not all, of the relevant underlying homeostatic mechanisms may be present.

(vii) In such cases, the relative importance of the various properties in F and of the various mechanisms in determining whether the thing falls under t—if it can be determined at all—is a theoretical rather than a conceptual issue.

(viii) Moreover, there will be many cases of extensional vagueness that are such that they are not resolvable even given all the relevant facts and all the true theories. There will be things that display some, but not all, of the properties in F (and/or in which some, but not all, of the relevant homeostatic mechanisms operate), such that no rational considerations dictate whether or not they are to be classed under t, assuming that a dichotomous choice is to be made.

(ix) The causal importance of the homeostatic property cluster F, together with the relevant underlying homeostatic mechanisms, is such that the kind or property denoted by t is a natural kind (see sect. 2.1.3).

(x) No refinement of usage that replaces t by a significantly less extensionally vague term will preserve the naturalness of the kind referred to. Any such refinement would either require that we treat as important distinctions that are irrelevant to causal explanation or to induction, or that we ignore similarities that are important in just these ways.

(ix) The homeostatic property cluster that serves to define t is not individuated extensionally. Instead, the property cluster is individuated like a (type or token) historical object or process: certain changes over time (or in space) in the property cluster or in the underlying homeostatic mechanisms preserve the identity of the defining cluster. In consequence, the properties that determine the conditions for falling under t may vary over time (or space), *while* t *continues to have the same definition*. The historicity of the individuation criterion for the definitional property cluster reflects the explanatory or inductive significance (for the relevant branches of theoretical or practical enquiry) of the historical development of the property cluster and of the causal factors that produce it, and considerations of explanatory and inductive significance determine the appropriate standards of individuation for the property cluster itself. The historicity of the individuation conditions for the property cluster is thus essential for the naturalness of the kind to which t refers.

The paradigm cases of natural kinds—biological species—are examples of homeostatic cluster kinds. The appropriateness of any particular biological species for induction and explanation in biology depends upon the imperfectly shared and homeostatically related morphological, physiological, and behavioural features that characterize its members. The definitional role of mechanisms of homeostasis is reflected in the role of interbreeding in the modern species concept; for sexually reproducing species, the exchange of genetic material between populations is thought by some evolutionary biologists to be essential to the homeostatic unity of the other properties characteristic of the species, and it is thus reflected in the species definition that they propose (see Mayr 1970). The *necessary* indeterminacy in extension of species terms is a consequence of evolutionary theory, as Darwin observed: speciation depends on the existence of populations that are intermediate between the parent species and the emerging one. Any 'refinement' of classification that artificially eliminated the resulting indeterminacy in classification would obscure the central fact about speciation upon which the cogency of evolutionary theory depends.

Similarly, the property-cluster and homeostatic mechanisms that define a species must be individuated non-extensionally as a process-like historical entity. It is universally recognized that selection for characters that enhance reproductive isolation from related species is a significant factor in phyletic evolution, and it is one which necessarily alters over time the species' defining property-cluster and homeostatic mechanisms (Mayr 1970).

It follows that a consistently developed scientific realism *predicts* indeterminacy of those natural kind or property terms that refer to complex homeostatic phenomena; such indeterminacy is a necessary consequence of 'cutting the world at its (largely mind-independent) joints' (contrast e.g. Putnam 1983 on 'metaphysical realism' and vagueness). Realists' accounts of approximation need not honour bivalence even when partial denotation is not at issue. Similarly, scientific realism predicts the existence of non-extensionally individuated definitional clusters for at least some natural kinds, and thus it treats as legitimate vehicles for the growth of approximate knowledge linguistic practices that would, from a more traditional empiricist perspective, look like diachronic inconsistencies in the standards for the application of such natural kind terms.

Moreover, the homeostatic cluster conception of definitions may permit a more perspicuous formulation of the central explanatory thesis of scientific realism. I have argued elsewhere (Boyd 1979, 1982, 1983) for an understanding of knowledge and of reference according to which (although I did not use this terminology) the relations 'x knows that y' and 'x refers to y' possess homeostatic property-cluster definitions. I will suggest in section 3.7 that scientific rationality has a homeostatic property-cluster definition, and that the realist's explanation for the reliability of scientific methods is best understood as the crucial component in an explanation of the homeostatic unity of scientific rationality.

Not all challenges to realism that arise from considerations about bivalence require in rebuttal an appeal to the possibility of actual bivalence failure. For example, the measurement problem in quantum mechanics is sometimes put by saying that quantum systems lack determinate values of classical magnitudes prior to measurement, and the problem is to characterize the interactions that relieve the indeterminacy with respect to a particular magnitude. Sometimes the alleged indeterminacy prior to measurement is seen as an indication of the failure of realism. Realism is seen as predicting determinacy for (pre-measurement) values of classical magnitudes.

In response, the realist need not appeal to the possibility of a realist explanation for failures of bivalence. There are two ways of understanding the claim about a physical system that it possesses a determinate value of a classical magnitude, a determinate component of orbital angular momentum, for example. On the first understanding, that claim is understood to incorporate the classical *mis*conception of the logical status of statements about angular momentum, in which case the statement is always false, in however many respects special cases of such statements may also have been usefully approximately true. Alternatively, the statement may be interpreted as attributing to the system an eigenstate of the relevant operator, in which case it need not be false, but it has, depending on the system in question, some determinate truth-value. On careful analysis there is no bivalence failure here to explain.

3. APPROXIMATE TRUTH AND PARAMETRIC SPECIFICATION: THE REALIST'S EXPLANATION AS ORDINARY SCIENCE

3.1. The Status of the Realist's Explanation

Recall that the argumentative strategy proposed in section 1.3 calls for us to first assess the evidence for the realist's explanation for the instrumental reliability of scientific methods considered as an ordinary scientific hypothesis. If the realist's explanation appears well confirmed, then there will remain the further and more distinctly philosophical task of determining whether or not, with respect to the realist's explanation itself, it is legiti-

REALISM, APPROXIMATE TRUTH, AND METHOD 239

mate to adopt the realist interpretation without which no defence of a realist position in the philosophy of science is forthcoming.

This approach presupposes that the realist's explanation has the form of an ordinary causal explanation in science subject to confirmation or disconfirmation by ordinary scientific standards. Two considerations might suggest that it does not. First, some philosophical explanations of epistemic matters seem non-causal; this is true, for example, of some transcendental explanations and of some 'ordinary language' analyses of notions like 'evidence', 'reliable', 'justification', and the like. Secondly, there are ways of thinking of the notions of truth and approximate truth (e.g. disquotational analyses) that make them non-causal.

The realist's conception of the epistemology and semantics of scientific theories does not raise any of these problems. Truth is definable from 'primitive denotation' (Field 1974), and denotation, on the realist's account, is an epistemic and thus a causal matter; truth is correspondence truth, and correspondence is a matter of complex causal interactions. Similarly, to talk of respects of approximation to the truth is to talk of respects of similarity and difference between actual causal situations and certain possible ones. It is philosophically challenging to give a general account of the nature of such comparisons with counterfactual possibilities, but such comparisons are so routine a feature of ordinary causal reasoning in science (including reasoning about the reliability of particular methods) that there is no reason to suppose that they raise difficulties in the present context.

Likewise, the explanatory claims of the realist are perfectly ordinary causal claims. Under certain sorts of historical and social circumstances, individually and socially held beliefs are said to exhibit a particular causal power—a tendency to generate methods that are (causally) conducive to the establishment of approximate knowledge—when they are in causally relevant ways approximately true. However controversial, this is an ordinary causal thesis about the interactions of scientific communities and the rest of the world. We may reasonably enquire about how it fares by ordinary scientific standards of evidence. It is to this issue that we now turn our attention.

3.2. Does the History of Science Immediately Refute the Realist's Explanation?

According to the historical objection, the realist's explanation for the reliability of scientific method is refuted by the fact that there have been episodes in the history of science during which methodological practices were successful, but during which the relevant background theories were not, by contemporary standards, approximately true as the realist's explanation requires. The realist's response comes in two parts.

First, the realist's explanation does not require that scientists, even during periods of mature enquiry, be especially good at doing metaphysics. The realist need not necessarily show about any episode in the history of science that the relevant background theories are close to the truth on metaphysical matters. The realist's position is not compromised by any respects of error in earlier background theories that do not undermine her appeal to the specific respects of approximation regarding unobservable phenomena that are crucial to her explanation of the reliability of methods during that episode.

Second, the realist's account of the methods of science predicts that there will be early stages in the history of *any* currently mature science in which the relevant background theories will have been too far from the truth to ensure the sort of reliability of methods that is characteristic of mature sciences. This conclusion is a consequence of the radical contingency in epistemology dictated by the realist explanation for the reliability of scientific methods and, in particular, of the claim that it is, in an epistemically important sense, accidental that the earliest relevantly approximately true theories arise within any scientific discipline. Of course, I do not mean that no historical explanations are possible for particular early successes, but only that, according to the realist, the explanation cannot involve appeal to the operation of rational methods with anything like the reliability of the methods of (what from the contemporary point of view are) theoretically more mature stages in the same sciences.

In sum, the realist's explanation is vulnerable to straightforward refutation by the phenomenon of successful science guided by deeply false background theories only if (a) the relevant historical episodes involve the operation of methods that exhibit the profound and routine reliability of judgements of projectability and related matters characteristic of the most mature sciences in the twentieth century, and (b) the respects of falsity in the relevant background theories are not merely deep, but such as to preclude an explanation of that reliability by appeal to the respects in which those theories are approximately true. The tendency in recent empiricist philosophy of science towards realism reflects precisely the opposite conception: philosophers were tempted by realism precisely because they thought they could see how to offer a realist explanation of the reliability of methodological practices in highly successful science, and they lost their confidence in alternative empiricist 'rational reconstructions' of those methods. In any event, what I envision as the realist's reply to the historical objection is simply that there aren't cases satisfying (a) and (b). Realism is, after all, supposed to be an empirical thesis, and here is one of the empirical claims upon which it rests.

3.3. Triviality, Contrivance, and the Methodology of Parametric Specification

Against the charge of immediate historical falsification, the realist replies by insisting, as the logic of her explanations dictates anyway, that her thesis is that background theories in mature sciences must be seen as approximately true in relevant respects. As we saw in section 1.2, the realist now faces the challenge that her explanations are trivial: that any consistent theory is true is some respects, and that she has offered no general theory of the relative importance of respects of approximate truth. Here the reply is the obvious one that the respects of approximation that are important are those that are required to sustain the realist's distinctive explanation of the reliability of scientific methods, and that it is with respect to these that approximations to the truth are claimed. The reply is successful just in case the charge of contrivance can be met: just in case, that is, the realist can argue that, even in the absence of a general context- and episode-neutral account of degrees of approximation, her appeal to respects of approximation appropriate to her own theoretical project does not constitute an *ad hoc* and thus methodologically inappropriate contrivance.

In order to assess the prospects for the realist's explanations, we need to know what distinguishes such contrivances from methodologically appropriate appeals to context-specific specifications of causal variables. Fortunately the question is not esoteric; frequently, especially in the context of historical explanations, we confirm theories that appeal to context-dependent specifications of causal parameters, and the methodology for avoiding *ad hoc* theorizing is well understood. Consider, for example, explanations in evolutionary theory. There are a variety of possible evolutionary mechanisms—individual selection, kin selection, genetic drift, selection for pleiotropically linked traits, etc.—for no one of which does evolutionary theory provide a context-independent prediction of its influence in any particular evolutionary episode. Moreover, in particular evolutionary episodes several of these factors may operate, and there is not context-independent way of predicting their relative influence. Still, the modern evolutionary explanation for the diversity of life is well confirmed. What methodological principles permit us to treat the explanations provided by evolutionary theory as appropriate rather than *ad hoc*, and as appropriate for 'inductive inference to the best explanation'?

The answer is pretty clear. What we require of the various individual evolutionary explanations for particular features of living organisms is that they cohere not only with each other but with the independent results of enquiry in the related scientific disciplines: geology, genetics, developmental biology, animal behaviour, atmospheric sciences, oceanography, anthropology, etc. This requirement of integration of the various particular explanations into the broader framework of scientific knowledge constitutes our methodological safeguard against the possibility that the apparent explanatory successes of evolutionary theory are reflections of mere contrivance. This pattern is quite general: particular explanations provide evidence for a broader theory whose explanatory resources they exploit just in case theory-dependent evidential standards, including requirements of theoretical integration, dictate the acceptance of the particular explanations, and just in case the success of those individual explanations lends inductive support for the causal claims of the broader theory (Boyd 1985c).

Exactly the same standards apply, of course, to the realist's broad explanation for the reliability of scientific methods. The charge of contrivance is met just in case the realist's explanations for the reliability of methods in particular episodes, including the context-dependent specifications of respects of approximation they contain, are independently supported by scientific evidence, and in particular that they pass the test of coherence with the rest of established scientific theory, and (this is the easier part) just in case these particular realist explanations lend inductive support to the broader realist explanatory picture of scientific epistemology.

3.4. The Local Coherence of Realism

Are the individual realist explanations for the reliability of specific scientific methods well confirmed, and do they, in particular, cohere appropriately with the rest of science? Do they inductively support the general realist conception of the growth of approximate knowledge? At an important level of analysis the answer to both questions must be 'Obviously yes'.

The particular realist explanations of the reliability of methods fall roughly into two categories. In the first category are the theoretical explanations for the reliability of particular measurement and computational procedures and for the reliability of various sorts of controls and other features of the design of experimental and observational studies. In the second category are the theoretical explanations for the reliability of the judgements of projectability which determine the broader outlines of ra-

tional experimental and observational enquiry. Explanations in either category may be either static or dialectical. By a static explanation, I understand an explanation that explains the reliability of some piece of
methodology by appeal to the approximate truth of some theories that
have been long established at the time of the relevant methodological
judgements; dialectical explanations explain the reliability of some
novel feature of methodology or of some revision of a previously established methodological practice by appealing to changes in theoretical
outlook that bring about a closer approximation to the truth along relevant
lines.

At any given time in the history of recent science, individual realist
explanations in the first category, both static and dialectical, look just like
well-established pieces of boringly normal science: they are the sorts of
claims that are routinely made explicit in the methods sections of papers in
the empirical sciences, in which scientists explain the appropriateness of
research design. Most explanations of the static sort and almost all of the
dialectical ones will embody reference to context-specific degrees and
respects of approximation in the current theoretical conception or its
immediate predecessors. Those explicit pieces of scientific theorizing
are not produced in service of any philosophical or historical project,
realist or otherwise. In the better-established sciences they are apparently
as well confirmed as anything gets; certainly there is no evidence that they
fail to cohere with the rest of established science. That, after all, is what
made such pieces of ordinary science so disturbing to empiricists. The
prospect that they are vulnerable to the contrivance objection is vanishingly remote.

Scientists seem rarely to investigate explicitly the causal question of the
reliability of particular projectability judgements *under that description*.
They do, however, offer justifications for their own methodological judgements, critiques of such judgements by others, and proposals for changes
in such judgements. Such justifications are made explicit in published
papers, in referees' reports, in grant proposals, in the introductory parts of
experimental papers, and in theoretical papers and books, and the judgements they justify are *in fact* judgements of projectability of the sort
to which the realist explanation refers. It is all but the consensus position
among students of the logic of scientific inference (e.g. Hanson 1958,
Kuhn 1970, Quine 1969b, van Fraassen 1980) that ordinary scientific standards of reasoning treat these projectability judgements as inductive
inferences from background theories, just as realism requires. Here again,
the justifications in question routinely appeal to context-specific respects
of approximation, especially in cases in which they mirror realist explan-

ations of the dialectical sort. There is again no prospect that scientists' reasoning in such cases is contrived to serve a philosophical purpose, nor is there any reason to hold that the requirement of coherence with the rest of science is not honoured in such reasoning—indeed, it is in reasoning of this sort that the requirement of coherence finds its expression in ordinary science!

Here, then, is the phenomenon of *local coherence*: the explicit and near-explicit findings of ordinary science examined synchronically seem to strongly confirm, if only tacitly, the particular explanations for the reliability of projectability judgements on which the realist's explanatory enterprise rests, and they appear to do so in a way that subjects the context-dependent judgements of relevant respects of approximation which they contain to the appropriate requirement of coherence.

Do the particular realist explanations we are considering, taken together, inductively support the realist conception of scientific epistemology developed in part 2? Here we cannot defer to any particular science except philosophy, but we can observe that the whole tendency to take realism seriously as an alternative to logical empiricism from the mid-1950s on reflects the extremely widespread judgement among philosophers of science that the actual practices of science *appear* to require a realist explanation. I conclude that, if we examine the question *pre*-philosophically, there appears to be very good reason to hold that the realist's explanation for the reliability of scientific methods is well confirmed as a scientific hypothesis, and, in particular, that there is no reason to think that the realist's approach to the problem of parametric specification is any more in doubt than, say, that of the evolutionary biologists who must also rely on specifications not given antecedently by a context-independent formula.

We turn now to the question of what the distinctly philosophical dimension is to the confirmation of the realist's explanatory hypothesis. The elaborate machinery rehearsed in part 2 indicates that a lot is going on philosophically. Some of it is relevant only to the question of circularity, but much is relevant also to a defence of realism as a scientific thesis in the methodological climate created by the philosophical disputes over realism.

3.5. What's Distinctly Philosophical? I: Diachronic Patterns of Inference and of Language Use

Central to the realist's abductive argument for realism is the claim that no alternative exists that adequately explains the reliability of scientific methods or justifies their use. It is possible to *imagine* that a case along these lines for realism—or, at any rate, against the verificationist insistence that knowledge of unobservables is impossible—could be made by the synchronic examination of only a few episodes in the history of science for which only realist explanations and justifications seem available. Nevertheless, the deep plausibility of empiricist epistemological principles, especially the evidential indistinguishability thesis, is so great that it is doubtful that realism about a few isolated cases would, even as a scientific hypothesis, be rationally acceptable. Instead, the plausibility of any individual realist explanation seems to rest upon diachronic considerations that provide additional and crucial support for the general realist explanation of the reliability of scientific methods. In effect, the role of these diachronic considerations is to establish that the individual synchronic-realist explanations can be coherently integrated into a scientifically acceptable historical conception of the reliability of scientific methodology.

In particular, there are two patterns in the history of science whose recognition is a distinctive contribution of philosophers and historians in making the case for the realist's explanations. In the first place, there is the utterly commonplace phenomenon of *mutual ratification* between consecutive stages in the development of scientific disciplines. It is routine in the case of theoretical innovations that (a) the new and innovative theoretical proposal is such that the only justification scientists have for accepting it, given the relevant evidence, is that it resolves some scientific problem or question *while preserving certain key features of the earlier theoretical conceptions*; and (b) the new proposal ratifies the earlier conception as approximately true in just those respects that justify their role in its own acceptance. Moreover, the patterns of mutual ratification are characteristically seen to be *retrospectively sustained*: although later theoretical innovations typically require a revision in our estimates of the degrees and respects of approximation of both the earlier innovative proposals and their predecessors, the initially discernible relation of mutual ratification is typically sustained as a very good first approximation to the evidentially and methodologically important relations between the innovation and its predecessor theories. It is the ubiquity of this sort of *retrospectively sustained mutual ratification* and the difficulty in 'rationally reconstructing' it away with respect to the justification of theoretical innovations that has made the case for realism so plausible.

A second pattern concerns the use of scientific language. The realist conception of projectability requires that the categories that scientists employ in formulating general laws and causal claims typically reflect underlying causal structures rather than conventionally specified nominal essences, and many of the changes in classificatory practice for which individual realist explanations are forthcoming seem to indicate an attempt to obtain a fit between categories and causal structure. It is essential to the case for realism that this pattern in scientific language use be sustained: that the diachronic linguistic behaviour of scientists involves an apparent disposition to take the definitions of scientific kinds, relations, magnitudes, etc. to be revisable in the light of new data and new theoretical developments. Thus the identification of just such a pattern of *apparent essentialism* in the actual linguistic practices in scientific communities is an important, distinctly philosophical contribution to the case for the realist's explanation of the reliability of scientific method.

3.6. What's Distinctly Philosophical? II: Epistemological, Metaphysical, and Semantic Underpinnings

The ubiquitous patterns of retrospectively sustained mutual ratification and apparent essentialism constitute philosophical reasons to accept the realist's explanation, and the recognition of those patterns was a central factor in the emergence of contemporary scientific realism. Still, their effect would not have been so significant were it not for more theoretical attempts to understand their philosophical import. The obvious examples here are causal theories of reference and associated naturalistic conceptions of definition. Had it not proved possible to articulate these distinctly philosophical theories, then it might have been rational to hold that the apparently rational theory-and-evidence-driven revision of definitions in science was only apparent, or only apparently rational. The initial case for the realist explanation would have been crucially undermined.

Analogous considerations hold for the epistemological dimension. Both the realist explanations for the reliability of scientific methods in particular cases and the view that the ubiquity of the pattern of mutual ratification supports the broader realist explanation entail that evidential considerations in science are deeply theory-dependent. Were it not possible to provide a realist epistemological framework that incorporates this conclusion—and, in particular, were it not possible to articulate that framework so as to refute the evidential indistinguishability thesis and make palatable the consequent abandonment of foundationalism—then it would have not been rational to take either the particular explanations or the pattern of mutual ratification as significant support for the realist explanation. Thus the development of a non-foundationalist realist treatment of projectability judgements and the incorporation of that treatment into an independently developing tradition of non-foundationalist naturalism in epistemology proves to have been essential for the rational acceptance of the realist explanation.

On to metaphysics. The causal theory of reference and the naturalistic conceptions within epistemology with which realist anti-foundationalism can be profitably assimilated all appear to reflect a distinctly non-Humean conception of causal relations. The cogency of these fundamental elements in the defence of the realist's explanation depends therefore (at least prima facie) on the successful articulation of a non-Humean conception of causation (e.g. Boyd 1985c, Mackie 1974, Shoemaker 1980).

Acceptance of the realist's explanation as a scientific theory does not entail the acceptance of scientific realism, since the realist's explanation might itself be interpreted non-realistically. What I have been suggesting is that, nevertheless, the realist's explanation is sufficiently novel in its apparent epistemological, semantic, and metaphysical implications that the articulation of just the sort of broader realistic and naturalistic conceptions of (scientific and other) knowledge, of language, and of metaphysics indicated in part 2 is essential for the defence of that explanation.

I think that the picture just presented captures the current case that the realist's explanation for the reliability of scientific methods is a well-confirmed scientific theory, context-dependent specifications of respects of approximation notwithstanding. An even broader philosophical setting for that case is available if we exploit the distinctly naturalistic conception of homeostatic property-cluster definitions outlined in section 2.5.

3.7. Realism and the Homeostatic Character of Scientific Rationality

I argued in section 2.5 that lots of natural kinds, properties, etc. possess homeostatic property-cluster definitions, and I suggested that knowledge and reference are among them. I want now to suggest a similar homeostatic cluster treatment of scientific rationality itself. Ordinarily we think of scientific rationality as being exhibited in two different features of the practice of science: the high level of deliberative rationality in the reasoning of researchers, and the spectacular successes of scientific research in understanding and predicting natural phenomena. If foundationalism is mistaken, as it surely seems to be, then the first of these features does not logically entail the second, and the realist explanation may be thought of as explaining why (and when) they reliably co-occur. Here is a kind of homeostasis of the two distinct components of scientific rationality.

Once it is recognized that this co-occurrence is a causal matter, then it is easy to see that at a finer level of analysis there is a family of similar sorts of co-occurrences requiring explanation. The methodological norms in a

particular subdiscipline are set not only by the background theoretical
findings in that subdiscipline, but as well by findings from other subdisci-
plines and from quite different disciplines altogether. That the methodological norms determined by such a wide range of theories should be
unified enough to be a practical guide to successful scientific research
requires explanation. Why aren't the resulting methodological norms characteristically irreconcilably conflicting, for instance?

Similarly, scientists working largely independently within different disciplines frequently converge on the same solution to a problem they may not
have recognized that they have in common. Why should this happen?
Likewise, it often happens that largely independently developing disciplines become ripe for interdisciplinary work, and their largely independently developed theories and methodologies prove (with some difficult
but not impossible negotiation) to be integrable. Why is this so frequently
possible?

What I propose is that we think of scientific rationality as being defined
by the homeostasis of all of these various components of scientific practice,
and that we should think of the realist explanation of the coincidence of
deliberative rationality and theoretical and empirical success as the first
step toward a more general realist explanation of the relevant homeostasis.
It is even possible that this project could be extended fruitfully to incorporate a naturalistic conception of moral rationality (Boyd 1988; Brink 1984,
1989; Miller 1984b; Railton 1986; Sturgeon 1984a,b).

If the proposal of the present section were to prove successful, it would
prima facie provide further support for the realist explanation and for the
philosophical naturalism that underwrites it. However, we still need to
know whether the realist's explanation should itself be understood realistically or whether, instead, as the circularity objection suggests, that would
simply beg the question against the anti-realist.

4. MEETING THE CIRCULARITY OBJECTION

4.1. Circularity and Philosophical Packages

According to the circularity objection, the realist's explanation for the
success of scientific methods, even if well confirmed, cannot without begging the question be interpreted realistically, and thus cannot without
circularity be treated as confirming scientific realism. The problem posed
by this objection is faced not only by the particular defence of realism
under consideration, but by almost any plausible defence of scientific
realism.

The reason is simple: in all but the most trivial cases the defence of realism regarding one or more theories or traditions will require the defence of a theory of epistemic contact that spells out the sort of epistemically relevant causal relations that are supposed to obtain between the subject-matter of the theories or traditions and the behaviour of the relevant enquirers. Because the realist thesis and the theory of epistemic contact that supports it are causal theses, their confirmation will always depend upon the confirmation of theories (or, for very simple cases, commonplaces) about the causal powers of the entities that are the putative subject-matter of the theory or tradition in question. The confirmation of specific theories of epistemic contact will, in turn, depend in part upon theoretical considerations grounded in the best available theories of the relevant subject-matter. Such theories will be a vital background assumption against which the evidence for the realist thesis is judged. As we have seen, the theory of epistemic contact, and (thus) the theories upon which its confirmation in turn depends, will themselves have to be understood realistically if they are to help to validate the realist thesis itself. But of course, these theories will, in any plausible case, be subject to the same anti-realist assessments as the theory or tradition about which realism is initially in question. Indeed, if that theory is a well-established contemporary theory, it may *itself* provide the foundations for the relevant theory of epistemic contact! Is this not a point at which the defence of realism begs the question against anti-realists?

Here the answer is 'no'. If theories of epistemic contact by themselves constituted the sole argument of the realist against anti-realism; if, for example, the *sole* argument in favour of realism in atomic theory consisted of the articulation of an apparently well-confirmed theory of epistemic contact between scientists and atoms, their properties and their constituent parts, then the question would indeed be begged by the assumption that that theory itself should be understood realistically. The actual role of theories of epistemic contact is quite different.

The issue of realism arises in the form we have been discussing only in the case of a theory or tradition of enquiry about which there is a prima-facie case that it possesses a theory-independent (even if unobservable) subject-matter. The prima-facie case for realism will thus rest upon the apparent confirmation of a (realistically understood) theory of epistemic contact. In the special case of realism defended along the lines proposed here, that theory of contact is the one embodied in the realist's explanation for the reliability of scientific methods. The defence of realism, however, depends not upon the theory of epistemic contact *alone*, but upon the ability of realists to incorporate suitably elaborated versions of it into an epistemological, semantic, and metaphysical conception of the theory or tradition in question (a *philosophical package*) that is superior to those available to defenders of the various anti-realist conceptions.

Thus, for example, the defence of realism regarding the tradition of atomic theory depends upon the best-confirmed atomic theories providing the basis for an apparently realistic theory of epistemic contact; but it depends as well upon additional, more explicitly philosophical considerations, which legitimize the realist treatment of such a theory. On the version of scientific realism presented here, these additional considerations are of two sorts. First, it is argued that only on a realist construal of atomic theory generally, and of the relevant theory of epistemic contact in particular, is it possible to avoid scepticism about the possibility of purely instrumental knowledge in physics and chemistry: knowledge of a sort acknowledged by empiricists and constructivists, as well as by realists. Secondly, it is argued that the picture that emerges from a realist treatment of atomic theory is consonant in its departures from foundationalism and in its treatment of scientific language with other quite independently defensible developments in epistemology and semantic theory.

In such a dialectical setting, the dependence of the realist theories of epistemic contact upon a realist understanding of the theory or tradition in question (or of some closely related theory or tradition) need not constitute begging the question against the anti-realist. Fairness to the case for realism requires that realism be understood in a context provided by a realist interpretation of the apparently best-confirmed realist theories of epistemic contact and of the apparently best-confirmed substantive theories of the alleged (theory-independent) subject-matter in question.

Importantly, just the same understanding of the issue is required by fairness to the case *against* realism. Both the empiricists' and the constructivists' anti-realist arguments depend upon the assumption that the realist accepts the prevailing theoretical conception and its associated methodology. The realist is understood to take the properties of the putative socially unconstructed referents of the terms of a theory or tradition to be, at least approximately, those required by (a realist understanding of) the apparently best-confirmed theories of the presumed subject-matter and to accept the methodology dictated by them as approximately reliable. On those assumptions (but not without them), the empiricist can reason that the realist's position commits her to the possibility of investigating the properties of unobservable phenomena, and thus to an epistemological position against which the empiricist has very powerful arguments.

The constructivist anti-realist similarly assumes that the realist accepts a realist interpretation of the prevailing theoretical and methodological conceptions. Only on such an understanding is it clear that the realist is committed to the possibility of investigating a theory-independent reality using theory-dependent methods—just the possibility that the constructivist critique of realism rejects. Thus, an adequate treatment of the controversy between realists and either of their standard opponents requires that we accept that the philosophical package offered in defence of realism contains the apparently best-confirmed theories of the alleged subject-matter, realistically understood, and in particular that it be understood as incorporating an associated realistically understood conception of epistemic contact.

Once it is seen that no question is begged against the anti-realist by adopting a realist interpretation of the realist's explanation for the reliability of scientific methodology, we are left with the question: Suppose that the realist's explanation is well confirmed, then why would a realist philosophical package incorporating a realist version of that explanation be superior to an empiricist package incorporating the explanation instrumentally interpreted or to a constructivist package incorporating the realist's explanation understood as a piece of social construction? My main aim in the present essay is to show that the realist's appeal to a distinctively realist explanation for the growth of approximate knowledge, incorporating an appropriate context-and-episode-dependent account of relevant respects of approximation, does not involve any triviality, contrivance, or begging of the question—not to finish once and for all the task of defending realism. I will therefore indicate only briefly the outlines of the considerations that seem to me to justify a preference for the realist package over the two alternatives in question.

4.2. Against the Empiricist Package

The key argument for scientific realism according to the programme presented here is that realism as a scientific hypothesis presents the only scientifically acceptable explanation for the reliability of scientific methods. The empiricist might be unimpressed by the demand for explanation in this case (Fine 1984 (Ch. 1, this volume); van Fraassen 1980). Still, the realist can also argue that accepting the realist explanation provides as well the only justification we have for accepting the instrumental findings of science (Boyd 1983, 1985a). One possible empiricist response is that we can justify accepting the inductive deliverances of an apparently realistic scientific method as a result of the second-order induction about induction whose conclusion is that reasoning like a realist in science is instrumentally reliable. Since this conclusion is only about observables, the empiricist can accept it and employ it to justify accepting currently accepted theories as empirically adequate.

Against this rebuttal I have argued (Boyd 1983, 1985a) that the induction in question is demonstrably just as theory-dependent as any other in science, and is thus unavailable to the empiricist who is adopting the proposed strategy. Here is a possible reply. We justify the second-order induction by a third-order induction about inductions about induction, the third-order induction by appeal to a fourth-order induction, etc. For the nth case the justification for the relevant projectability judgements is provided not by apparently realistic theoretical considerations, but by the n + first-order induction.

If I am right, this last response is what the incorporation of the realist's explanation into an empiricist philosophical package would require if that package were to provide any even remotely plausible account of the justification of (instrumental) scientific knowledge. I claim that the resulting philosophical package would prove to be only remotely plausible in consequence. Here we have not just infinite regress but infinite ascent: each level of inductive inference is justified by appeal to a more abstract and problematical level of inductive inference. Given that the realist's package already incorporates an alternative, less speculative, and independently justified naturalistic epistemology, I predict that it will prove superior.

4.3. Against the Constructivist Package

Response to the sort of constructivist philosophical package that might be constructed so as to include the realist's explanation for the reliability of scientific methods is substantially more difficult. Constructivism is a richer philosophical programme than empiricism, and at the same time it incorporates features (often just the ones that add to its richness) whose consistency is disputable. Rather than even beginning to sort out all of the issues that a thoroughgoing realist response to constructivism would have to address, I will just indicate briefly how two quite standard objections to constructivism might be brought to bear on the proposed package.

In the first place, any adequate philosophical package will have to incorporate versions of most of the apparently best-established scientific and methodological findings. The suggestion outlined in section 2.4, that the establishment of social institutions and linguistic conventions does not contribute non-causally to the causal powers of the objects studied by participants in those institutions and conventions, has very deep roots in quite diverse features of our understanding both of causation and of social phenomena. Thus, any constructivist philosophical package will be prima facie vulnerable at any point at which it incorporates a distinctly constructivist conception of the social construction of causal relations. The proposed constructivist package would incorporate this doubtful feature into its version of the naturalistic account of the reliability of scientific methods, and thus into the very centre of its basic epistemology. It is doubtful, therefore, that the proposed package will afford as satisfactory a treatment of absolutely central epistemological issues as its realist rivals.

A second standard objection to constructivism is that the historical fact of anomalies indicates that the world scientists study does not have a structure logically, socially, or conceptually determined by the paradigms or theories they accept. It is beyond the scope of this essay to examine the variants on this objection and the range of possible replies. It cannot be doubted, however, that it does pose a serious challenge to the acceptability of any constructivist package. Since there are anomalies in methodological matters that exactly parallel those in theoretical matters, the incorporation of a doctrine of social construction of the reliability of scientific method seems hardly to strengthen the constructivist philosophical package.

I conclude that the resources exist for a spirited defence of a realist philosophical package against empiricist and constructivist alternatives, and in particular that the incorporation of a realist interpretation of the realist's explanation of the reliability of scientific methodology strengthens rather than (as the circularity challenge suggests) weakens the realist package.

REFERENCES

Armstrong, D. M. (1973). *Belief, Truth and Knowledge.* Cambridge: Cambridge University Press.

Boyd, R. (1972). 'Determinism, Laws and Predictability in Principle.' *Philosophy of Science*, 39: 431–50.

——(1973). 'Realism, Underdetermination and a Causal Theory of Evidence.' *Noûs* 7: 1–12.

——(1979) 'Metaphor and Theory Change.' In A. Ortony (ed.), *Metaphor and Thought*. Cambridge: Cambridge University Press.

——(1980). 'Materialism without Reductionism: What Physicalism does not Entail.' In N. Block (ed.), *Readings in Philosophy of Psychology*, i. 67–106. Cambridge, Mass.: Harvard University Press.

——(1982). 'Scientific Realism and Naturalistic Epistemology.' In P. D. Asquith and R. N. Giere (eds.), *PSA 1980, vol. 2*, 613–62. East Lansing, Mich.: Philosophy of Science Association.

254   RICHARD BOYD

—— (1983). 'On the Current Status of the Issue of Scientific Realism.' *Erkenntnis*, 19: 45–90.
—— (1985a). 'Lex Orandi est Lex Credendi.' In P. Churchland and C. Hooker (eds.), *Images of Science: Scientific Realism versus Constructive Empiricism*, 3–34. Chicago: University of Chicago Press.
—— (1985b). 'The Logician's Dilemma.' *Erkenntnis*, 22: 197–252.
—— (1985c). 'Observations, Explanatory Power, and Simplicity.' In P. Achinstein and O. Hannaway (eds.), *Observation, Experiment, and Hypothesis in Modern Physical Science*, 45–94. Cambridge, Mass.: MIT Press.
—— (1988). 'How to be a Moral Realist.' In G. Sayre McCord (ed.), *Moral Realism*. Ithaca, NY: Cornell University Press.
Brink, D. (1984). 'Moral Realism and the Skeptical Arguments from Disagreement and Queerness.' *Australasian Journal of Philosophy*, 62/2: 111–25.
—— (1989). *Moral Realism and the Foundations of Ethics*. Cambridge: Cambridge University Press.
Byerly, H. and Lazara, V. (1973). 'Realist Foundations of Measurement.' *Philosophy of Science*, 40: 10–28.
Carnap, R. (1934). *The Unity of Science*, tr. M. Black. London: Kegan Paul.
Feigl, H. (1956). 'Some Major Issues and Developments in the Philosophy of Science of Logical Empiricism.' In H. Feigl and M. Scriven (eds.), *The Foundations of Science and the Concepts of Psychology and Psychoanalysis*, Minnesota Studies in the Philosophy of Science, vol. 1. 3–37. Minneapolis: University of Minnesota Press.
Field, H. (1973). 'Theory Change and the Indeterminacy of Reference.' *Journal of Philosophy*, 70: 462–81.
—— (1974). 'Tarski's Theory of Truth.' *Journal of Philosophy*, 69: 347–75.
Fine, A. (1984). 'The Natural Ontological Attitude.' In J. Leplin (ed.), *Scientific Realism*, 83–107. Berkeley: University of California Press. Reprinted as Ch. I, this volume.
Goldman, A. (1967). 'A Causal Theory of Knowing.' *Journal of Philosophy*, 64: 357–72.
—— (1976). 'Discrimination and Perceptual Knowledge.' *Journal of Philosophy*, 73: 771–91.
Goodman, N. (1973). *Fact, Fiction and Forecast*, 3rd edn. Indianapolis and New York: Bobbs-Merrill.
Hanson, N. R. (1958). *Patterns of Discovery*. Cambridge: Cambridge University Press.
Hardin, C., and Rosenberg, A. (1982). 'In Defense of Convergent Realism.' *Philosophy of Science*, 49: 604–15.
Kripke, S. A. (1971). 'Identity and Necessity.' In M. K. Munitz (ed.), *Identity and Individuation*, 135–64. New York: New York University Press.
—— (1972). 'Naming and Necessity.' In D. Davidson and G. Harman (eds.), *The Semantics of Natural Language*, 253–355. Dordrecht: Reidel.
Kuhn, T. (1970). *The Structure of Scientific Revolution*, 2nd edn. Chicago: University of Chicago Press.
Laudan, L. (1981). 'A Confutation of Convergent Realism.' *Philosophy of Science*, 48: 218–49. Reprinted as Ch. VI, this volume.
Mackie, J. L. (1974). *The Cement of the Universe*. Oxford: Oxford University Press.
Mayr, E. (1970). *Populations, Species and Evolution*. Cambridge, Mass.: Harvard University Press.

McMullin, E. (1984). 'A Case for Scientific Realism.' In J. Leplin (ed.), *Scientific Realism*, 8–40. Berkeley: University of California Press.

Miller, R. (1984*a*). *Analyzing Marx*. Princeton: Princeton University Press.

——(1984*b*). 'Ways of Moral Learning.' *Philosophical Review*, 94: 507–56.

Nagel, E. (1961). *The Structure of Science*. New York: Harcourt Brace.

Putnam, H. (1962). 'The Analytic and the Synthetic.' In H. Feigl and G. Maxwell (eds.), *Realism and Reason*, Minnesota Studies in the Philosophy of Science, vol. 3, 358–97. Minneapolis: University of Minnesota Press.

——(1972). 'Explanation and Reference.' In G. Pearce and P. Maynard (eds.), *Conceptual Change*, Dordrecht: Reidel.

——(1975*a*). 'Language and Reality.' In Putnam (ed.), *Mind, Language and Reality*, 272–90. Cambridge: Cambridge University Press.

——(1975*b*). 'The Meaning of Meaning.' In Putnam (ed.), *Mind, Language and Reality*, 215–41. Cambridge: Cambridge University Press.

——(1983). 'Vagueness and Alternative Logic.' In *Realism and Reason*, 271–86. Cambridge: Cambridge University Press.

Quine, W. V. O. (1969*a*). 'Epistemology Naturalized.' In *Ontological Relativity and Other Essays*, 69–90. New York: Columbia University Press.

——(1969*b*). 'Natural Kinds.' In *Ontological Relativity and Other Essays*, 114–38. New York: Columbia University Press.

Railton, P. (1986). 'Moral Realism.' *Philosophical Review*, 95: 163–207.

Rawls, J. (1971). *A Theory of Justice*. Cambridge, Mass.: Harvard University Press.

Shoemaker, S. (1980). 'Causality and Properties.' In P. van Inwagen (ed.), *Time and Cause*, 109–36. Dordrecht: Reidel.

Sturgeon, N. (1984*a*). 'Moral Explanations.' In D. Copp and D. Zimmerman (eds.), *Morality, Reason and Truth*, 49–78. Totowa, NJ: Rowman and Allanheld.

——(1984*b*). Review of *Moral Relativism* and *Virtues and Vices*, by P. Foot. *Journal of Philosophy*, 81: 326–33.

Tarski, A. (1951). 'The Concept of Truth in Formalized Languages.' In *Logic, Semantics and Metamathematics*, 152–278. New York: Oxford University Press.

Van Fraassen, B. (1980). *The Scientific Image*. Oxford: Clarendon Press.

XI

RATIONALITY AND OBJECTIVITY IN SCIENCE *OR* TOM KUHN MEETS TOM BAYES

WESLEY C. SALMON

Twenty-five years ago, as of this writing, Thomas S. Kuhn published *The Structure of Scientific Revolutions* (1962).[1] It has been an extraordinarily influential book. Coming at the height of the hegemony of *logical empiri-cism*—as espoused by such figures as R. B. Braithwaite, Rudolf Carnap, Herbert Feigl, Carl G. Hempel, and Hans Reichenbach—it posed a severe challenge to the logistic approach that they practised.[2] It also served as an unparalleled source of inspiration to philosophers with a historical bent. For a quarter of a century there has been a deep division between the logical empiricists and those who adopt the historical approach, and Kuhn's book was undoubtedly a key document in the production and preservation of this gulf.

At a 1983 meeting of the American Philosophical Association (Eastern Division), Kuhn and Hempel—the most distinguished living advocates for their respective viewpoints—shared the platform in a symposium (1983) devoted to Hempel's philosophy. I had the honour to participate in this symposium. On that occasion Kuhn chose to address certain issues pertaining to the rationality of science that he and Hempel had been discussing for several years. It struck me that a bridge could be built between the differ-

Reprinted from C. Wade Savage (ed.), *Scientific Theories*, Minnesota Studies in the Philosophy of Science, vol. 14 (Minneapolis: University of Minnesota Press, 1990), 175–204, by permission of the University of Minnesota Press.

[1] Kuhn 2nd edn. 1970. 'Postscript—1969', added to 2nd edn., contains discussions of some of the major topics that are treated in the present essay.

[2] Such philosophers are often characterized by their opponents as *logical positivists*, but this is an egregious historical inaccuracy. Although some of them had been members of, or closely associated with, the Vienna Circle in their earlier years, none of them retained the early positivistic commitment to phenomenalism and/or instrumentalism in their more mature writings. Reichenbach and Feigl, e.g., were outspoken realists, and Carnap regarded physicalism as a tenable philosophical framework. Richenbach never associated himself with positivism; indeed, he regarded his 1938 book as a refutation of logical positivism. I could go on and on . . .

ing views of Kuhn and Hempel if Bayes's theorem were invoked to explicate the concept of scientific confirmation.[3] At the time it seemed to me
that this manoeuvre could remove a large part of the dispute between
standard logical empiricism and the historical approach to philosophy of
science on this fundamental issue.

I still believe that we have the basis for a new consensus regarding the
choice among scientific theories. Although such a consensus, if achieved,
would not amount to total agreement on every problem, it would represent
a major rapprochement on an extremely fundamental issue. The purpose
of the present essay is to develop this approach more fully. As it turns out,
the project is much more complex than I thought in 1983.

1. KUHN ON SCIENTIFIC RATIONALITY

A central part of Kuhn's challenge to the logical empiricist philosophy of
science concerns the nature of theory choice in science. The choice
between two fundamental theories (or paradigms), he maintains, raises
issues that 'cannot be resolved by proof'. To see how they are resolved,
we must talk about 'techniques of persuasion', or about 'argument
and counterargument in a situation in which there can be no proof'. Such
choices involve the exercise of the kind of judgement that cannot
be rendered logically explicit and precise. Such statements, along with
many others that are similar in spirit, led a number of critics to attribute
to Kuhn the view that science is fundamentally irrational and lacking
in objectivity.

Kuhn was astonished by this response, which he regarded as a serious
misinterpretation. In his 'Postscript—1969' in the second edition of *The
Structure of Scientific Revolutions* and in 'Objectivity, Value Judgment,
and Theory Choice' (1977: 320–39)[4] he replies to these charges. What he
had intended to convey was the claim that the decision by the community
of trained scientists *constitutes* the best criterion of objectivity and rationality we can have. In order better to understand the nature of such objective and rational methods, we need to look in more detail at the
considerations that are actually brought to bear by scientists when they
endeavour to make comparative evaluations of competing theories.

For purposes of illustration, Kuhn offers a (non-exhaustive) list of
characteristics of good scientific theories that are, he claims:

---

[3] I had offered a similar suggestion in my 1970.

[4] The response is given in greater detail in this article than it is in the Postscript.

individually important and collectively sufficiently varied to indicate what is at stake. . . . These five characteristics—accuracy, consistency, scope, simplicity, and fruitfulness—are all standard criteria for evaluating the adequacy of a theory. . . . Together with others of much the same sort, they provide the shared basis for theory choice. (1977: 321–2)

Two sorts of problems arise when one attempts to use them.

Individually the criteria are imprecise: individuals may legitimately differ about their applicability to concrete cases. In addition, when deployed together, they repeatedly prove to conflict with one another; accuracy may, for example, dictate the choice of one theory, scope the choice of its competitor. (1977: 322)

For reasons of these sorts—and others as well—individual scientists may, at a given moment, differ regarding a particular choice of theories. In the course of time, however, the interactions among individual members of the community of scientists produce a consensus for the group. Individual choices inevitably depend upon idiosyncratic and subjective factors; only the outcome of the group activity can be considered objective and fully rational.

One of Kuhn's major claims seems to be that observation and experiment, in conjunction with hypothetico-deductive reasoning, do not adequately account for the choice of scientific theories. This has led some philosophers to believe that theory choice is not rational. Kuhn, in contrast, has tried to locate the additional factors that are involved. These additional factors constitute a crucial aspect of scientific rationality.

2. BAYES'S THEOREM

The first step in coming to grips with the problem of evaluating and choosing scientific hypotheses or theories[5] is the recognition of the inadequacy of the traditional hypothetico-deductive (H-D) schema as a characterization of the logic of science. According to this schema, we confirm a scientific hypothesis by deducing from it, in conjunction with suitable initial conditions and auxiliary hypotheses, an observational prediction that turns out to be true. The H-D method has a number of well-known shortcomings. (1) It does not take account of alternative hypotheses that might be invoked to explain the same prediction. (2) It makes no reference to the initial plausibility of the hypothesis being evaluated. (3) It cannot accommodate cases, such as the testing of statistical hypotheses, in which

[5] Throughout this paper I shall use the terms 'hypothesis' and 'theory' more or less interchangeably. Kuhn tends to prefer 'theory', while I tend to prefer 'hypothesis', but nothing of importance hinges on this usage here.

the observed outcome is not deducible from the hypothesis (in conjunction with the pertinent initial conditions and auxiliary hypotheses), but only rendered more or less probable.

In view of these and other considerations, many logical empiricists agreed with Kuhn regarding the inadequacy of hypothetico-deductive confirmation. A number—including Carnap and Reichenbach—appealed to Bayes's theorem, which may be written in the following form:

$$P(T/E.B) = \frac{P(T/B)P(E/B.T)}{P(T/B)P(E/B.T) + P(\sim T/B)P(E/B.\sim T)}. \qquad (1)$$

Let 'T' stand for the theory or hypothesis being tested, 'B' for our background information, and 'E' for some new evidence we have just acquired. Then the expression on the left-hand side of the equation represents the probability of our hypothesis on the basis of the background information and the new evidence. This is known as the *posterior probability*. The right-hand side of the equation contains four probability expressions. Two of these, $P(T/B)$ and $P(\sim T/B)$, are called *prior probabilities*; they represent the probability, on the basis of background information alone, without taking account of the new evidence E, that our hypothesis is true or false respectively. Obviously the two prior probabilities must add up to one; if the value of one of them is known, the value of the other can be inferred immediately. The remaining two probabilities, $P(E/T.B)$ and $P(E/\sim T.B)$, are known as *likelihoods*; they are, respectively, the probability that the new evidence would occur if our hypothesis were true and the probability that it would occur if our hypothesis were false. The two likelihoods, in contrast to the two prior probabilities, must be established independently; the value of one does not automatically determine the value of the other. To calculate the posterior probability of our hypothesis, then, we need three separate probability values to plug into the right-hand side of Bayes's theorem—a prior probability and two likelihoods.

Before attempting to resolve any important issues concerning the nature of scientific reasoning, let us look at a simple and non-controversial application of Bayes's theorem. Consider a factory that produces can-openers at the rate of 6,000 per day. This factory has two machines, a new one that produces 5,000 can-openers per day and an old one that produces 1,000 per day. Among the can-openers produced by the new machine 1 per cent are defective; among those produced by the old machine 3 per cent are defective. We pick one can-opener at random from today's production and find it defective. What is the probability that it was produced by the new machine?

We can get the answer to this question via Bayes's theorem. If we let 'B' stand for the class of can-openers produced in this factory today, 'T' for the class of can-openers produced by the new machine, and 'E' for a can-opener that is defective, then the probability we seek is the posterior probability P(T/B.E)—the probability that a defective can-opener from today's production was produced by the new machine. The values of the prior probabilities and likelihoods have been given, namely:

$$P(T/B) = 5/6 \qquad\qquad P(\sim T/B) = 1/6$$
$$P(E/T.B) = 1/100 \qquad\qquad P(E/\sim T.B) = 3/100.$$

Plugging these values into equation (1) immediately yields $P(T/B.E) = 5/8$. Notice that the old machine has a greater probability of producing a defective can-opener than does the new, but the probability that a defective can-opener was produced by the new machine is greater than that it was produced by the old one. This results, obviously, from the fact that the new machine produces so many more can-openers overall than does the old one.

One way to look at this example is to consider the hypothesis T that a given can-opener was produced by the new machine. This is a causal hypothesis. Our background information B is simply that the can-opener is part of today's production at this factory. On the basis of this prior information, we can evaluate the prior probability of T; it is 5/6. Now we add to our knowledge about this can-opener the information E that it is defective. This knowledge is relevant to the hypothesis that it was produced by the new machine; the posterior probability is 5/8. Although one does not *need* to appeal to Bayes's theorem to establish this result,[6] the highly artificial example shows clearly just how Bayes's theorem can be used to ascertain the posterior probability of a simple causal hypothesis.

When we come to more realistic scientific cases, it is not so easy to see how to apply Bayes's theorem; the prior probabilities may seem particularly difficult. I believe that, in fact, they reflect the plausibility arguments scientists often bring to bear in their deliberations about scientific hypotheses. I shall discuss this issue in section 4; indeed, in subsequent sections we shall have to take a close look at all of the probabilities that enter into Bayes's theorem.

---

[6] As Adolf Grünbaum pointed out to me, if we assume that in a given day the actual frequency of defective can-openers produced by the two machines matches precisely the respective probabilities, we can calculate the result as follows. The new machine produces 50 defective can-openers and the old machine produces 30, so that 50 out of a total of 80 are produced by the new machine. However, it would be *incorrect* to assume that the frequencies match the probabilities each day; in fact, the probability of an exact match is quite small.

In this section I have been concerned to present Bayes's theorem and to make a few preliminary remarks about its application to the problem of evaluating scientific hypotheses. In the next section I shall try to spell out the connections between Bayes's theorem and Kuhn's views on the nature of theory choice. Before moving on to that discussion, however, I want to present two other useful forms in which Bayes's theorem can be given. In the first place, because of the theorem on total probability:

$$P(E/B) = P(T/B)P(E/T.B) + P(\sim T/B)P(E/\sim T.B), \qquad (2)$$

equation (1) can obviously be rewritten as:

$$P(T/E.B) = \frac{P(T/B) \times P(E/B.T)}{P(E/B)}. \qquad (3)$$

In the second place, equation (1) can be generalized to handle several alternative hypotheses, instead of just one hypothesis and its negation, as follows:

$$P(T_i/B.E) = \frac{P(T_i/B) \times P(E/T_i.B)}{\sum_{j=1}^{k} \left[P(T_j/B) \times P(E/T_j.B) \right]}, \qquad (4)$$

where T_1–T_k are mutually exclusive and exhaustive alternative hypotheses and $1 \le i \le k$.

Strictly speaking, (4) is the form that is needed for realistic historical examples—such as the corpuscular (T_1) and wave (T_2) theories of light in the nineteenth century. In that case, although we could construe T_1 and T_2 as mutually exclusive, we could not legitimately consider them exhaustive, for we cannot be sure that one or the other is true. Therefore, we would have to introduce T_3—what Abner Shimony has called the *catch-all hypothesis*—which says that T_1 and T_2 are both false. T_1–T_3 thus constitute a mutually exclusive and exhaustive set of hypotheses. This is the sort of situation that obtains when scientists are attempting to choose a correct hypothesis from among two or more serious candidates.

3. KUHN AND BAYES

For purposes of discussion, Kuhn is willing to admit that 'each scientist chooses between competing theories by deploying some Bayesian algorithm which permits him to compute a value for P(T/E), i.e., for the probability of the theory T on the evidence E available both to him and the

other members of his professional group at a particular period of time'
(1977: 328). He then formulates the crucial issue in terms of the question of
whether there is one unique algorithm used by all rational scientists, yielding a unique value for P, or whether different scientists, though fully
rational, may use different algorithms yielding different values of P. I want
to suggest a third possibility to account for the phenomena of theory
choice—namely, that many different scientists might use the same algorithm, but nevertheless arrive at different values of P.

When one speaks of a Bayesian algorithm, the first thought that comes
to mind is Bayes's theorem itself, as embodied in any of the equations (1),
(3), or (4). We have, for instance:

$$P(T/E.B) = \frac{P(T/B) \times P(E/B.T)}{P(E/B)}, \qquad (3)$$

which constitutes an algorithm in the most straightforward sense of the
term. Let us call P(E/B) the *expectedness* of the evidence. Given values for
the prior probability, likelihood, and expectedness, the value of the posterior probability can be computed by trivial arithmetical operations.[7]

If we propose to use equation (3) as an algorithm, the obvious question
is how to get values for the expressions on the right-hand side. Several
answers are possible in principle, depending on what interpretation of the
probability concept is espoused. If one adopts a Carnapian approach to
inductive logic and confirmation theory, all of the probabilities that appear
in Bayes's theorem can be derived a priori from the structure of the
descriptive language and the definition of degree of confirmation. Since it
is extremely difficult to see how any genuine scientific case could be
handled by means of the highly restricted apparatus available within that
approach, not many philosophers are tempted to follow this line. Moreover, even if a rich descriptive language were available, it is not philosophically tempting to suppose that the probabilities associated with serious
scientific theories are a priori semantic truths.

Two major alternatives remain. First, one might maintain that the probabilities on the right-hand side of (3)—especially the prior probability
P(T/B)—are objective and empirical. I have attempted to defend the view
that they refer, at bottom, to the frequencies with which various kinds of
hypotheses or theories have been found successful (Salmon 1967: ch. 7).

---

[7] I remarked above that three probabilities are required to calculate the posterior probability—a prior probability and two likelihoods. Obviously, in view of (2), the theorem on total
probability, if we have a prior probability, one of the likelihoods, and the expectedness, we can
compute the other likelihood; likewise, if we have one prior probability and both likelihoods, we
can compute the expectedness.

Clearly, enormous difficulties are involved in working out that alternative; I shall return to the issue below. In the meantime, let us consider the other—far more popular—alternative. The remaining alternative approach involves the use of personal probabilities. Personal probabilities are subjective in character; they represent subjective degrees of conviction on the part of the individual who has them, provided that they fulfil the condition of coherence.[8] Consider a somewhat idealized situation. Suppose that, in the presence of background knowledge B (which may include initial conditions, boundary conditions, auxiliary hypotheses), theory T deductively entails evidence E. This is the situation to which the hypothetico-deductive method appears to be applicable. In this case, P(E/T.B) must equal 1, and equation (3) reduces to:

$$P(T/E.B) = P(T/B)/P(E/B). \qquad (5)$$

One might then ask a particular scientist for his or her plausibility rating of the theory T on background knowledge B alone, quite irrespective of whether evidence E obtains or not. Likewise, the same individual may be queried regarding the degree to which evidence E is to be expected irrespective of the truth or falsity of T. According to the personalist, it should be possible—by direct questioning or by some less direct method—to elicit such *psychological* facts regarding a scientist involved in investigations concerning the theory in question. This information is sufficient to determine the degree of belief this individual should have in the theory T, given the background knowledge B and the evidence E, namely, the posterior probability P(T/E.B).

In the more general case, when T and B do not deductively entail E, the procedure is the same, except that the value of P(E/T.B) must also be ascertained. In many contexts, where statistical significance tests can be applied, a value of the likelihood P(E/T.B) can be calculated, and the personal probability will coincide with the value thus derived. In any case, whether statistical tests apply or not, there is no *new* problem in principle involved in procuring the needed degree of confidence. This reflects the standard Bayesian approach in which all of the probabilities are taken to be personal probabilities.

In any case, whether one adopts an objective or a personalistic interpretation of probability, equation (3)—or some other version of Bayes's theorem—can be taken as an algorithm for evaluating scientific hypotheses or theories. Individual scientists, using the same algorithm, may arrive at

[8] A set of degrees of conviction is coherent provided that its members do not violate any of the conditions embodied in the mathematical calculus of probability.

different evaluations of the same hypothesis because they plug in different values for the probabilities. If the probabilities are construed as objective, different individuals may well have different estimates of these objective values. If the probabilities are construed as personal, different individuals may well have different subjective assessments of them. Bayes's theorem provides a mechanical algorithm, but the judgement of individual scientists are involved in procuring the values that are to be fed into it. This is a general feature of algorithms; they are not responsible for the data they are given.

4. PRIOR PROBABILITIES

In section 2 I remarked that the prior probabilities in Bayes's theorem can best be seen as embodying the kinds of plausibility judgements that scientists regularly make regarding the hypotheses with which they are concerned. Einstein, who was clearly aware of this consideration, contrasted two points of view from which a theory can be criticized or evaluated:

The first point of view is obvious: the theory must not contradict empirical facts. . . . [it] is concerned with the confirmation of the theoretical foundation by the available empirical facts. The second point of view is not concerned with the relation of the material of observation but with the premises of the theory itself, with what may briefly but vaguely be characterized as the 'naturalness' or 'logical simplicity' of the premises. . . . The second point of view may briefly be characterized as concerning itself with the 'inner perfection' of a theory, whereas the first point of view refers to the 'external confirmation.' (1949: 21–2)

Einstein's second point of view is the sort of thing I have in mind in referring to plausibility arguments or judgements concerning prior probabilities.

Plausibility considerations are pervasive in the sciences; they play a significant—indeed, *indispensable*—role. This fact provides the initial reason for appealing to Bayes's theorem as an aid to understanding the logic of evaluating scientific hypotheses. Plausibility arguments serve to enhance or diminish the probability of a given hypothesis prior to—that is, without reference to—the outcome of a particular observation or experiment. They are designed to answer the question 'Is this the kind of hypothesis that is likely to succeed in the scientific situation in which the scientist finds himself or herself?' On the basis of their training and experience, scientists are qualified to make such judgements.

This point can best be explained, I believe, in terms of concrete examples. Since before the time of Newton, for instance, a well-known plausibility argument for the inverse-square character of gravitational forces has been around. It is natural to think of the gravitational force emanating from a particle of matter as one that spreads spherically from it in a uniform manner. In the seventeenth and eighteenth centuries all competent physical scientists believed that physical space has a three-dimensional Euclidean structure. Since the surface of a Euclidean sphere increases as the square of the radius, it is reasonable to suppose that the force of gravity is diluted in just the same way, for the farther one goes from the particle, the greater the spherical surface over which the force must be spread.

A famous Canadian study of the effects of the consumption of large doses of saccharin provides another example.[9] A statistically significant association between heavy saccharin consumption and bladder cancer in a controlled experiment with rats lends considerable plausibility to the hypothesis that use of saccharin as an artificial sweetener in diet soft drinks increases the risk of bladder cancer in humans. This example, unlike the preceding one, is inherently statistical, and does not have even the prima-facie appearance of a hypothetico-deductive inference.

In order to come to a clearer understanding of the nature of prior probabilities, it will be necessary to look at them from the point of view of the personalist and that of the objectivist (frequency or propensity theorist).[10] The frightening thing about pure unadulterated personalism is that nothing prevents prior probabilities (and other probabilities as well) from being determined by all sorts of idiosyncratic and objectively irrelevant considerations. A given hypothesis might get an extremely low prior probability because the scientist considering it has a hangover, has had a recent fight with his or her lover, is in passionate disagreement with the politics of the scientist who first advanced the hypothesis, harbours deep prejudices against the ethnic group to which the originator of the hypothesis belongs, etc. What we want to demand is that the investigator make every effort to bring all of his or her *relevant* experience in evaluating hypotheses to bear on the question of whether the hypothesis under consideration is of a type likely to succeed, and to leave aside emotional irrelevancies.

It is rather easy to construct really perverse systems of belief that do not violate the coherence requirement. But we need to keep in mind the

---

[9] This example is discussed in Giere 1984: 274–6.

[10] I reject the so-called propensity interpretation of probability because, as Paul Humphreys pointed out, the probability calculus accommodates inverse probabilities of the type that occur in Bayes's theorem, but the corresponding inverse propensities do not exist. In the example of the can-opener factory, each machine has a certain propensity to produce defective can-openers, but it does not make sense to speak of the propensity of a given defective can-opener to have been produced by the new machine.

objectives of science. When we have a long series of events, such as tosses of fair or biased coins, or radioactive decays of unstable nuclei, we want our subjective degrees of conviction to match what either a frequency theorist or a propensity theorist would regard as the objective probability. Carnap was profoundly correct in his notion that inductive or logical or epistemic probabilities should be reasonable estimates of relative frequencies.

A sensible personalist, I would suggest, is someone who wants his or her personal probabilities to reflect objective fact. Betting on a sequence of tosses of a coin, a personalist wants not only to avoid Dutch books,[11] but also to stand a reasonable chance of winning (or of not losing too much too fast). As I read it, the whole point of F. P. Ramsey's famous article on degrees of belief (1950) is to consider what you get if your subjective degrees of belief match the relevant frequencies. One of the facts recognized by the sensible personalist is that whether the coin lands heads or tails is not affected by on which side of the bed he or she got out that morning. If we grant that the personalist's aim is to do as well as possible in betting on heads and tails, it would obviously be counterproductive to allow the betting odds to be affected by such irrelevancies.

The same general sort of consideration should be brought to bear on the assignment of probabilities to hypotheses. Whether a particular scientist is dyspeptic on a given morning is irrelevant to the question of whether a physical hypothesis that is under consideration is correct or not. Much more troubling, of course, is the fact that any given scientist may be inadvertently influenced by ideological or metaphysical prejudices. It is obvious that an unconscious commitment to capitalism or racism might seriously affect theorizing in the behavioural sciences.

Similar situations may arise in the physical sciences as well; another historical example will illustrate the point. In 1800 Alessandro Volta invented the battery, thereby providing scientists with a way of producing steady electrical currents. It was not until 1820 that Hans Christian Oersted discovered the effect of an electrical current on a magnetic needle. Why was there such a delay? One reason was the previously established fact that a static electric charge has no effect on a magnetic needle. Another reason that has been mentioned is the fact that, contrary to the expectation if there were such an effect, it aligns the needle perpendicular to the current carrying wire. As Holton and Brush remark, 'But even if one has currents and compass needles available, one does not observe the effect unless the compass is placed in the right position so that the needle can

[11] A so-called Dutch book is a combination of bets such that, no matter what the outcome of the event upon which the wagers are made, the subject is bound to suffer a net loss.

respond to a force that seems to act in a direction *around* the current rather than *toward* it' (1973: 416, emphasis original). I found it amusing when, on one occasion, a colleague set up the demonstration with the magnetic needle oriented at right angles to the wire to show why the experiment fails if one begins with the needle in that position. When the current was turned on, the needle rotated through 180 degrees; he had neglected to take account of polarity. How many times, between 1800 and 1820, had the experiment been performed without reversing the polarity? Not many. The experiment had apparently not been tried by others because of Cartesian metaphysical commitments. It was undertaken by Oersted as a result of his proclivities toward *Naturphilosophie*.

How should scientists go about evaluating the prior probabilities of hypotheses? In elaborating a view he calls *tempered personalism*—a view that goes beyond standard Bayesian personalism by placing further constraints on personal probabilities—Shimony point out (1970) that experience shows that the hypotheses seriously advanced by serious scientists stand some chance of being successful. Science has, in fact, made considerable progress over the past four or five centuries, which constitutes strong empirical evidence that the probability of success among members of this class is non-vanishing. Likewise, hard experience has also taught us to reject claims of scientific infallibility. Thus, we have good reasons for avoiding the assignment of extreme values to the priors of the hypotheses with which we are seriously concerned. Moreover, Shimony reminds us, experience has taught that science is difficult and frustrating; consequently, we ought to assign fairly low prior probabilities to the hypotheses that have been explicitly advanced, allowing a fairly high prior for the catch-all hypothesis—the hypothesis that we have not yet thought of the correct hypothesis. The history of science abounds with situations of choice among theories in which the successful candidate has not even been conceived at the time.

In *The Foundations of Scientific Inference*, I proposed that the problem of prior probabilities be approached in terms of an objective interpretation of probability, in particular, the frequency interpretation. I suggested three sorts of criteria that can be brought to bear in assessing the prior probabilities of hypotheses: formal, material, and pragmatic.

Pragmatic criteria have to do with the circumstances in which a new hypothesis originates. We have already seen an example of a pragmatic criterion in Shimony's observation that hypotheses advocated by serious scientists have non-vanishing chances of success. The opposite side of the same coin is provided by Martin Gardner, who offers an enlightening characterization of scientific cranks (1957: 7–15). Since it is doubtful that a single useful scientific suggestion has ever been originated by anyone in that category, hypotheses advanced by people of that ilk have negligible chances of being correct. I recall when L. Ron Hubbard's *Dianetics* was first published. A psychologist friend, asked what he thought of it, said, 'I can't condemn this book before reading it, but after I have read it, I will.' When competent scientists offer hypotheses outside of their areas of specialization, we have a right to wonder whether appreciable plausibility accrues to such suggestions. Hubbard was, incidentally, an engineer with no training in psychology.

The *formal criteria* have to do not only with matters of internal consistency of a new hypothesis, but also with relations of entailment or incompatibility of the new hypothesis with accepted laws and theories. The fact that Immanuel Velikovski's *Worlds in Collision* (1950) contradicts many of the accepted basic laws of physics—for example, the law of conservation of angular momentum—renders his 'explanations' of such biblically reported incidents as the parting of the waters of the Red Sea and the brief interruption of the rotation of the earth (the sun standing still) utterly implausible.

It should be recalled that among his five considerations for the evaluation of scientific theories—mentioned above—Kuhn includes consistency of the sort we are discussing. I take this as a powerful hint that one of the main issues Kuhn has raised about scientific theory choice involves the use of prior probabilities and plausibility judgements.

The *material criteria* have to do with the actual structure and content of the hypothesis or theory under consideration. The most obvious example is simplicity—another of Kuhn's five items. Simplicity strikes me as singularly important, for it has often been treated by scientists and philosophers as an a priori criterion. It has been suggested, for example, that the hypothesis that quarks are fundamental constitutents of matter loses plausibility as the number of different types of quarks increases, since it becomes less simple as a result (see Harari 1983). It has also been advocated as a universal methodological maxim: *Search for the simplest possible hypothesis.* Only if the simpler hypotheses do not stand up under testing should one resort to more complex hypotheses.

Although simplicity has obviously been an important consideration in the physical sciences, its applicability in the social/behavioural sciences is problematic. In a recent article, 'Slips of the Tongue', Michael T. Motley criticizes Freud's theory for being too simple—an oversimplification.

Further still, the categorical nature of Freud's claim that all slips have hidden meanings makes it rather unattractive. It is difficult to imagine, for example, that my six-year-old daughter's mealtime request to 'help cut up my meef' was the result of repressed anxieties or anything of that kind. It seems more likely that she simply merged 'meat' and 'beet' into 'meef'. Similarly, about the only meaning one can easily read into someone's saying 'room mock' instead of 'moon rock' is that the m and r got switched. Even so, how does it happen that words can merge or sounds can be switched in the course of speech production? And in the case of my 'pleased to beat you' error [to a competitor for a job], might Freud have been right?' (1985: 116)[12]

The most reasonable way to look at simplicity, I think, is to regard it as a highly relevant characteristic, but one whose applicability varies from one scientific context to another. Specialists in any given branch of science make judgements about the degree of simplicity or complexity that is appropriate to the context at hand, and they do so on the basis of extensive experience in that particular area of scientific investigation. Since there is no precise measure of simplicity as applied to scientific hypotheses and theories, scientists must use their judgement concerning the degree of simplicity a given hypothesis or theory possesses and concerning the degree of simplicity that is desirable in the given context. The kind of judgement to which I refer is not spooky; it is the kind of judgement that arises on the basis of training and experience. This experience is far too rich to be the sort of thing that can be spelled out explicitly. As Patrick Suppes has pointed out (1966: 202–3), the assignment of prior probability by the Bayesian can be regarded as the best estimate of the chances of success of the hypothesis or theory on the basis of all relevant experience in that particular scientific domain. The personal probability represents, not an effort to contaminate science with subjective irrelevancies, but rather an attempt to facilitate the inclusion of all relevant evidence.

Simplicity is only one among many material criteria. Another closely related criterion—frequently employed in contemporary physics—is symmetry. Perhaps the most striking historical example is de Broglie's hypothesis regarding matter waves. Since light exhibits both particle and wave behaviour, which are linked in terms of linear momentum, he suggested, why should not material particles, which obviously possess linear momentum, also have such wave characteristics as wavelength and frequency? Unbeknownst to de Broglie, experimental work by Davisson was, at that very time, providing positive evidence of wave-like behaviour of electrons.

A third widely used material criterion is analogy, as illustrated by the saccharin study. The physiological analogy between rats and humans is

[12] Grünbaum 1984: 202–4 criticizes Motley's account of Freud's theory; he considers Motley's version a distortion, and points out that Freud's motivational explanations were explicitly confined to a very circumscribed set of slips. He defends Freud against Motley's criticism on the grounds that Freud's actual account has greater complexity than Motley gives it credit for.

sufficiently strong to lend considerable plausibility to the hypothesis that saccharin can cause bladder cancer in humans. I suspect that the use of arguments by analogy in science is almost always aimed at establishing prior probabilities. The formal criteria enable us to take account of the ways in which a given hypothesis fits *deductively* with what else we know. Analogy helps us to assess the degree to which a given hypothesis fits *inductively* with what else we know.

The moral I would draw concerning prior probabilities is that they can be understood as our best estimates of the frequencies with which certain kinds of hypotheses succeed. These estimates are rough and inexact; some philosophers might prefer to think of them in terms of intervals. If, however, one wants to construe them as personal probabilities, there is no harm in it, as long as we attribute to the subject who has them the aim of bringing to bear all his or her experience that is relevant to the success or failure of hypotheses similar to that being considered. The personalist and the frequentist need not be in any serious disagreement over the construal of prior probabilities.[13]

One point is apt to be immediately troublesome. If we are to use Bayes's theorem to compute values of posterior probabilities, it would appear that we must be prepared to furnish numerical values for the prior probabilities. Unfortunately, it seems preposterous to suppose that plausibility arguments of the kind we have considered could yield exact numerical values. The usual answer is that, because of a phenomenon known as 'washing out of the priors' or 'swamping of the priors', even very crude estimates of the prior probabilities will suffice for the kinds of scientific judgements we are concerned to make. Obviously, however, this sort of convergence depends upon agreement regarding the likelihoods.

5. THE EXPECTEDNESS

The term '$P(E/B)$' occurring in the denominator of equation (3) is called the *expectedness* because it is the opposite of surprisingness. The smaller the value of $P(E/B)$, the more surprising E is; the larger the value of $P(E/B)$, the less surprising, and hence, the more expected E is. Since the expectedness occurs in the denominator, a smaller value tends to increase the value of the fraction. This conforms to a widely held intuition that the more surprising the predictions a theory can make, the greater is their evidential value when they come true.

[13] I have discussed the relations between personal probabilities and objective probabilities in my 1988.

A classic example of a surprising prediction that came true is the Poisson bright spot. If we ask someone who is completely naïve about theories of light how probable it is that a bright spot appears in the centre of the shadow of a brightly illuminated circular object (ball or disk), we would certainly anticipate the response that it is very improbable indeed. There is a good inductive basis for this answer. In our everyday lives we have all observed many shadows of opaque objects, and they do not contain bright spots at their centres. Once, when I demonstrated the Poisson bright spot to an introductory class, one student carefully scrutinized the ball-bearing that cast the shadow, because he strongly suspected that it had a hole through it.

Another striking example, to my mind, is the Cavendish torsion-balance experiment. If we ask someone who is totally ignorant of Newton's theory of universal gravitation how strongly they expect to find a force of attraction between a lead ball and a pith ball in a laboratory, I should think the answer, again, would be that it is very unlikely. There is, in this example as well, a sound inductive basis for the response. We are all familiar with the gravitational attraction of ordinary-size objects to the earth, but we do not have everyday experience of an attraction between two such relatively small (electrically neutral and unmagnetized) objects as those Cavendish used to perform his experiment. Newton's theory predicts, of course, that there will be a gravitational attraction between any two material objects. The trick was to figure out how to measure it.

As the foregoing two examples show, there is a possible basis for assigning a low value to the expectedness; it was made plausible by assuming that the subject was completely naïve concerning the relevant physical theory. The trouble with this approach is that a person who wants to use Bayes's theorem—in the form of equation (3), say—cannot be totally innocent of the theory T that is to be evaluated, since the other terms in the equation refer explicitly to T. Consequently, we have to recognize the relationship between $P(E/B)$ and the prior probabilities and likelihoods that appear on the right-hand side in the theorem on total probability:

$$P(E/B) = P(T/B)P(E/T.B) + P(\sim T/B)P(E/\sim T.B), \qquad (2)$$

Suppose that the prior probability of T is not negligible, and that T, in conjunction with suitable initial conditions, entails E. Under these circumstances, E cannot be totally surprising; the expectedness cannot be vanishingly small. Moreover, to evaluate the expectedness of E, we must also consider its probability if T is false. By focusing on the expectedness, we cannot really avoid dealing with likelihoods.

There is a further difficulty. Suppose, for example, that the wave theory of light is true. It is surely *true enough* in the context of the Poisson bright spot experiment. If we want to evaluate P(E/B), we must include in B the initial conditions of the experiment—the circular object illuminated by a bright light in such a way that the shadow falls upon a screen. Given the truth of the wave theory, the *objective probability* of the bright spot is 1, for whenever those initial conditions are realized, the bright spot appears. It makes no difference whether we know that the wave theory is true, or believe it, or reject it, or have ever thought of it. Under the conditions specified in B, the bright spot invariably occurs. Interpreted either as a frequency or a propensity, P(E/B) = 1. If we are to avoid trivialization in many important cases, the expectedness must be treated as a personal probability. To anyone who, like me, wants to base scientific theory preference or choice on objective considerations, this result poses a serious problem.

The net result is a twofold problem. First, by focusing on the expectedness, we do *not* escape the need to deal explicitly with the likelihoods. In section 6 I shall discuss the difficulties that arise when we focus on the likelihoods, especially the problem of the likelihood on the catch-all hypothesis. Second, the expectedness defies interpretation as an objective probability. In section 7 I shall propose a strategy for avoiding involvement with either the expectedness or the likelihood on the catch-all. That manœuvre will, I hope, keep open the possibility of an objective basis for the evaluation of scientific hypotheses.

6. LIKELIHOODS

Equations (1), (3), and (4) are different forms of Bayes's theorem, and each of them contains a likelihood, P(E/T.B), in the numerator. Two trivial cases can be noted at the outset. First, if the conjunction of theory T and background knowledge B are logically incompatible with evidence E, the likelihood equals 0, and the posterior probability, P(T/E.B), automatically becomes 0.[14] Second, as we have already noticed, if T.B entails E, that likelihood equals 1, and consequently drops out, as in equation (5).

Another easy case occurs when the hypothesis T involves various kinds of randomness assumptions, for example, the independence of a series of

[14] As Duhem has made abundantly clear, in such cases we may be led to re-examine our background knowledge B, which normally involves auxiliary hypotheses, to see whether it remains acceptable in the light of the negative outcome E. Consequently, refutation of T is not usually as automatic as it appears in the simplified account just given. Nevertheless, the probability relation just stated is correct.

trials on a chance set-up.[15] Consider, for example, the case of a coin that has been tossed 100 times, with the result that heads showed in 63 cases and tails in 37. We assume that the tosses are independent, but we are concerned whether the system consisting of the coin and tossing mechanism is biased. Calculation shows that the probability, given an unbiased coin and tossing mechanism, of the actual frequency of heads differing from 1/2 by 20 per cent or more on 100 tosses (i.e. falling outside the range 40 to 60) is about 0.05. Thus, the likelihood of the outcome on the hypothesis that the coin and mechanism are fair is less than 0.05. On the hypothesis that the coin has a 60 to 40 bias for heads, by contrast, the probability that the number of heads in 100 trials differs from 6/10 by less than 20 per cent (i.e. lies within the 48 to 72 range) is well above 0.95. These are the kinds of likelihoods that would be used to compare the *null hypothesis* that the coin is fair with the hypothesis that it has a certain bias.[16] This example typifies a wide variety of cases, including the above-mentioned controlled experiment on rats and saccharin, in which statistical significance tests are applied. These yield a comparison between the probability of the observed result if the hypothesis is correct and the probability of the same result on a null hypothesis.

In still another kind of situation the likelihood P(E/T.B) is straightforward. Consider, for example, the case in which a physician takes an X-ray for diagnostic purposes. Let T be the hypothesis that the patient has a particular disease, and let E be a certain appearance on the film. From long medical experience it may be known that E occurs in 90 per cent of all cases in which that disease is present. In many cases, as this example suggests, there may be accumulated frequency data from which the value of P(E/T.B) can be derived.

Unfortunately, life with likelihoods is not always as simple as the foregoing cases suggest. Consider an important case, which I will present in a highly unhistorical way. In comparing the Copernican and Ptolemaic cosmologies, it is easy to see that the phases of Venus are critical. According to the Copernican system, Venus should exhibit a broad set of phases from a narrow crescent to an almost full disc. According to the Ptolemaic system, Venus should always present nearly the same crescent-shaped appearance. One of Galileo's celebrated telescopic observations was of the phases of Venus. The likelihood of such evidence on the Copernican

---

[15] Exchangeability is the personalist's surrogate for randomness; it means that the subject would draw the same conclusion regardless of the order in which the members of an observed sample occurred.

[16] Note that, in order to get the posterior probability—the probability that the observed results were produced by a biased device—the prior probabilities have to be taken into account.

system is unity; on the Ptolemaic it is zero. This is the decisive sort of case that we cherish.

The Copernican system did, however, face one serious obstacle. On the Ptolemaic system, because the earth does not move, the fixed stars should not appear to change their positions. On the Copernican system, because the earth makes an annual trip around the sun, the fixed stars should appear to change their positions in the course of the year. The very best astronomical observations, including those of Tycho Brahe, failed to reveal any observable stellar parallax.[17] However, it was realized that, if the fixed stars are at a very great distance from the earth, stellar parallax, though real, would be too small to be observed. Consequently, the likelihood P(E/T.B), where T is the Copernican system and E the absence of observable stellar parallax, is not zero. At the time of the scientific revolution, prior to the advent of Newtonian mechanics, there seemed no reasonable way to evaluate this likelihood. The assumption that the fixed stars are almost unimaginably distant from the earth was a highly *ad hoc*, and consequently implausible, auxiliary hypothesis to adopt just to save the Copernican system. Among other things, Christians did not like the idea that heaven was so very far away.

The most reasonable resolution of this anomaly was offered by Tycho Brahe, whose cosmology placed the earth at rest, with the sun and moon moving in orbits around the earth, but with all of the other planets moving in orbits around the sun. In this way both the observed phases of Venus and the absence of observable stellar parallax could be accommodated. Until Newton's dynamics came upon the scene, it seems to me, Tycho's system was clearly the best available theory.

In section 2 I suggested that the following form of Bayes's theorem is the most appropriate for use in actual scientific cases in which more than one hypothesis is available for serious consideration:

$$P(T_i/B.E) = \frac{P(T_i/B) \times P(E/T_i.B)}{\sum_{j=1}^{k} \left[P(T_j/B) \times P(E/T_j.B) \right]}, \qquad (4)$$

It certainy fits the foregoing example in which we compared the Ptolemaic, Copernican, and Tychonic systems. This equation involves a mutually exclusive and exhaustive set of hypotheses $T_1, \ldots, T_{k-1}, T_k$, where T_1–T_{k-1} are seriously entertained, and T_k is the catch-all. Thus, the scientist who wants to calculate the posterior probability of one particular hypothesis T_i on the basis of evidence E must ascertain likelihoods of three types: (1) the probability of evidence E, given T_i; (2) the probability of that evidence on

---

[17] Indeed, stellar parallax was not detected until the nineteenth century.

each of the other seriously considered alternatives T_j ($j \neq i, j \neq k$); and (3) the probability of that evidence on the catch-all T_k.

In considering the foregoing example, I suggested that, although likelihoods in the first two categories are sometimes straightforward, there are cases in which they turn out to be quite problematic. We shall look at more examples in which they present difficulties as our discussion proceeds. But the point to be emphasized right now is the utter intractability of the likelihood on the catch-all. The reason for this difficulty is easy to see. Whereas the seriously considered candidates are bona fide hypotheses, the catch-all is a hypothesis only in a Pickwickian sense. It refers to all of the hypotheses we are *not* taking seriously, including all those that have not been thought of as yet; indeed, the catch-all is logically equivalent to their disjunction. These will often include brilliant discoveries in the future history of science that will eventually solve our most perplexing problems.

Among the hypotheses hidden in the catch-all are some that, in conjunction with present available background information, entail the present evidence E. On such as-yet-undiscovered hypotheses the likelihood is 1. Obviously, however, the fact that its probability on one particular hypothesis is unity does not entail anything about its probability on some disjunction containing that hypothesis as one of its disjuncts. These considerations suggest to me that the likelihood on the catch-all is totally intractable. To try to evaluate the likelihood on the catch-all involves, it seems to me, an attempt to guess the future history of science. That is something we cannot do with any reliability.

In any situation in which a small number of theories are competing for ascendancy, it is tempting, though quite illegitimate, simply to ignore the likelihood on the catch-all. In the nineteenth century, for instance, scientists asked what the probability of a given phenomenon is on the wave theory of light and what it is on the corpuscular theory. They did not seriously consider its probability if neither of these theories is correct. Yet we see, from the various forms in which Bayes's theorem is written, that either the expectedness or the likelihood on the catch-all is an indispensable ingredient. In the next section I shall offer a *legitimate* way of eliminating those probabilities from our consideration.

7. CHOOSING BETWEEN THEORIES

Kuhn has often maintained that in actual science the problem is never to evaluate one particular hypothesis or theory in isolation; it is always a matter of choosing from among two or more viable alternatives. He has emphasized that an old theory is never completely abandoned unless there is currently available a rival to take its place. Given that circumstance, it is a matter of choosing between the old and the new. On this point I think that Kuhn is quite right, especially as regards reasonably mature sciences. And this insight provides a useful clue on how to use Bayes's theorem to explicate the logic of scientific confirmation.

Suppose that we are trying to choose between T_1 and T_2, where there may or may not be other serious alternatives in addition to the catch-all. By letting $i = 1$ and $i = 2$, we can proceed to write equation (4) for each of these candidates. Noting that the denominators of the two are identical, we can form their ratio as follows:

$$\frac{P(T_1/E.B)}{P(T_2/E.B)} = \frac{P(T_1/B)P(E/T_1.B)}{P(T_2/B)P(E/T_2.B)}. \tag{6}$$

No reference to the catch-all hypothesis appears in this equation. Since the catch-all is not a bona fide hypothesis, it is not a contender, and we need not try to calculate its posterior probability. The use of equation (6) frees us from the need to deal either with the expectedness of E or with its probability on the catch-all.

Equation (6) yields a relation that can be regarded as a *Bayesian algorithm for theory preference*. Suppose that, prior to the emergence of evidence E, you prefer T_1 to T_2; that is, $P(T_1/B) > P(T_2/B)$. Then E becomes available. You should change your preference in the light of E if and only if $P(T_2/E.B) > P(T_1/E.B)$. From (6) it follows that

$$P(T_2/E.B) > P(T_1/E.B) \text{ iff } P(E/T_2.B)/(E/T_1.B) > (T_1/B)/P(T_2/B). \tag{7}$$

In other words, you should change your preference to T_2 if the ratio of the likelihoods is greater than the reciprocal of the ratio of the respective prior probabilities. A corollary is that, if both $T_1.B$ and $T_2.B$ entail E, so that:

$$P(E/T_1.B) = P(E/T_2.B) = 1,$$

the occurrence of E can never change the preference rating between the two competing theories.

At the end of section 4 I made reference to the well-known phenomenon of washing out of priors in connection with the use of Bayes's theorem. One might well ask what happens to this swamping when we switch from Bayes's theorem to the ratio embodied in equation (6).[18] The best answer, I believe, is this. If we are dealing with two hypotheses that are serious

---

[18] This question was, in fact, raised by Adolf Grünbaum in a private communication.

contenders in the sense that they do not differ too greatly in plausibility, the ratio of the priors will be of the order of unity. If, as the observational evidence accumulates, the likelihoods come to differ greatly, the ratio of the likelihoods will swamp the ratio of the priors. Recall the example of the tossed coin. Suppose we consider the prior probability of a fair device to be ten times as large as that of a biased device. If about the same proportion of heads occurs in 500 tosses as occurred in the aforementioned 100, the likelihood on the null hypothesis would be virtually zero, and the likelihood on the hypothesis that the device has a bias approximating the observed frequency would be essentially indistinguishable from unity. The ratio of prior probabilities would obviously be completely dominated by the likelihood ratio.

8. PLAUSIBLE SCENARIOS

Although, by appealing to equation (6), we have eliminated the need to deal with the expectedness or the likelihood on the catch-all, we cannot claim to have dealt adequately with the likelihoods on the hypotheses we are seriously considering, for their values are not always straightforwardly ascertainable. We have already mentioned one example, namely, the probability of absence of observable stellar parallax on the Copernican hypothesis. We noted that, by adding an auxiliary hypothesis to the effect that the fixed stars are located an enormous distance from the Earth, we could augment the Copernican hypothesis in such a way that the likelihood on this augmented hypothesis is 1. But, for many reasons, this auxiliary assumption could hardly be considered plausible in that historical context. By now, of course, we have measured the parallax of relatively nearby stars, and from those values have calculated these distances. They are extremely far from us in comparison to the familiar objects in our solar system.

Consider another well-known example. During the seventeenth and eighteenth centuries the wave and corpuscular theories of light received considerable scientific attention. Each was able to explain certain important optical phenomena, and each faced fundamental difficulties. The corpuscular hypothesis easily explained how light could travel vast distances through empty space, and it readily explained sharp shadows. The theory of light as a longitudinal wave explained various kinds of diffraction phenomena, but failed to deal adequately with polarization. When, early in the nineteenth century, light was conceived as a transverse wave, the wave theory explained polarization as well as diffraction quite straightforwardly.

And Huygens had long since shown how the wave theory could handle rectilinear propagation and sharp shadows. For most of the nineteenth century the wave theory dominated optics. The proponent of the particle theory could still raise a serious objection. What is the likelihood of a wave propagating in empty space? Lacking a medium, the answer is zero. So wave theorists augmented their theory with the auxiliary assumption that all of space is filled with a peculiar substance known as the *luminiferous ether*. This substance was postulated to have precisely the properties required to transmit light waves.

The process I have been describing can appropriately be regarded as the discovery and introduction of *plausible scenarios*. A theory is confronted with an *anomaly*—a phenomenon that appears to have a small, possibly zero, likelihood, given that theory. Proponents of the theory search for some auxiliary hypothesis that, if conjoined to the theory, renders the likelihood high, possibly unity. This move shifts the burden of the argument to the plausibility of the new auxiliary hypothesis. I mentioned two instances involved in the wave theory of light. The first was the auxiliary assumption that the wave is transverse. This modification of the theory was sufficiently plausible to be incorporated as an integral part of the theory. The second was the luminiferous ether. The plausibility of this auxiliary hypothesis was debated throughout the nineteenth, and into the twentieth, century. The ether had to be dense enough to transmit transverse waves (which require a denser medium than do longitudinal waves) and thin enough to allow astronomical bodies to move through it without noticeable diminution of speed. Attempts to detect the motion of the earth relative to the ether were unsuccessful. The Lorentz–Fitzgerald contraction hypothesis was an attempt to save the ether theory—that is, another attempt at a plausible scenario—but it was, of course, abandoned in favour of special relativity.

I am calling these auxiliaries *scenarios* because they are stories about how something could have happened, and *plausible* because they must have some degree of acceptability if they are to be of any help in handling problematic phenomena. The wave theory could handle the Poisson bright spot by deducing it from the theory. There seemed to be no plausible scenario available to the particle theory that could deal with this phenomenon. The same has been said with respect to Foucault's demonstration that the velocity of light is greater in air than it is in water.[19]

One nineteenth century optician of considerable importance who did not adopt the wave theory, but remained committed to the Newtonian

[19] See e.g. Holton and Brush 1973: 392–3.

emission theory, was David Brewster.[20] In a 'Report on the Present State of Physical Optics', presented to the British Association for the Advancement of Science in 1831, he maintained that the undulatory theory is 'still burthened with difficulties and cannot claim our implicit assent' (quoted in Worrall 1990: 321). Brewster freely admitted the unparalleled explanatory and predictive success of the wave theory; nevertheless, he considered it false.

Among the difficulties Brewster found with the wave theory, two might be mentioned. First, he considered the wave theory implausible, for the reason that it required 'an *ether* invisible, intangible, imponderable, inseparable from all bodies, and extending from our own eye to the remotest verge of the starry heavens' (quoted in Worrall 1990: 322). History has certainly vindicated him on that issue. Second, he found the wave theory incapable of explaining a phenomenon that he had discovered himself, namely, *selective absorption*—dark lines in the spectrum of sunlight that has passed through certain gases. Brewster points out that a gas may be opaque to light of one particular index of refraction in flint glass, while transmitting freely light whose refractive indices in the same glass are only the tiniest bit higher or lower. Brewster maintained that there was no plausible scenario the wave theorists could devise that would explain why the ether permeating the gas transmits two waves of very nearly the same wavelength, but does not transmit light of a very precise wavelength lying in between:

There is no fact analogous to this in the phenomena of sound, and I can form no conception of a simple elastic medium so modified by the particles of the body which contains it, as to make such an extraordinary selection of the undulations which it stops or transmits. (Quoted in Worrall 1990: 323)

Brewster never found a plausible scenario by means of which the Newtonian theory he favoured could cope with absorption lines, nor could proponents of the wave theory find one to bolster their viewpoint. Dark absorption lines remained anomalous for both the wave and particle theories; neither could see a way to furnish them with high likelihood.

With hindsight we can say that the catch-all hypothesis was looking very strong at this point. We recognize that the dark absorption lines in the spectrum of sunlight are closely related to the discrete lines in the emission spectra of gases, and that they, in turn, are intimately bound up with the problem of the stability of atoms. These phenomena played a major role in the overthrow of classical physics at the turn of the twentieth century.

---

[20] An excellent account of Brewster's position can be found in Worrall 1990.

I have introduced the notion of a plausible scenario to deal with problematic likelihoods. Likelihoods can cause trouble for a scientific theory for either of two reasons. First, if you have a pet theory that confers an extremely small—for all practical purposes zero—likelihood on some observed phenomenon, that is a problem for that favoured theory. You try to come up with a plausible scenario according to which the likelihood will be larger—ideally, unity. Second, if there seems to be no way to evaluate the likelihood of a piece of evidence with respect to some hypothesis of interest, that is another sort of problem. In this case, we search for a plausible scenario that will make the likelihood manageable, whether this involves assigning it a high, medium, or low value.

What does this mean in terms of the Bayesian approach I am advocating? Let us return to:

$$\frac{P(T_1/E.B)}{P(T_2/E.B)} = \frac{P(T_1/B)P(E/T_1.B)}{P(T_2/B)P(E/T_2.B)}, \tag{6}$$

which contains two likelihoods. Suppose, as in nineteenth-century optics, that both likelihoods are problematic. As we have seen, we search for plausible scenarios A_1 and A_2 to augment T_1 and T_2 respectively. If the search has been successful, we can assess the likelihoods of E with respect to the augmented theories $A_1.T_1$ and $A_2.T_2$. Consequently, we can modify (6) so as to yield

$$\frac{P(A_1.T_1/E.B)}{P(A_2.T_2/E.B)} = \frac{P(A_1.T_1/B)P(E_1/A_1.T_1.B)}{P(A_2.T_2/B)P(E/A_2.T_2.B)}. \tag{8}$$

In order to use this equation to compare the posterior probabilities of the two augmented theories, we must assess the plausibilities of the scenarios, for the prior probabilities of both augmented theories—$A_1.T_1$ and $A_2.T_2$—appear in it. In section 4 I tried to explain how prior probabilities can be handled—that is, how we can obtain at least rough estimates of their values. If, as suggested, the plausible scenarios have made the likelihoods ascertainable, then we can use them in conjunction with our determinations of the prior probabilities to assess the ratio of the posterior probabilities. We have, thereby, handled the central issue raised by Kuhn, namely, what is the basis for preference between two theories.[21] Equation (8) is a Bayesian algorithm.

If either augmented theory, in conjunction with background knowledge

---

[21] If more than two theories are serious candidates, the pairwise comparison can be repeated as many times as necessary.

B, entails E, then the corresponding likelihood is one, and it drops out of (8). If both likelihoods drop out, we have the special case in which:

$$\frac{P(A_1.T_1/E.B)}{P(A_2.T_2/E.B)} = \frac{P(A_1.T_1/B)}{P(A_2.T_2/B)}, \tag{9}$$

thereby placing the *whole* burden on the prior probabilities—the plausibility considerations. Equation (9) represents a simplified Bayesian algorithm that is applicable in this type of special case.

Another type of special case was mentioned above. If, as in our coin-tossing example, the values of the prior probabilities do not differ drastically from one another, but the likelihoods become widely divergent as the observational evidence accumulates, there will be a washing out of the priors. In this case, the ratio of the posterior probabilities equals, for practical purposes, the ratio of the likelihoods.

The use of either (8) or (9) as an algorithm for theory choice does not imply that all scientists will agree on the numerical values or prefer the same theory. The evaluation of prior probabilities clearly demands the kind of scientific judgement whose importance Kuhn has rightly insisted upon. It should also be clearly remembered that these formulae provide no evaluations of individual theories; they furnish only comparative evaluations. Thus, instead of yielding a prediction regarding the chances of one particular theory being a component of 'completed science', they compare existing theories with regard to their present merits.

9. KUHN'S CRITERIA

Early in this paper I quoted five criteria that Kuhn mentioned in connection with his views on the rationality and objectivity of science. The time has come to relate them explicitly to the Bayesian approach I have been attempting to elaborate. In order to appreciate the significance of these criteria, it is important to distinguish three aspects of scientific theories that may be called *informational virtues*, *confirmational virtues*, and *economic virtues*. Up to this point we have concerned ourselves almost exclusively with confirmation, for our use of Bayes's theorem is germane only to the confirmational virtues. But since Kuhn's criteria patently refer to the other virtues as well, we must also say a little about them.

Consider, for example, the matter of *scope*. Newton's three laws of motion and his law of universal gravitation obviously have greater scope than the conjunction of Galileo's law of falling bodies and Kepler's three laws of planetary motion. This means, simply, that Newtonian mechanics contains more information than the laws of Kepler and Galileo taken together. Given a situation of this sort, we prefer the more informative theory because it is a basic goal of science to increase our knowledge as much as possible. We might, of course, hesitate to choose a highly informative theory if the evidence for it were extremely limited or shaky, because the desire to be right might overrule the desire to have more information content. But in the case at hand that consideration does not arise.

In spite of its intuitive attraction, however, the appeal to scope is not altogether unproblematic. There are two ways in which we might construe the Galileo–Kepler–Newton example of the preceding paragraph. First, we might ignore the small corrections mandated by Newton's theory in the laws of Galileo and Kepler. In that case we can clearly claim greater scope for Newton's laws than for the conjunction of Galileo's and Kepler's laws, since the latter is entailed by the former but not conversely. Where an entailment relation holds, we can make good sense of comparative scope.

Kuhn, however, along with most of the historically oriented philosophers, has been at pains to deny that science progresses by finding more general theories that include earlier theories as special cases. Theory choice or preference involves *competing* theories that are *mutually incompatible* or *mutually incommensurable*. To the best of my knowledge, Kuhn has not offered any precise characterization of scope; Karl Popper, in contrast, has made serious attempts to do so. In response to Popper's efforts, Adolf Grünbaum (1976) has effectively argued that none of the Popperian measures can be usefully applied to make comparisons of scope among mutually incompatible competing theories. Consequently, the concept of scope requires fundamental clarification if we are to use it to understand preferences among competing theories. However, since scope refers to information rather than confirmation, it plays no role in the Bayesian programme I have been endeavouring to explicate. We can thus put aside the problem of explicating that difficult concept.

Another of Kuhn's criteria is *accuracy*. It can, I think, be construed in two different ways. The first has to do with informational virtues, the second with economic. On the one hand, two theories might both make true predictions regarding the same phenomena, but one of them might give us precise predictions where the other gives only predictions that are less exact. If, for example, one theory enables us to predict that there will be a solar eclipse on a given day, and that its path of totality will cross North America, it may well be furnishing correct information about the eclipse. If another theory gives not only the day, but also the time, and not only the continent, but also the precise boundaries, the second provides much more information, at least with respect to this particular occurrence. It is not that either is incorrect; rather, the second yields more knowledge than the first. However, it should be clearly noted—as it was in the case of scope—that these theories are not incompatible or incommensurable competitors (at least with respect to this eclipse), and hence do not illustrate the interesting type of theory preference with which Kuhn is primarily concerned.

On the other hand, one theory may yield predictions that are nearly, but not quite, correct, while another theory yields predictions that are entirely correct—or, at least, more nearly correct. Newtonian astrophysics does well in ascertaining the orbit of the earth, but general relativity introduces a correction of 3.8 seconds of arc per century in the precession of its perihelion (Weinberg 1972: 198).[22] Although the Newtonian theory is literally false, it is used in contexts of this sort because its inaccuracy is small, and the economic gain involved in using it instead of general relativity (the saving in computational effort) is enormous.

The remaining three criteria are *simplicity*, *consistency*, and *fruitfulness*; all of them have direct bearing upon the confirmational virtues. In the treatment of prior probabilities in section 4, I briefly mentioned simplicity as a factor having a significant bearing upon the plausibility of theories. More examples could be added, but I think the point is clear.

In the same section I also made passing reference to consistency, but more can profitably be said on that topic. Consistency has two aspects, internal consistency of a theory and its compatibility with other accepted theories. While scientists may be fully justified in *entertaining* collections of statements that contain contradictions, the goal of science is surely to accept only logically consistent theories (see Smith 1987). The discovery of an internal inconsistency has a distinctly adverse effect on the prior probability of that theory, to wit, it must go straight to zero.

When we consider the relationships of a given theory to other accepted theories we again find two aspects. There are *deductive* relations of entailment and incompatibility, and there are *inductive* relations of fittingness and incongruity. The deductive relations are quite straightforward. Incompatibility with an accepted theory makes for implausibility; being a logical consequence of an accepted theory makes for a high prior probability. Although deductive subsumption of narrower theories under broader theories is probably something of an oversimplification of actual cases,

---

[22] Note that this correction is smaller by an order of magnitude than the correction of 43 seconds of arc per century for Mercury.

nevertheless, the ability of an overarching theory to deductively unify diverse domains furnishes a strong plausibility argument.

When it comes to the inductive relations among theories, analogy is, I think, the chief consideration. I have already mentioned the use of analogy in inductively transferring results of experiments from rats to humans. In archaeology, the method of ethnographic analogy, which exploits similarities between extant primitive societies and prehistoric societies, is widely used. In physics, the analogy between the inverse-square law of electrostatics and the inverse-square law of gravitation provides an example of an important plausibility consideration.

Kuhn's criteria of consistency (broadly construed) and simplicity seem clearly to pertain to assessments of the prior probabilities of theories. They cry out for a Bayesian interpretation.

The final criterion in Kuhn's list is *fruitfulness*; it has many aspects. Some theories prove fruitful by unifying a great many apparently different phenomena in terms of a few simple principles. The Newtonian synthesis is, perhaps, the outstanding example; Maxwellian electrodynamics is also an excellent case. As I suggested above, this ability to accommodate a wide variety of facts tends to enhance the prior probability of a given theory. To attribute diverse success to happenstance, rather than basic correctness, is implausible.

Another sort of fertility involves the predictability of theretofore unknown phenomena. We might mention as familiar illustrations the prediction of the Poisson bright spot by the wave theory of light and the prediction of time dilation by special relativity. These are the kinds of instances in which, in an important sense, the expectedness is low. As we have noted, a small expectedness tends to increase the posterior probability of a hypothesis.

A further type of fertility relates directly to plausible scenarios; a theory is fruitful in this way if it successfully copes with difficulties with the aid of suitable auxiliary assumptions. Newtonian mechanics again provides an excellent example. The perturbations of Uranus were explained by postulating Neptune. The perturbations of Neptune were explained by postulating Pluto.[23] The motions of stars within galaxies and of galaxies within clusters are explained in terms of *dark matter*, concerning which there are many current theories. A theory that readily gives rise to plausible scenarios to deal with problematic likelihoods can boast this sort of fertility.

---

[23] Unfortunately, recent evidence regarding the mass of Pluto strongly suggests that Pluto is not sufficiently massive to explain the perturbations of Neptune. A different plausible scenario is needed, but I do not know of any serious candidates that have been offered.

The discussion of Kuhn's criteria in this section is intended to show how adequately they can be understood within a Bayesian framework—in so far as they are germane to confirmation. If it is sound, we have constructed a fairly substantial bridge connecting Kuhn's views on theory choice with those of the logical empiricists—at least, those who find in Bayes's theorem a suitable schema for characterizing the confirmation of hypotheses and theories.

10. RATIONALITY VERSUS OBJECTIVITY

In the title of this essay I have used both the concept of *rationality* and that of *objectivity*. It is time to say something about their relationship. Perhaps the best way to approach the distinction between them is to enumerate various grades of rationality. In a certain sense one can be rational without paying any heed at all to objectivity. It is essentialiy a matter of good housekeeping as far as one's beliefs and degrees of confidence are concerned. As Bayesians have often emphasized, it is important to avoid logical contradictions in one's beliefs and to avoid probabilistic incoherence in one's degrees of conviction. If contradiction or incoherence are discovered, they must somehow be eliminated; the presence of either constitutes a form of irrationality. But the removal of such elements of irrationality can be accomplished without any appeal to facts outside of the subject's corpus of beliefs and degrees of confidence. To achieve this sort of rationality is to achieve a minimal standard that I have elsewhere called *static* rationality.[24]

One way in which additional facts may enter the picture is via Bayes's theorem. We have a theory T in which we have a particular degree of confidence. A new piece of evidence turns up—some objective fact E of which we were previously unaware—and we use Bayes's theorem to calculate a posterior probability of T. To accept this value of the posterior probability as one's degree of confidence in T is known as *Bayesian conditionalization*. Use of Bayes's theorem does not, however, guarantee objectivity. If the resulting posterior probability of T is one we are not willing to accept, we can make adjustments elsewhere to avoid incoherence. After all, the prior probabilities and likelihoods are simply personal probabilities, so they can be adjusted to achieve the desired result. If, however, the requirement of *Bayesian conditionalization* is added to those

---

[24] See Salmon 1988: 5–12 for a more detailed discussion of various grades of rationality. The term 'static' was chosen to indicate the lack of any principled method for changing personal probabilities in the face of inconsistency or incoherence.

of static rationality, we have a stronger type of rationality that I have called *kinematic* (Salmon 1988: 11–12).

The highest grade of rationality—what I have called *dynamic rationality*—requires much fuller reference to objective fact than is demanded by advocates of personalism. The most obvious way to inject a substantial degree of objectivity into our deliberations regarding choices of scientific theories is to provide an objective interpretation of the probabilities in Bayes's theorem. Throughout this discussion I have adopted that approach as thoroughly as possible. For instance, I have argued that prior probabilities can be given an objective interpretation in terms of frequencies of success. I have tried to show how likelihoods could be objective— by virtue of entailment relations, tests of statistical significance, or observed frequencies. When the likelihoods created major difficulties, I appealed to plausible scenarios. The result was that an intractable likelihood could be exchanged for a tractable prior probability—namely, the prior probability of a theory in conjunction with an auxiliary assumption.

We noted that the denominators of the right-hand sides of the various versions of Bayes's theorem—equations (1), (3), and (4)—contain either an expectedness or a likelihood on the catch-all. It seems to me futile to try to construe either of these probabilities objectively. Consequently, in section 7 I introduced equation (6), which involves a ratio of two instances of Bayes's theorem, and from which the expectedness and the likelihood on the catch-all drop out. Confining our attention, as Kuhn recommends, to comparing the merits of competing theories, rather than offering absolute evaluations of individual theories, we were able to eliminate the probabilities that most seriously defy objective interpretation.

11. CONCLUSIONS

For many years I have been convinced that plausibility arguments in science have constituted a major stumbling-block to an understanding of the logic of scientific inference. Kuhn was not alone, I believe, in recognizing that considerations of plausibility constitute an essential aspect of scientific reasoning, without seeing where they fit into the logic of science. If one sees confirmation solely in terms of the crude hypothetico-deductive method, there is no place for them. There is, consequently, an obvious incentive for relegating plausibility considerations to heuristics. If one accepts the traditional distinction between the *context of discovery* and the *context of justification*, it is tempting to place them in the former context. But Kuhn recognized, I think, that plausibility arguments enter into the justifications of choices of theories, with the result that he became sceptical of the value of that distinction. If, as I believe, plausibility considerations are simply evaluations of prior probabilities of hypotheses or theories, then it becomes apparent via Bayes's theorem that they play an indispensable role in the context of justification. We do not need to give up that important distinction.

At several places in this paper I have spoken of Bayesian algorithms, mainly because Kuhn introduced that notion into the discussion. I have claimed that such algorithms exist—and attempted to exhibit them—but I accord *very little* significance to that claim. The algorithms are trivial; what is important is the scientific judgement involved in assessing the probabilities that are fed into the equations. The algorithms give frameworks in terms of which to understand the role of the sort of judgement upon which Kuhn rightly placed great emphasis.

The history of science chronicles the successes and failures of attempts at scientific theorizing. If the Bayesian analysis I have been offering is at all sound, history of science—in addition to contemporary scientific experience, of course—provides a rich source of information relevant to the prior probabilities of the theories among which we are at present concerned to make objective and rational choices. This viewpoint captures, I believe, the point Kuhn made at the beginning of his first book:

But an age as dominated by science as our own does need a perspective from which to examine the scientific beliefs which it takes so much for granted, and history provides one important source of such perspective. If we can discover the origins of some modern scientific concepts and the way in which they supplanted the concepts of an earlier age, we are more likely to evaluate intelligently their chances for survival. (1957: 3–4)

I suggested at the outset that an appeal to Bayesian principles could provide some aid in bridging the gap between Hempel's logical-empiricist approach and Kuhn's historical approach. I hope I have offered a convincing case. However that may be, there remain many unresolved issues. For instance, I have not even broached the problem of incommensurability of paradigms or theories. This is a major issue. For another example, I have assumed uncritically throughout the discussion that the various parties to disputes about theories share a common body B of background knowledge. It is by no means obvious that this is a tenable assumption. No doubt other points for controversy remain. I do not for a moment maintain that complete consensus would be in the offing even if both camps were to buy the Bayesian line I have been peddling. But I do hope that some areas of misunderstanding have been clarified.[25]

REFERENCES

Einstein, A. (1949). 'Autobiographical Notes.' In P. A. Schilpp (ed.), *Albert Einstein: Philosopher-Scientist*, 1–95. Evanston, Ill.: Library of Living Philosophers.
Gardner, M. (1957). *Fads and Fallacies in the Name of Science*. New York: Dover.
Giere, R. N. (1984). *Understanding Scientific Reasoning*, 2nd edn. New York: Holt, Rinehart, and Winston.
Grünbaum, A. (1976). 'Can a Theory Answer More Questions than One of its Rivals?' *British Journal for Philosophy of Science*, 27: 1–23.
——(1984). *The Foundations of Psychoanalysis*. Berkeley: University of California Press.
Harari, H. (1983). 'The Structure of Quarks and Leptons.' *Scientific American*, 248: 56–68.
Holton, G., and Brush, S. G. (1973). *Introduction to Concepts and Theories in Physical Science*, 2nd edn. Reading, Mass.: Addison-Wesley.
Kuhn, T. S. (1957). *The Copernican Revolution*. Cambridge, Mass.: Harvard University Press.
——(1962). *The Structure of Scientific Revolutions*. Chicago: University of Chicago Press; 2nd edn. 1970.
——(1977). *The Essential Tension*. Chicago: University of Chicago Press.
Motley, M. T. (1985). 'Slips of the Tongue.' *Scientific American*, 253: 116.
Ramsey, F. P. (1950). 'Truth and Probability.' In R. B. Braithwaite (ed.), *The Foundations of Mathematics*, 156–98. New York: Humanities Press.
Reichenbach, H. (1938). *Experience and Prediction*. Chicago: University of Chicago Press.
Salmon, W. C. (1967). *The Foundations of Scientific Inference*. Pittsburgh: University of Pittsburgh Press.
——(1970). 'Bayes's Theorem and the History of Science.' In R. Stuewer (ed.), *Historical and Philosophical Perspectives of Science*, Minnesota Studies in the Philosophy of Science, vol. 5, 68–86.
——(1988). 'Dynamic Rationality: Propensity, Probability, and Credence.' In J. H. Fetzer (ed.), *Probability and Causality*, 3–40. Dordrecht: Reidel.
Shimony, A. (1970). 'Scientific Inference.' In R. G. Colodny (ed.), *The Nature and Function of Scientific Theories*, 79–172. Pittsburgh: University of Pittsburgh Press.
Smith, J. (1987). 'The Status of Inconsistent Statements in Scientific Inquiry.' Ph.D. diss., University of Pittsburgh.
Suppes, P. (1966). 'A Bayesian Approach to the Paradoxes of Confirmation.' In J. Hintikka and P. Suppes (eds.), *Aspects of Inductive Logic*, 198–218. Amsterdam: North Holland.
Symposium (1983). 'Symposium: The Philosophy of Carl G. Hempel.' *Journal of Philosophy*, 80/10: 555–72.

[25] I should like to express my deepest gratitude to Adolf Grünbaum and Philip Kitcher for important criticism and valuable suggestions with respect to an earlier version of this paper.

Velikovski, I. (1950). *Worlds in Collision*. New York: Doubleday.
Weinberg, S. (1972). *Gravitation and Cosmology*. New York: Wiley and Sons.
Worrall, J. (1990). 'Scientific Revolutions and Scientific Rationality: The Case of the Elderly Holdout.' In C. Savage (ed.), *Scientific Theories*, Minnesota Studies in the Philosophy of Science, vol. 14, 319–36. Minneapolis: University of Minnesota Press.

XII

WHY I AM NOT A BAYESIAN*

CLARK GLYMOUR

The aim of confirmation theory is to provide a true account of the principles that guide scientific argument in so far as that argument is not, and does not purport to be, of a deductive kind. A confirmation theory should serve as a critical and explanatory instrument quite as much as do theories of deductive inference. Any successful confirmation theory should, for example, reveal the structure and fallacies, if any, in Newton's argument for universal gravitation, in nineteenth-century arguments for and against the atomic theory, in Freud's arguments for psychoanalytic generalizations. Where scientific judgements are widely shared, and sociological factors cannot explain their ubiquity, and analysis through the lens provided by confirmation theory reveals no good explicit arguments for the judgements, confirmation theory ought at least sometimes to suggest some good arguments that may have been lurking misperceived. Theories of deductive inference do that much for scientific reasoning in so far as that reasoning is supposed to be demonstrative. We can apply quantification theory to assess the validity of scientific arguments, and although we must almost always treat such arguments as enthymematic, the premises we interpolate are not arbitrary; in many cases, as when the same subject-matter is under discussion, there is a common set of suppressed premises. Again, there may be differences about the correct logical form of scientific claims; differences of this kind result in (or from) different formalizations, for example, of classical mechanics. But such differences often make no difference for the assessment of validity in actual arguments. Confirmation theory should do as well in its own domain. If it fails, then it may still be of interest for many purposes, but not for the purpose of understanding scientific reasoning.

The aim of confirmation theory ought not to be simply to provide precise replacements for informal methodological notions, that is, expli-

Reprinted from Clark Glymour, *Theory and Evidence* (Chicago: University of Chicago Press, 1981), 63–93, by permission.
* Who cares whether a pig-farmer is a Bayesian?—R. C. Jeffrey.

cations of them. It ought to do more; in particular, confirmation theory ought to *explain* both methodological truisms and particular judgements that have occurred within the history of science. By 'explain' I mean at least that confirmation theory ought to provide a rationale for methodological truisms, and ought to reveal some systematic connections among them and, further, ought, without arbitrary or question-begging assumptions, to reveal particular historical judgements as in conformity with its principles.

Almost everyone interested in confirmation theory today believes that confirmation relations ought to be analysed in terms of *probability* relations. Confirmation theory is the theory of probability plus introductions and appendices. Moreover, almost everyone believes that confirmation proceeds through the formation of conditional probabilities of hypotheses on evidence. The basic tasks facing confirmation theory are thus just those of explicating and showing how to determine the probabilities that confirmation involves, developing explications of such meta-scientific notions as 'confirmation', 'explanatory power', 'simplicity', and so on in terms of functions of probabilities and conditional probabilities, and showing that the canons and patterns of scientific inference result. It was not always so. Probabilistic accounts of confirmation really became dominant only after the publication of Carnap's *Logical Foundations of Probability* (1950), although of course many probabilistic accounts had preceded Carnap's. An eminent contemporary philosopher (Putnam 1967) has compared Carnap's achievement in inductive logic with Frege's in deductive logic: just as before Frege there was only a small and theoretically uninteresting collection of principles of deductive inference, but after him the foundation of a systematic and profound theory of demonstrative reasoning, so with Carnap and inductive reasoning. After Carnap's *Logical Foundations*, debates over confirmation theory seem to have focused chiefly on the interpretation of probability and on the appropriate probabilistic explications of various meta-scientific notions. The meta-scientific notions remain controversial, as does the interpretation of probability, although, increasingly, logical interpretations of probability are giving way to the doctrine that probability is degree of belief.[1] In very recent years a few philosophers have attempted to apply probabilistic analyses to derive and to explain particular methodological practices and precepts, and even to elucidate some historical cases.

I believe these efforts, ingenious and admirable as many of them are, are none the less misguided. For one thing, probabilistic analyses remain at too

[1] A third view, that probabilities are to be understood exclusively as frequencies, has been most ably defended by Wesley Salmon (1969).

great a distance from the history of scientific practice to be really informative about that practice, and in part they do so exactly because they are probabilistic. Although considerations of probability have played an important part in the history of science, until very recently, explicit probabilistic arguments for the confirmation of various theories, or probabilistic analyses of data, have been great rarities in the history of science. In the physical sciences at any rate, probabilistic arguments have rarely occurred. Copernicus, Newton, Kepler, none of them give probabilistic arguments for their theories; nor does Maxwell or Kelvin or Lavoisier or Dalton or Einstein or Schrödinger or.... There are exceptions. Jon Dorling has discussed a seventeenth-century Ptolemaic astronomer who apparently made an extended comparison of Ptolemaic and Copernican theories in probabilistic terms; Laplace, of course, gave Bayesian arguments for astronomical theories. And there are people—Maxwell, for example—who scarcely give a probabilistic argument when making a case for or against scientific hypotheses but who discuss *methodology* in probabilistic terms. This is not to deny that there are many areas of contemporary physical science where probability figures large in confirmation; regression analysis is not uncommon in discussions of the origins of cosmic rays, correlation and analysis of variance in experimental searches for gravitational waves, and so on. It *is* to say that, explicitly, probability is a distinctly minor note in the history of scientific argument.

The rarity of probability considerations in the history of science is more an embarrassment for some accounts of probability than for others. Logical theories, whether Carnap's or those developed by Hintikka and his students, seem to lie at a great distance from the history of science. Still, some of the people working in this tradition have made interesting steps towards accounting for methodological truisms. My own inclination is to believe that the interest such investigations have stems more from the insights they obtain into syntactic versions of structural connections among evidence and hypotheses than to the probability measures they mesh with these insights. Frequency interpretations suppose that for each hypothesis to be assessed there is an appropriate reference class of hypotheses to which to assign it, and the prior probability of the hypothesis is the frequency of true hypotheses in this reference class. The same is true for statements of evidence, whether they be singular or general. The matter of how such reference classes are to be determined, and determined so that the frequencies involved do not come out to be zero, is a question that has only been touched upon by frequentist writers. More to the point, for many of the suggested features that might determine reference classes, we have no statistics, and cannot plausibly imagine those who figure in the history of our sciences to have had them. So conceived, the history of scientific argument must turn out to be largely a history of fanciful guesses. Further, some of the properties that seem natural candidates for determining reference classes for hypotheses—simplicity, for example—seem likely to give perverse results. We prefer hypotheses that posit simple relations among observed quantities, and so on a frequentist view should give them high prior probabilities. Yet simple hypotheses, although often very useful approximations, have most often turned out to be literally false.

At present, perhaps the most philosophically influential view of probability understands it to be degree of belief. The subjectivist Bayesian (hereafter, for brevity, simply Bayesian) view of probability has a growing number of advocates who understand it to provide a general framework for understanding scientific reasoning. They are singularly unembarrassed by the rarity of explicit probabilistic arguments in the history of science, for scientific reasoning need not be explicitly probabilistic in order to be probabilistic in the Bayesian sense. Indeed, a number of Bayesians have discussed historical cases within their framework. Because of its influence and its apparent applicability, in what follows it is to the subjective Bayesian account that I shall give my full attention.

My thesis is several-fold. First, there are a number of attempts to demonstrate a priori the rationality of the restrictions on belief and inference that Bayesians advocate. These arguments are altogether admirable, but ought, I shall maintain, to be unconvincing. My thesis in this instance is not a new one, and I think many Bayesians do regard these a priori arguments as insufficient. Second, there are a variety of methodological notions that an account of confirmation ought to explicate and methodological truisms involving these notions that a confirmation theory ought to explain: for example, variety of evidence and why we desire it, *ad hoc* hypotheses and why we eschew them, what separates a hypothesis integral to a theory from one 'tacked on' to the theory, simplicity and why it is so often admired, why 'de-Occamized' theories are so often disdained, what determines when a piece of evidence is relevant to a hypothesis, and what, if anything, makes the confirmation of one bit of theory by one bit of evidence stronger than the confirmation of another bit of theory (or possibly the same bit) by another (or possibly the same) bit of evidence. Although there are plausible Bayesian explications of some of these notions, there are not plausible Bayesian explications of others. Bayesian accounts of methodological truisms and of particular historical cases are of one of two kinds: either they depend on general principles restricting prior probabilities, or they don't. My claim is that many of the principles proposed by the first kind of Bayesian are either implausible or incoherent, and that, for want of such principles, the explanations the second kind of Bayesians provide for particular historical cases and for truisms of method are chimeras. Finally, I claim that there are elementary but perfectly common features of the relation of theory and evidence that the Bayesian scheme cannot capture at all without serious—and perhaps not very plausible—revision.

It is not that I think the Bayesian scheme or related probabilistic accounts capture nothing. On the contrary, they are clearly pertinent where the reasoning involved is explicitly statistical. Further, the accounts developed by Carnap, his predecessors, and his successors are impressive systematizations and generalizations, in a probabilistic framework, of certain principles of ordinary reasoning. But so far as understanding scientific reasoning goes, I think it is very wrong to consider our situation to be analogous to that of post-Fregean logicians, our subject-matter transformed from a hotchpotch of principles by a powerful theory whose outlines are clear. We flatter ourselves that we possess even the hotchpotch. My opinions are outlandish, I know; few of the arguments I shall present in their favour are new, and perhaps none of them is decisive. Even so, they seem sufficient to warrant taking seriously entirely different approaches to the analysis of scientific reasoning.

The theories I shall consider share the following framework, more or less. There is a class of sentences that express all hypotheses and all actual or possible evidence of interest; the class is closed under Boolean operations. For each ideally rational agent, there is a function defined on all sentences such that, under the relation of logical equivalence, the function is a probability measure on the collection of equivalence classes. The probability of any proposition represents the agent's degree of belief in that proposition. As new evidence accumulates, the probability of a proposition changes according to Bayes's rule: the posterior probability of a hypothesis on the new evidence is equal to the prior conditional probability of the hypothesis on the evidence. This is a scheme shared by diverse accounts of confirmation. I call such theories 'Bayesian', or sometimes 'personalist'.

We certainly have *grades* of belief. Some claims I more or less believe, some I find plausible and tend to believe, others I am agnostic about, some I find implausible and far-fetched, still others I regard as positively absurd. I think everyone admits some such gradations, although descriptions of them might be finer or cruder. The personalist school of probability theorists claim that we also have *degrees* of belief, degrees that can have any value between 0 and 1 and that ought, if we are rational, to be representable by a probability function. Presumably, the degrees of belief are to co-vary with everyday gradations of belief, so that one regards a proposition as preposterous and absurd just if his degree of belief in it is somewhere near zero, and he is agnostic just if his degree of belief is somewhere near a half, and so on. According to personalists, then, an ideally rational agent always has his degrees of belief distributed so as to satisfy the axioms of probability, and when he comes to accept a new belief, he also forms new *degrees* of belief by conditionalizing on the newly accepted belief. There are any number of refinements, of course; but that is the basic view.

Why should we think that we really do have *degrees* of belief? Personalists have an ingenious answer: people have them because we can measure the degrees of belief that people have. Assume that no one (rational) will accept a wager on which he expects a loss, but anyone (rational) will accept any wager on which he expects a gain. Then we can measure a person's degree of belief in proposition P by finding, for fixed amount v, the highest amount u such that the person will pay u in order to receive $u + v$ if P is true, but receive nothing if P is not true. If u is the greatest amount the agent is willing to pay for the wager, his expected gain on paying u must be zero. The agent's gain if P is the case is v; his gain if P is not the case is $-u$. Thus

$$v \cdot \text{prob}(P) + (-u) \cdot \text{prob}(\sim P) = 0.$$

Since prob $(\sim P) = 1 - \text{prob}(P)$, we have

$$\text{prob}(P) = u/(u+v).$$

The reasoning is clear: any sensible person will act so as to maximize his expected gain; thus, presented with a decision whether or not to purchase a bet, he will make the purchase just if his expected gain is greater than zero. So the betting odds he will accept determine his degree of belief.[2]

I think that this device really does provide evidence that we have, or can produce, degrees of belief, in at least some propositions, but at the same time it is evident that betting odds are not an unobjectionable device for the measurement of degrees of belief. Betting odds could fail to measure degrees of belief for a variety of reasons: the subject may not believe that

[2] More detailed accounts of means for determining degrees of belief may be found in Jeffrey 1965. It is a curious fact that the procedures that Bayesians use for determining subjective degrees of belief empirically are an instance of the general strategy described in Glymour 1981, ch. 5. Indeed, the strategy typically used to determine whether or not actual people behave as rational Bayesians involves the bootstrap strategy described in that chapter.

the bet will be paid off if he wins, or he may doubt that it is clear what constitutes winning, even though it is clear what constitutes losing. Things he values other than monetary gain (or whatever) may enter into his determination of the expected utility of purchasing the bet: for example, he may place either a positive or a negative value on risk itself. And the very fact that he is offered a wager on P may somehow change his degree of belief in P.

Let us suppose, then, that we do have degrees of belief in at least some propositions, and that in some cases they can be at least approximately measured on an interval from 0 to 1. There are two questions: why should we think that, for rationality, one's degrees of belief must satisfy the axioms of probability, and why should we think that, again for rationality, changes in degrees of belief ought to proceed by conditionalization? One question at a time. In using betting quotients to measure degrees of belief, it was assumed that the subject would act so as to maximize *expected* gain. The betting quotient determined the degree of belief by determining the coefficient by which the gain is multiplied in case that P is true in the expression for the expected gain. So the betting quotient determines a degree of belief, as it were, in the *role* of a probability. But why should the things, degrees of belief, that play this role be probabilities? Supposing that we do choose those actions that maximize the sum of the product of our degrees of belief in each possible outcome of the action and the gain (or loss) to us of that outcome. Why must the degrees of belief that enter into this sum be probabilities? Again, there is an ingenious argument: if one acts so as to maximize his expected gain using a degree-of-belief function that is not a probability function, and if for every proposition there were a possible wager (which, if it is offered, one believes will be paid off if it is accepted and won), then there is a circumstance, a combination of wagers, that one would enter into if they were offered, and in which one would suffer a net loss whatever the outcome. That is what the Dutch-book argument shows; what it counsels is prudence.

Some of the reasons why it is not clear that betting quotients are accurate measures of degrees of belief are also reasons why the Dutch-book argument is not conclusive: there are many cases of propositions in which we may have degrees of belief, but on which, we may be sure, no acceptable wager will be offered us; again, we may have values other than the value we place on the stakes, and these other values may enter into our determination whether or not to gamble; and we may not have adopted the policy of acting so as to maximize our expected gain or our expected utility: that is, we may save ourselves from having book made against us by refusing to make certain wagers, or combinations of wagers, even though we judge the odds to be in our favour.

The Dutch-book argument does not succeed in showing that in order to avoid absurd commitments, or even the possibility of such commitments, one must have degrees of belief that are probabilities. But it does provide a kind of justification for the personalist viewpoint, for it shows that if one's degrees of belief are probabilities, then a certain kind of absurdity is avoided. There are other ways of avoiding that kind of absurdity, but at least the personalist way is one such.[3]

One of the common objections to Bayesian theory is that it fails to provide any connection between what is inferred and what is the case. The Bayesian reply is that the method guarantees that, in the long run, everyone will agree on the truth. Suppose that B_i are a set of mutually exclusive, jointly exhaustive hypotheses, each with probability $B(i)$. Let $\bar{x}_r$ be a sequence of random variables with a finite set of values and conditional distribution given by $P(\bar{x}_r = x_r|B_i) = \varepsilon(x_r|B_i)$; then we can think of the values x_r as the outcomes of experiments, each hypothesis determining a likelihood for each outcome. Suppose that no two hypotheses have the same likelihood distribution; that is, for $i \neq j$ it is not the case that for all values x_r of $\bar{x}_r$, $\varepsilon(x_r|B_i) = \varepsilon(x_r|B_j)$, where the ε's are defined as above. Let $\bar{x}$ denote the first n of these variables, where x is a value of $\bar{x}$. Now imagine an observation of these n random variables. In Savage's words:

Before the observation, the probability that the probability given x of whichever element of the partition actually obtains will be greater than α is

$$\sum_i B(i)P\Big(P\big(B_i|x\big) > \alpha|B_i\Big),$$

where summation is confined to those i's for which $B(i) \neq 0$. (1972: 49)

In the limit as n approaches infinity, the probability that the probability given x of whichever element of the partition actually obtains is greater than α is 1. That is the theorem. What is its significance? According to Savage, 'With the observation of an abundance of relevant data, the person is almost certain to become highly convinced of the truth, and it has also been shown that he himself knows this to be the case' (p. 50). That is a little misleading. The result involves second-order probabilities, but these too, according to personalists, are degrees of belief. So what has been shown seems to be this: in the limit as n approaches infinity, an ideally rational Bayesian has degree of belief 1 that an ideally rational Bayesian (with degrees of belief as in the theorem) has degree of belief, given x, greater than α in whichever element of the partition actually

---

[3] For further criticisms of the Dutch-book argument see Kyburg, 1978.

obtains. The theorem does not tell us that in the limit any rational Bayesian will assign probability 1 to the true hypothesis and probability 0 to the rest; it only tells us that rational Bayesians are certain that he will. It may reassure those who are already Bayesians, but it is hardly grounds for conversion. Even the reassurance is slim. Mary Hesse points out (1974: 117–19), entirely correctly I believe, that the assumptions of the theorem do not seem to apply even approximately in actual scientific contexts. Finally, some of the assumptions of stable estimation theorems can be dispensed with if one assumes instead that all of the initial distributions considered must agree regarding which evidence is relevant to which hypotheses. But there is no evident a priori reason why there should be such agreement.

I think relatively few Bayesians are actually persuaded of the correctness of Bayesian doctrine by Dutch-book arguments, stable estimation theorems, or other a priori arguments. Their frailty is too palpable. I think that the appeal of Bayesian doctrine derives from two other features. First, with only very weak or very natural assumptions about prior probabilities, or none at all, the Bayesian scheme generates principles that seem to accord well with common sense. Thus, with minor restrictions, one obtains the principle that hypotheses are confirmed by positive instances of them; and, again, one obtains the result that if an event that actually occurs is, on some hypothesis, very unlikely to occur, then that occurrence renders the hypothesis less likely than it would otherwise have been. These principles, and others, can claim something like the authority of common sense, and Bayesian doctrine provides a systematic explication of them. Second, the restrictions placed a priori on rational degrees of belief are so mild, and the device of probability theory at once so precise and so flexible, that Bayesian philosophers of science may reasonably hope to explain the subtleties and vagaries of scientific reasoning and inference by applying their scheme together with plausible assumptions about the distribution of degrees of belief. This seems, for instance, to be Professor Hesse's line of argument. After admitting the insufficiency of the standard arguments for Bayesianism, she sets out to show that the view can account for a host of alleged features of scientific reasoning and inference. My own view is different: particular *inferences* can almost always be brought into accord with the Bayesian scheme by assigning degrees of belief more or less *ad hoc*, but we learn nothing from this agreement. What we want is an explanation of scientific argument; what the Bayesians give us is a theory of learning—indeed, a theory of personal learning. But arguments are more or less impersonal; I make an argument to persuade anyone informed of the premises, and in doing so I am not reporting any bit of autobiography. To ascribe to me degrees of belief that make my slide from my premises to my conclusion a plausible one fails to explain anything, not only because the ascription may be arbitrary, but also because, even if it is a correct assignment of my degrees of belief, it does not explain why what I am doing is *arguing*—why, that is, what I say should have the least influence on others, or why I might hope that it should. Now, Bayesians might bridge the gap between personal inference and argument in either of two ways. In the first place, one might give arguments in order to change others' beliefs because of the respect they have for his opinion. This is not very plausible; if that were the point of giving arguments, one would not bother with them, but would simply state one's opinion. Alternatively, and more hopefully, Bayesians may suggest that we give arguments exactly because there are general principles restricting belief, principles that are widely subscribed to, and in giving arguments we are attempting to show that, supposing our audience has certain beliefs, they must in view of these principles have other beliefs, those we are trying to establish. There is nothing controversial about this suggestion, and I endorse it. What is controversial is that the general principles required for argument can best be understood as conditions restricting prior probabilities in a Bayesian framework. Sometimes they can, perhaps; but I think that when arguments turn on relating evidence to theory, it is very difficult to explicate them in a plausible way within the Bayesian framework. At any rate, it is worth seeing in more detail what the difficulties may be.

There is very little Bayesian literature about the hotchpotch of claims and notions that are usually canonized as scientific method; very little seems to have been written, from a Bayesian point of view, about what makes a hypothesis *ad hoc*, about what makes one body of evidence more various than another body of evidence, and why we should prefer a variety of evidence, about why, in some circumstances, we should prefer simpler theories, and what it is that we are preferring when we do. And so on. There is little to nothing of this in Carnap, and more recent, and more personalist, statements of the Bayesian position are almost as disappointing. In a lengthy discussion of what he calls 'tempered personalism', Abner Shimony (1970) discusses only how his version of Bayesianism generalizes and qualifies hypothetico-deductive arguments. (Shimony does discuss simplicity, but only to argue that it is overvalued.) Mary Hesse devotes the later chapters of her book to an attempt to show that certain features of scientific method do result when the Bayesian scheme is supplemented with a postulate that restricts assignments of prior probabilities. Unfortunately, as we shall see, her restrictive principle is incoherent.[4]

One aspect of the demand for a variety of evidence arises when there is

---

[4] Moreover, I believe that much of her discussion of methodological principles has only the loosest relation to Bayesian principles.

some definite set of alternative hypotheses between which we are trying to decide. In such cases we naturally prefer the body of evidence that will be most helpful in eliminating false competitors. This aspect of variety is an easy and natural one for Bayesians to take account of, and within an account such as Shimony's it is taken care of so directly as hardly to require comment. But there is more to variety. In some situations we have some reason to suspect that if a theory is false, its falsity will show up when evidence of certain kinds is obtained and compared. For example, given the tradition of Aristotelian distinctions, there was some reason to demand both terrestrial and celestial evidence for seventeenth-century theories of motion that subjected all matter to the same dynamical laws. Once again, I see no special reason why this kind of demand for a variety of evidence cannot be fitted into the Bayesian scheme. But there is still more. A complex theory may contain a great many logically independent hypotheses, and particular bodies of evidence may provide grounds for some of those hypotheses but not for others. Surely part of the demand for a variety of evidence, and an important part, derives from a desire to see to it that the various independent parts of our theories are tested. Taking account of this aspect of the demand for a variety of evidence is just taking account of the relevance of evidence to pieces of theory. How Bayesians may do this we shall consider later.

Simplicity is another feature of scientific method for which some Bayesians have attempted to account. There is one aspect of the scientific preference for the simple that seems beyond Bayesian capacities, and that is the disdain for 'de-Occamized' hypotheses, for theories that postulate the operation of a number of properties, determinable only in combination, when a single property would do. Such theories can be generated by taking any ordinary theory and replacing some single quantity, wherever it occurs in the statement of the theory, by an algebraic combination of new quantities. If the original quantity was not one that occurs in the statement of some body of evidence for the theory, then the new, de-Occamized theory will have the same entailment relations with that body of evidence as did the original theory. If the old theory entailed the evidence, so will the new, de-Occamized one. Now, it follows from Bayesian principles that if two theories both entail e, then (provided the prior probability of each hypothesis is neither 1 nor 0), if e confirms one of them, it confirms the other. How then is the fact (for so I take it to be) that pieces of evidence just don't seem to *count* for de-Occamized theories to be explained? Not by supposing that de-Occamized theories have lower prior probabilities than un-de-Occamized theories, for being 'de-Occamized' is a feature that a theory has only with respect to a certain body of evidence, and it is not hard to imagine artificially restricted bodies of evidence with respect to which perfectly good theories might count as de-Occamized. Having extra wheels is a feature a theory has only in relation to a body of evidence; the only Bayesian relation that appears available and relevant to scientific preference is the likelihood of the evidence on the theory, and unfortunately the likelihood is the same for a theory and for its de-Occamized counterparts whenever the theory entails the evidence.

It is common practice in fitting curves to experimental data, in the absence of an established theory relating the quantities measured, to choose the 'simplest' curve that will fit the data. Thus linear relations are preferred to polynomial relations of higher degree, and exponential functions of measured quantities are preferred to exponential functions of algebraic combinations of measured quantities, and so on. The problem is to account for this preference. Harold Jeffreys, a Bayesian of sorts, offered an explanation (1979) along the following lines. Algebraic and differential equations may be ordered by simplicity; the simpler the hypothetical relation between two or more quantities, the greater is its prior probability. If measurement error has a known probability distribution, we can then compute the likelihood of any set of measurement results given an equation relating the measured quantities. It should be clear, then, that with these priors and likelihoods, ratios of posterior probabilities may be computed from measurement results. Jeffreys constructed a Bayesian significance test for the introduction of higher-degree terms in the equation relating the measured quantities. Roughly, if one's equation fits the data *too* well, then the equation has too many terms and too many arbitrary parameters; and if the equation does not fit the data well enough, then one has not included enough terms and parameters in the equation. The whole business depends, of course, entirely on the ordering of prior probabilities. In his *Theory of Probability* Jeffreys (1967) proposed that the prior probability of a hypothesis decreases as the number of arbitrary parameters increases, but hypotheses having the same number of arbitrary parameters have the same prior probability. This leads immediately to the conclusion that the prior probability of every hypothesis is zero. Earlier, Jeffreys proposed a slightly more complex assignment of priors that did not suffer from this difficulty. The problem is not really one of finding a way to assign finite probabilities to an infinite number of incompatible hypotheses, for there are plenty of ways to do that. The trouble is that it is just very implausible that scientists typically have their prior degrees of belief distributed according to any plausible simplicity ordering, and still less plausible that they would be rational to do so. I can think of very few simple relations between experimentally determined quantities that have withstood continued investigation, and often simple relations are replaced by relations that are infinitely complex: consider the fate of Kepler's laws. Surely it would be naïve for anyone to suppose that a set of newly measured quantities will truly stand in a simple relation, especially in the absence of a well-confirmed theory of the matter. Jeffreys' strategy requires that we proceed in ignorance of our scientific experience, and that can hardly be a rational requirement.

Consider another Bayesian attempt, this one due to Mary Hesse. Hesse puts a 'clustering' constraint on prior probabilities: for any positive r, the conjunction of $r + 1$ positive instances of a hypothesis is more probable than a conjunction of r positive instances with one negative instance. This postulate, she claims, will lead us to choose, *ceteris paribus*, the most economical, the simplest, hypotheses compatible with the evidence. Here is the argument:

Consider first evidence consisting of individuals $a_1, a_2, \ldots, a_n$, all of which have properties P and Q. Now consider an individual a_{n+1} with property P. Does a_{n+1} have Q or not? If nothing else is known, the clustering postulate will direct us to predict Q_{a+1} since, *ceteris paribus*, the universe is to be postulated to be as homogeneous as possible consistently with the data. . . . But this is also the prediction that would be made by taking the most economical general law which is both confirmed by the data and of sufficient content to make a prediction about the application of Q to a_{n+1}. For h = 'All P are Q' is certainly more economical than the 'grubfied' conflicting hypothesis of equal content h': 'All x up to a_n that are P and Q, and all other x that are P are $\sim Q$.'

If follows in the [case] considered that if a rule is adopted to choose the prediction resulting from the most probable hypothesis on grounds of content, or, in case of a tie in content, the most economical hypothesis on those of equal content, this rule will yield the same predictions as the clustering postulate.

Here is the argument applied to curve-fitting:

Let f be the assertion that two data points $(x_1, . y_1)$, (x_2, y_2) are obtained from experiments. . . . The two points are consistent with the hypothesis $y = a + bx$, and also of course with an indefinite number of other hypotheses of the form $y = a_0 + a_1 + \ldots + a_n x$, where the values of $a_0, \ldots, a_n$ are not determined by (x_1, y_1), (x_2, y_2). What is the most economical prediction of the y-value of a further point g, where the x-value of g is x_3? Clearly it is the prediction which uses only the information already contained in f, that is, the calculable values of a, b rather than a prediction which assigns arbitrary values to the parameters of a higher-order hypothesis. Hence the most economical prediction is about the point $g = (x_3, a + bx_3)$, which is also the prediction given by the 'simplest' hypothesis on almost all accounts of the simplicity of curves. Translated into probabilistic language, this is to say that to conform to intuitions about economy we should assign higher initial probability to the assertion that points $(x_1, a + bx_1)$, $(x_2, a + bx_2)$, $(x_3, a + bx_3)$ are satisfied by the experiment than to that in which the third point is inexpressible in terms of a and b alone. In this formulation economy is a function of finite descriptive lists of points rather than general hypotheses, and the relevant initial probability is that of a universe containing these particular points rather than that of a universe in which the corresponding general law is true. . . . Description in terms of a minimum number of parameters may therefore be regarded as another aspect of homogeneity or clustering of the universe. (Hesse 1974: 230–2)

Hesse's clustering postulate applies directly to the curve-fitting case, for her clustering postulate then requires that if two paired values of x and y satisfy the predicate $y = ax + b$, then it is more probable than not that a third pair of values will satisfy the predicate. So the preference for the linear hypothesis in the next instance results from Hesse's clustering postulate and the probability axioms. Unfortunately, with trivial additional assumptions, everything results. For, surely, if $y = a + bx$ is a legitimate predicate, then so is $y = a_1 + b_1x^2$, for any definite values of a_1 and b_1. Now Hesse's first two data points can be equally well described by $(x_1, a_1 + b_1x_1^2)$ and $(x_2, a_1 + b_1x_2^2)$, where

$$b_1 = \frac{y_1 - y_2}{x_1^2 - x_2^2} \qquad a_1 = y_1 - x_1^2\left(\frac{y_1 - y_2}{x_1^2 - x_2^2}\right).$$

Hence her first two data points satisfy both the predicate $y = a + bx$ and the predicate $y = a_1 + b_1x^2$. So, by the clustering postulate, the probability that the third point satisfies the quadratic expression must be greater than one-half, and the probability that the third point satisfies the linear expression must also be greater than one-half, which is impossible.

Another Bayesian account of our preference for simple theories has recently been offered by Roger Rosencrantz (1976). Suppose that we have some criterion for 'goodness of fit' of a hypothesis to data—for example, confidence regions based on the χ^2 distribution for categorical data, or in curve-fitting perhaps that the average sum of squared deviations is less than some figure. Where the number of possible outcomes is finite, we can compare the number of such possible outcomes that meet the goodness-of-fit criterion with the number that do not. This ratio Rosencrantz calls the 'observed sample coverage' of the hypothesis. Where the possible outcomes are infinite, if the region of possible outcomes meeting the goodness-of-fit criterion is always bounded for all relevant hypotheses, we can compare the volumes of such regions for different hypotheses, and thus obtain a measure of comparative sample coverage.

It seems plausible enough that the smaller the observed sample coverage of a hypothesis, the more severely it is tested by observing outcomes. Rosencrantz's first proposal is this: the smaller the observed sample coverage, the simpler the hypothesis. But further, he proves the following for hypotheses about categorical data: if H_1 and H_2 are hypotheses with parameters, and H_1 is a special case of H_2 obtained by letting a free parameter in H_2 take its maximum likelihood value, then if we average the likelihood of getting evidence that fits each hypothesis well enough over all the possible parameter values, the average likelihood of H_1 will be greater than the average likelihood of H_2. The conclusion Rosencrantz suggests is that the simpler the theory, the greater the average likelihood of data that fit it sufficiently well. Hence, even if a simple theory has a lower prior probability than more complex theories, because the average likelihood is higher for the simple theory, its posterior probability will increase more rapidly than that of more complex theories. When sufficient evidence has accumulated, the simple theory will be preferred. Rosencrantz proposes to identify average likelihood with support.

Rosencrantz's approach has many virtues; I shall concentrate on its vices. First, observed sample coverage does not correlate neatly with simplicity. If H is a hypothesis, T another utterly irrelevant to H and to the phenomena about which H makes predictions, then H & T will have the same observed sample coverage as does H. Further, if H^* is a de-Occamization of H, then H^* and H will have the same observed sample coverage. Second, Rosencrantz's theorem does not establish nearly enough. It does not establish, for example, that in curve-fitting the average likelihood of a linear hypothesis is greater than the average likelihood of a quadratic or higher-degree hypothesis. We cannot explicate support in terms of average likelihood unless we are willing to allow that evidence supports a de-Occamized hypothesis as much as un-de-Occamized ones, and a hypothesis with tacked-on parts as much as one without such superfluous parts.

Finally, we come to the question of the relevance of evidence to theory. When does a piece of evidence confirm a hypothesis according to the Bayesian scheme of things? The natural answer is that it does so when the posterior probability of the hypothesis is greater than its prior probability, that is, if the conditional probability of the hypothesis on the evidence is greater than the probability of the hypothesis. That is what the condition of positive relevance requires, and that condition is the one most commonly advanced by philosophical Bayesians. The picture is a kinematic one: a Bayesian agent moves along in time having at each moment a coherent set of degrees of belief; at discrete intervals he learns new facts, and each time he learns a new fact, e, he revises his degrees of belief by conditionalizing on e. The discovery that e is the case has confirmed those hypotheses whose probability after the discovery is higher than their probability before. For several reasons, I think this account is unsatisfactory; moreover, I doubt that its difficulties are remediable without considerable changes in the theory.

The first difficulty is a familiar one. Let us suppose that we can divide the consequences of a theory into sentences consisting of reports of actual or possible observations, and simple generalizations of such observations, on the one hand; and on the other hand, sentences that are theoretical. Then the collection of 'observational' consequences of the theory will always be at least as probable as the theory itself; generally, the theory will be less probable than its observational consequences. A theory is never any better established than is the collection of its observational consequences. Why, then, should we entertain theories at all? On the probabilist view, it seems, they are a gratuitous risk. The natural answer is that theories have some special function that their collection of observational consequences cannot serve; the function most frequently suggested is explanation—theories explain; their collection of observational consequences do not. But however sage this suggestion may be, it only makes more vivid the difficulty of the Bayesian why of seeing things. For whatever explanatory power may be, we should certainly expect that goodness of explanation will go hand in hand with warrant for belief; yet, if theories explain, and their observational consequences do not, the Bayesian must deny the linkage. The difficulty has to do both with the assumption that rational degrees of belief are generated by probability measures and with the Bayesian account of evidential relevance. Making degrees of belief probability measures in the Bayesian way already guarantees that a theory can be no more credible than any collection of its consequences. The Bayesian account of confirmation makes it impossible for a piece of evidence to give us more total credence in a theory than in its observational consequences. The Bayesian way of setting things up is a natural one, but it is not inevitable, and wherever a distinction between theory and evidence is plausible, it leads to trouble.

A second difficulty has to do with how praise and blame are distributed among the hypotheses of a theory. Recall the case of Kepler's laws (discussed in Glymour 1981, ch. 2). It seems that observations of a single planet (and, of course, the sun) might provide evidence for or against Kepler's first law (all planets move on ellipses) and for or against Kepler's second law (all planets move according to the area rule), but no observations of a single planet would constitute evidence for or against Kepler's third law (for any two planets, the ratio of their periods equals the $3/2$ power of the ratio of their distances). Earlier [in Ch. 2 of Glymour's *Theory and Evidence*] we saw that hypothetico-deductive accounts of confirmation have great difficulty explaining this elementary judgement. Can the Bayesians do any better? One thing that Bayesians can say (and some have said) is that our degrees of belief are distributed—and historically were

distributed—so that conditionalizing on evidence about one planet may
change our degrees of belief in the first and second laws, but not our degree
of belief in the third law.[5] I don't see that this is an explanation for our
intuition at all; on the contrary, it seems merely to restate (with some
additional claims) what it is that we want to be explained. Are there any
reasons why people had their degrees of belief so distributed? If their
beliefs had been different, would it have been equally rational for them to
view observations of Mars as a test of the third law, but not of the first? It
seems to me that we never succeed in explaining a widely shared judgement about the relevance or irrelevance of some piece of evidence merely
by asserting that degrees of belief happened to be so distributed as to
generate those judgements according to the Bayesian scheme. Bayesians
may instead try to explain the case by appeal to some structural difference
among the hypotheses; the only gadget that appears to be available is the
likelihood of the evidence about a single planet on various combinations of
hypotheses. If it is supposed that the observations are such that Kepler's
first and second laws entail their description, but Kepler's third law does
not, then it follows that the likelihood of the evidence on the first and
second laws—that is, the conditional probability of the evidence given
those hypotheses—is unity, but the likelihood of the evidence on the third
law may be less than unity. But any attempt to found an account of the case
on these facts alone is simply an attempt at a hypothetico-deductive account. The problem is reduced to one already unsolved. What is needed to
provide a genuine Bayesian explanation of the case in question (as well as
of many others that could be adduced) is a *general* principle restricting
conditional probabilities and having the effect that the distinctions about
the bearing of evidence that have been noted here do result. Presumably,
any such principles will have to make use of relations of content or structure between evidence and hypothesis. The case does nothing to establish
that no such principles exist; it does, I believe, make it plain that without
them the Bayesian scheme does not *explain* even very elementary features
of the bearing of evidence on theory.

A third difficulty has to do with Bayesian kinematics. Scientists commonly argue for their theories from evidence known long before the
theories were introduced. Copernicus argued for his theory using observations made over the course of millennia, not on the basis of any startling
new predictions derived from the theory, and presumably it was on the
basis of such arguments that he won the adherence of his early disciples.
Newton argued for universal gravitation using Kepler's second and third

[5] This is the account suggested by Horwich 1978.

laws, established before the *Principia* was published. The argument that
Einstein gave in 1915 for his gravitational field equations was that they
explained the anomalous advance of the perihelion of Mercury, established more than half a century earlier. Other physicists found the argument enormously forceful, and it is a fair conjecture that without it the
British would not have mounted the famous eclipse expedition of 1919.
Old evidence can in fact confirm new theory, but according to Bayesian
kinematics, it cannot. For let us suppose that evidence e is known before
theory T is introduced at time t. Because e is known at t, $\text{prob}_t(e) = 1$.
Further, because $\text{prob}_t(e) = 1$, the likelihood of e given T, $\text{prob}_t(e, T)$, is also
1. We then have

$$\text{prob}_t(T, e) = \frac{\text{prob}_t(T) \times \text{prob}_t(e, T)}{\text{prob}_t(e)} = \text{prob}_t(T).$$

The conditional probability of T on e is therefore the same as the prior
probability of T: e cannot constitute evidence for T in virtue of the positive
relevance condition nor in virtue of the likelihood of e on T. None of the
Bayesian mechanisms apply, and if we are strictly limited to them, we have
the absurdity that old evidence cannot confirm new theory. The result is
fairly stable. If the probability of e is very high but not unity, $\text{prob}_t(e, T)$
will still be unity if T entails e, and so $\text{prob}_t(T, e)$ will be very close to
$\text{prob}_t(T)$. How might Bayesians deal with the old evidence/new theory
problem?[6] Red herrings abound. The prior probability of the evidence,
Bayesians may object, is not really unity; when the evidence is stated as
measured or observed values, the theory does not really entail that those
exact values obtain; an ideal Bayesian would never suffer the embarrassment of a novel theory. None of these replies will do: the acceptance of old
evidence may make the degree of belief in it as close to unity as our degree
of belief in some bit of evidence ever is; although the exact measured value
(of, e.g., the perihelion advance) may not be entailed by the theory and
known initial conditions, that the value of the measured quantity lies in a
certain interval may very well be entailed, and that is what is believed
anyway; and, finally, it is beside the point that an ideal Bayesian would
never face a novel theory, for the idea of Bayesian confirmation theory is
to explain scientific inference and argument by means of the assumption
that good scientists are, about science at least, approximately ideal

[6] All of the defences sketched below were suggested to me by one or another philosopher
sympathetic to the Bayesian view; I have not attributed the arguments to anyone for fear of
misrepresenting them. None the less, I thank Jon Dorling, Paul Teller, Daniel Garber, Ian
Hacking, Patrick Suppes, Richard Jeffrey, and Roger Rosencrantz for valuable discussions and
correspondence on the point at issue.

Bayesians, and we have before us a feature of scientific argument that seems incompatible with that assumption.

A natural line of defence lies through the introduction of counterfactual degrees of belief. When using Bayes's rule to determine the posterior probability of a new theory on old evidence, one ought not to use one's actual degree of belief in the old evidence, which is unity or nearly so; one ought instead to use the degree of belief one would have had in *e* if. . . . The problem is to fill in the blanks in such a way that it is both plausible that we have the needed counterfactual degrees of belief, and that they do serve to determine how old evidence bears on new theory. I tend to doubt that there is such a completion. We cannot merely throw *e* and whatever entails *e* out of the body of accepted beliefs; we need some rule for determining a counterfactual degree of belief in *e* and a counterfactual likelihood of *e* on *T*. To simplify, let us suppose that *T* does logically entail *e*, so that the likelihood is fixed.

If one flips a coin three times and it turns up heads twice and tails once, in using this evidence to confirm hypotheses (e.g. of the fairness of the coin), one does not take the probability of two heads and one tail to be what it is after the flipping—namely, unity—but what it was before the flipping. In this case there is an immediate and natural counterfactual degree of belief that is used in conditionalizing by Bayes's rule. The trouble with the scientific cases is that no such immediate and natural alternative distribution of degree of belief is available. Consider someone trying, in a Bayesian way, to determine in 1915 how much Einstein's derivation of the perihelion advance confirmed general relativity. There is no single event, like the coin flipping, that makes the perihelion anomaly virtually certain. Rather, Leverrier first computed the anomaly in the middle of the nineteenth century; Simon Newcomb calculated it again around 1890, using Leverrier's method but new values for planetary masses, and obtained a substantially higher value than had Leverrier. Both Newcomb and Leverrier had, in their calculations, approximated an infinite series by its first terms without any proof of convergence, thus leaving open the possibility that the entire anomaly was the result of a mathematical error. In 1912 Eric Doolittle calculated the anomaly by a wholly different method, free of any such assumption, and obtained virtually the same value as had Newcomb.[7] For actual historical cases, unlike the coin-flipping case, there is no single counterfactual degree of belief in the evidence ready to hand, for belief in the evidence sentence may have grown gradually—in some cases, it may have even waxed, waned, and waxed again. So

---

[7] The actual history is still more complicated. Newcomb and Doolittle obtained values for the anomaly differing by about 2 seconds of arc per century. Early in the 1920s. Grossmann discovered that Newcomb had made an error in calculation of about that magnitude.

the old evidence/new theory problem cannot be assimilated to coin flipping.

The suggestion that what is required is a counterfactual degree of belief is tempting, none the less; but there are other problems with it besides the absence of any unique historical degree of belief. A chief one is that various ways of manufacturing counterfactual degrees of belief in the evidence threaten us with incoherence. One suggestion, for example, is the following, used implicitly by some Bayesian writers. At about the time T is introduced, there will be a number of alternative competing theories available; call them $T_1, T_2, \ldots, T_k$, and suppose that they are mutually exclusive of T and of each other. Then $P(e)$ is equal to

$$P(T_1) P(e, T_1) + P(T_2) P(e, T_2) + \ldots + P(T_k) P(e, T_k)$$
$$+ P(\sim(T_1 \vee \ldots \vee T_k) P(e, T_1 \vee \ldots \vee T_k)),$$

and we may try to use this formula to evaluate the counterfactual degree of belief in e. The problem is with the last term. Of course, one could suggest that this term just be ignored when evaluating $P(e)$, but it is difficult to see within a Bayesian framework any rationale at all for doing so. For if one does ignore this term, then the collection of prior probabilities used to evaluate the posterior probability of T will not be coherent unless either the likelihood of e on T is zero or the prior probability of T is zero. One could remedy this objection by replacing the last term by

$$P(T) P(e, T),$$

but this will not do either, for if one's degree of belief in

$$P(T_1 \vee T_2 \vee \ldots \vee T_k \vee T)$$

is not unity, then the set of prior degrees of belief will still be incoherent. Moreover, not only will it be the case that if the actual degree of belief in e is replaced by a counterfactual degree of belief in e according to either of these proposals, then the resulting set of priors will be incoherent, it will further be the case that if we conditionalize on e the resulting conditional probabilities will be incoherent. For example, if we simply delete the last term, one readily calculates that

$$P(T_1 \vee \ldots \vee T_k, e) = \frac{P(T_1 \vee \ldots \vee T_k) P(e, T_1 \vee \ldots \vee T_k)}{P(e, T_1 \vee \ldots \vee T_k) P(T_1 \vee \ldots \vee T_k)} = 1,$$

and further that

$$P(T, e) = \frac{P(T) P(e, T)}{P(e, T_1 \vee \ldots \vee T_k) P(T_1 \vee \ldots \vee T_k)}.$$

But because T is supposed inconsistent with $T_1 \vee \ldots \vee T_k$ and $P(T, e)$ is not zero, this is incoherent.

Let us return to the proposal that when new theory confronts old evidence, we should look backwards to the time when the old evidence e had not yet been established and use for the prior probability of e whatever degree of belief we would have had at that time. We cannot just stick in such a counterfactual value for the prior probability of e and change nothing else without, as before, often making both prior and conditionalized probabilities incoherent. If we give all of our sentences the degree of belief they would have had in the relevant historical period (supposing we somehow know what period that is) and then conditionalize on e, incoherence presumably will not arise; but it is not at all clear how to combine the resulting completely counterfactual conditional probabilities with our actual degrees of belief. It does seem to me that the following rather elaborate procedure will work when a new theory is introduced. Starting with your actual degree of belief function P, consider the degree of belief you would have had in e in the relevant historical period, call it $H(e)$. Now change P by regarding $H(e)$ as an arbitrary change in degree of belief in e and using Richard Jeffrey's (1965) rule,

$$P'(S) = H(e) P(S, e) + (1 - H(e)) P(S, \sim e).$$

Jeffrey's rule guarantees that P' is a probability function. Finally, conditionalize on e:

$$P''(S) = P'(S, e),$$

and let P'' be your new actual degree of belief function. (Alternatively, P'' can be formed by using Jeffrey's rule a second time.)

There remain a number of objections to the historical proposal. It is not obvious that there are, for each of us, degrees of belief we personally would have had in some historical period. It is not at all clear which historical period is the relevant one. Suppose, for example, that the gravitational deflection of sunlight had been determined experimentally around 1900, well before the introduction of general relativity.[8] In trying to assess the confirmation of general relativity, how far back in time should a twen-

---

[8] Around 1900 is fanciful, before general relativity is not. In 1914 E. Freundlich mounted an expedition to Russia to photograph the eclipse of that year in order to determine the gravitational deflection of starlight. At that time, Einstein had predicted an angular deflection for light passing near the limb of the sun that was equal in value to that derived from Newtonian principles by Soldner in 1801. Einstein did not obtain the field equations that imply a value for the deflection equal to twice the Newtonian value until late in 1915. Freundlich was caught in Russia by the outbreak of World War I, and was interned there. Measurement of the deflection had to wait until 1919.

tieth-century physicist go under this supposition? If only to the nineteenth, then if he would have shared the theoretical prejudices of the period, gravitational deflection of light would have seemed quite probable. Where ought he to stop, and why? But laying aside these difficulties, it is implausible indeed that such a historical Bayesianism, however intriguing a proposal, is an accurate account of the principles by which scientific judgements of confirmation are made. For if it were, then we should have to condemn a great mass of scientific judgements on the grounds that those making them had not studied the history of science with sufficient closeness to make a judgement as to what their degrees of belief would have been in relevant historical periods. Combined with the delicacy that is required to make counterfactual degrees of belief fit coherently with actual ones, these considerations make me doubt that we should look to counterfactual degrees of belief for a plausible Bayesian account of how old evidence bears on new theory.

Finally, consider a quite different Bayesian response to the old evidence/new theory problem. Whereas the ideal Bayesian agent is a perfect logician, none of us are, and there are always consequences of our hypotheses that we do not know to be consequences. In the situation in which old evidence is taken to confirm a new theory, it may be argued that there is *something* new that is learned, and typically, what is learned is that the old evidence is entailed by the new theory. Some old anomalous result is lying about, and it is not this old result that confirms a new theory, but rather the new discovery that the new theory entails (and thus explains) the old anomaly. If we suppose that semi-rational agents have degrees of belief about the entailment relations among sentences in their language, and that

$$P(h \mid - e) = 1 \quad \text{implies} \quad P(e, h) = 1,$$

this makes a certain amount of sense. We imagine the semi-rational Bayesian changing his degree of belief in hypothesis h in light of his new discovery that h entails e by moving from his prior degree of belief in h to his conditional degree of belief in h given that e, that $h \mid - e$, and whatever background beliefs there may be. Old evidence can, in this vicarious way, confirm a new theory, then, provided that

$$P(h, b \& e \& (h \mid - e)) > P(h, b \& e).$$

Now, in a sense, I believe this solution to the old evidence/new theory problem to be the correct one; what matters is the discovery of a certain logical or structural connection between a piece of evidence and a piece of theory, and it is in virtue of that connection that the evidence, if believed to be true, is thought to be evidence for the bit of theory. What I do not believe is that the relation that matters is simply the entailment relation between the theory, on the one hand, and the evidence, on the other. The reasons that the relation cannot be simply that of entailment are exactly the reasons why the hypothetico-deductive account (see Glymour 1981, ch. 2) is inaccurate; but the suggestion is at least correct in sensing that our judgement of the relevance of evidence to theory depends on the perception of a structural connection between the two, and that degree of belief is, at best, epiphenomenal. In the determination of the bearing of evidence on theory, there seem to be mechanisms and stratagems that have no apparent connection with degrees of belief, which are shared alike by people advocating different theories. Save for the most radical innovations, scientists seem to be in close agreement regarding what would or would not be evidence relevant to a novel theory; claims as to the relevance to some hypothesis of some observation or experiment are frequently buttressed by detailed calculations and arguments. All of these features of the determination of evidential relevance suggest that that relation depends somehow on structural, objective features connecting statements of evidence and statements of theory. But if that is correct, what is really important and really interesting is what these structural features may be. The condition of positive relevance, even if it were correct, would simply be the least interesting part of what makes evidence relevant to theory.

None of these arguments is decisive against the Bayesian scheme of things, nor should they be; for in important respects that scheme is undoubtedly correct. But taken together, I think they do at least strongly suggest that there must be relations between evidence and hypotheses that are important to scientific argument and to confirmation but to which the Bayesian scheme has not yet penetrated.

REFERENCES

Carnap, R. (1950). *The Logical Foundations of Probability*. Chicago: University of Chicago Press.
Glymour, C. (1981). *Theory and Evidence*. Chicago: University of Chicago Press.
Hesse, M. (1974). *The Structure of Scientific Inference*. Berkeley: University of California Press.
Horwich, P. (1978). 'An Appraisal of Glymour's Confirmation Theory.' *Journal of Philosophy*, 75: 98–113.
Jeffrey, R. (1965). *The Logic of Decision*. New York: McGraw-Hill.
Jeffreys, H. (1967). *Theory of Probability*. Oxford: Clarendon Press.

——(1973). *Scientific Inference*. Cambridge: Cambridge University Press.
Kyburg, H. (1978). 'Subjective Probability: Criticisms, Reflections and Problems.' *Journal of Philosophical Logic*, 7: 157–80.
Putnam, H. (1967). 'Probability and Confirmation.' In S. Morgenbesser (ed.), *Philosophy of Science Today*. New York: Basic Books.
Rosencrantz, R. (1976). 'Simplicity.' In W. Harper and C. Hooker (eds.), *Foundations and Philosophy of Statistical Inference*. Boston Reidel
Salmon, W. C. (1969). *Foundations of Scientific Inference*. Pittsburgh: University of Pittsburgh Press.
Savage, L. (1972). *The Foundations of Statistics*. New York: Dover.
Shimony, A. (1970). 'Scientific Inference.' In R. G. Colodny (ed.), *The Nature and Function of Scientific Theories*, 79–179. Pittsburgh: University of Pittsburgh Press.

XIII

FUNDAMENTALISM vs THE PATCHWORK OF LAWS

NANCY CARTWRIGHT

I

For realism. A number of years ago I wrote *How the Laws of Physics Lie.* That book was generally perceived to be an attack on realism. Nowadays I think that I was deluded about the enemy: it is not *realism* but *fundamentalism* that we need to combat.

My advocacy of realism—local realism about a variety of different kinds of knowledge in a variety of different domains across a range of highly differentiated situations—is Kantian in structure. Kant frequently used what should be a puzzling argument form to establish quite abstruse philosophical positions (Ø): We have X—perceptual knowledge, freedom of the will, whatever. But without Ø (the transcendental unity of apperception, or the kingdom of ends) X would be impossible, or inconceivable. Hence Ø. The objectivity of local knowledge is my Ø; X is the possibility of planning, prediction, manipulation, control, and policy setting. Unless our claims about the expected consequences of our actions are reliable, our plans are for nought. Hence knowledge is possible.

What might be found puzzling about the Kantian argument form are the X's from which it starts. These are generally facts that appear in the clean and orderly world of pure reason as refugees with neither proper papers nor proper introductions, of suspect worth and suspicious origin. The facts that I take to ground objectivity are similarly alien in the clear, well-lighted streets of reason, where properties have exact boundaries, rules are unambiguous, and behaviour is precisely ordained. I know that I can get an oaktree from an acorn, but not from a pine-cone; that nurturing will make my child more secure; that feeding the hungry and housing the homeless will make for less misery; and that giving more smear tests will lessen the incidence of vaginal cancer. Getting closer to physics, which is ultimately

Reprinted from *Proceedings of the Aristotelian Society,* 93/2 (1994): 279–92, by permission of the Editor of the Aristotelian Society. © 1994.

our topic here, I also know that I can drop a pound coin from the upstairs window into the hands of my daughter below, but probably not a paper tissue; that I can head north by following my compass needle (so long as I am on foot and not in my car), that . . .

I know these facts, even though they are vague and imprecise, and I have no reason to assume that that can be improved on. Nor, in many cases, am I sure of the strength or frequency of the link between cause and effect, nor of the range of its reliability. And I certainly do not know in any of the cases which plans or policies would constitute an optimal strategy. But I want to insist that these items are items of knowledge. They are, of course, like all genuine items of knowledge (as opposed to fictional items like sense-data or the synthetic a priori) defeasible and open to revision in the light of further evidence and argument. If I do not know these things, what do I know, and how can I come to know anything?

Besides this odd assortment of inexact facts, we also have a great deal of very precise and exact knowledge, chiefly supplied by the natural sciences. I am not thinking here of abstract laws, which as an empiricist I take to be of considerable remove from the world to which they are supposed to apply, but rather of the precise behaviour of specific kinds of concrete systems, knowledge of, say, what happens when neutral K-mesons decay, which allows us to establish c–p violation, or of the behaviour of SQUIDS (superconducting quantum interference devices) in a shielded fluctuating magnetic field, which allows us to detect the victims of strokes. This knowledge is generally regimented within a highly articulated, highly abstract theoretical scheme.

One cannot do positive science without the use of induction, and where those concrete phenomena can be legitimately derived from an abstract scheme, they serve as a kind of inductive base for that scheme. *How the Laws of Physics Lie* challenged the soundness of these derivations and hence of the empirical support for the abstract laws. I still maintain that these derivations are generally shaky, but that is not the point I want to make here. So let us for the sake of argument assume the contrary: the derivations are deductively correct, and they use only true premises. Then, granting the validity of the appropriate inductions,[1] we have reason to be realists about the laws in question. But that does not give us reason to be fundamentalists. To grant that a law is true—even a law of 'basic' physics or a law about the so-called fundamental particles—is far from admitting that it is universal, that it holds everywhere and governs in all domains.

---

[1] These will depend on the circumstances and on our general understanding of the similarities and structures or kinds and essences that obtain in those circumstances.

II

Against fundamentalism. Return to my rough division of law-like items of knowledge into two categories: (1) those that are legitimately regimented into theoretical schemes, these generally, though not always, being facts about behaviour in highly structured, manufactured environments like a spark chamber; (2) those that are not. There is a tendency to think that all facts must belong to one grand scheme, and, moreover, that this is a scheme in which the facts in the first category have a special and privileged status. They are exemplary of the way nature is supposed to work. The others must be made to conform to them. This is the kind of fundamentalist doctrine that I think we must resist. Biologists are clearly already doing so on behalf of their own special items of knowledge. Reductionism has long been out of fashion in biology, and now emergentism is again a real possibility. But the long-debated relations between biology and physics are not good paradigms for the kind of anti-fundamentalism I urge. Biologists used to talk about how new laws emerge with the appearance of 'life'; nowadays they talk, not about life, but about levels of complexity and organization. Still, in both cases the relation in question is that between larger, richly endowed, complex systems, on the one hand, and fundamental laws of physics, on the other: it is the possibility of 'downwards' reduction that is at stake.

I want to go beyond this. Not only do I want to challenge the possibility of downwards reduction, but also the possibility of 'cross-wise reduction'. Do the laws of physics that are true of systems (literally true, we may imagine for the sake of argument) in the highly contrived environments of a laboratory or inside the housing of a modern technological device, do these laws carry across to systems, even systems of very much the same kind, in different and less regulated settings? Can our refugee facts always, with sufficient effort and attention, be remoulded into proper members of the physics community, behaving tidily in accord with the fundamental code? Or must—and should—they be admitted into the body of knowledge on their own merit?

In moving from the physics experiment to the facts of more everyday experience, we are not only changing from controlled to uncontrolled environments, but often from micro to macro as well. In order to keep separate the issues which arise from these two different shifts, I am going to choose for illustration a case from classical mechanics, and will try to keep the scale constant. Classical electricity and magnetism would serve as well. Moreover, in order to make my claims as clear as possible, I shall consider the simplest and most well-known example, that of Newton's third law and its application to falling bodies: $F = ma$. Most of us, brought up within the fundamentalist canon, read this with a universal quantifier in front: for any body in any situation, the acceleration it undergoes will be equal to the force exerted on it in that situation divided by its inertial mass. I want instead to read it, as indeed I believe we should read *all* nomologicals, as a *ceteris paribus* law: for any body in any situation, *if nothing interferes*, its acceleration will equal the force exerted on it divided by its mass. But what can interfere with a force in the production of motion other than another force? Surely there is no problem: the acceleration will always be equal to the *total* force divided by the mass. That is just what I want to question.

Think again about how we construct a theoretical treatment of a real situation. Before we can apply the abstract concepts of basic theory—assign a quantum field, a tensor, a Hamiltonian, or in the case of our discussion, write down a force function—we must first produce a model of the situation in terms the theory can handle. From that point the theory itself provides 'language-entry rules' for introducing the terms of its own abstract vocabulary, and thereby for bringing its laws into play. *How the Laws of Physics Lie* illustrated this for the case of the Hamiltonian—which is roughly the quantum analogue of the classical force function. Part of learning quantum mechanics is learning how to write the Hamiltonian for canonical models—for example, for systems in free motion, for a square well potential, for a linear harmonic oscillator, and so forth. Ronald Giere (1988) has made the same point for classical mechanics.

The basic strategy for treating a real situation is to piece together a model from these fixed components; and then to determine the prescribed composite Hamiltonian from the Hamiltonians for the parts. Questions of realism arise when the model is compared with the situation it is supposed to represent. *How the Laws of Physics Lie* argued that even in the best cases, the fit between the two is not very good. I concentrated there on the best cases, because I was trying to answer the question 'Do the explanatory successes of modern theories argue for their truth?' Here I want to focus on the multitude of 'bad' cases, where the models, if available at all, provide a very poor image of the situation. These are not cases that disconfirm the theory. You can't show that the predictions of a theory for a given situation are false until you have managed to describe the situation in the language of the theory. When the models are too bad a fit, the theory is not disconfirmed; it is just inapplicable.[2]

Now consider a falling object. Not Galileo's from the leaning tower, nor the pound coin I earlier described dropping from the upstairs window,

[2] Here I follow Alan Musgrave (1981: 381): 'We do not falsify a theory containing a domain assumption by showing that this assumption is not true of some situations ... ; we merely show that that assumption is not applicable to that situation in the first place.'

but rather something more vulnerable to non-gravitational influence. Otto Neurath has a nice example. My doctrine about the case is much like his.

In some cases a physicist is a worse prophet than a [behaviourist psychologist], as when he is supposed to specify where in St. Stephen's Square a thousand dollar bill swept away by the wind will land, whereas a [behaviourist] can specify the result of a conditioning experiment rather accurately. (1933: 13)

Mechanics provides no model for this situation. We have only a partial model, which describes the 1,000-dollar bill as an unsupported object in the vicinity of the earth, and thereby introduces the force exerted on it due to gravity. Is that the total force? The fundamentalist will say no: there is in principle (in God's completed theory?) a model in mechanics for the action of the wind, albeit probably a very complicated one that we may never succeed in constructing. This belief is essential for the fundamentalist. If there is no model for the 1,000-dollar bill in mechanics, then what happens to the note is not determined by its laws. Some falling objects, indeed a very great number, will be outside the domain of mechanics, or only partially affected by it. But what justifies this fundamentalist belief? The successes of mechanics in situations that it can model accurately do not support it, no matter how precise or surprising they are. They show only that the theory is true in its domain, not that its domain is universal. The alternative to fundamentalism that I want to propose supposes just that: mechanics is true, literally true we may grant, for all those motions whose causes can be adequately represented by the familiar models that get assigned force functions in mechanics. For these motions, mechanics is a powerful and precise tool for prediction. But for other motions, it is a tool of limited serviceability.

Let us set our problem of the 1,000-dollar bill in St Stephen's Square to an expert in fluid dynamics. The expert should immediately complain that the problem is ill defined. What exactly is the bill like: is it folded or flat? straight down the middle, or . . . ? is it crisp or crumpled? how long versus wide? and so forth and so forth and so forth. I do not doubt that when answers can be supplied, fluid dynamics can provide a practicable model. But I do doubt that for every real case, or even for the majority, fluid dynamics has enough of the 'right questions' to ask to allow it to model the full set of causes, or even the dominant ones. I am equally sceptical that the models that work will do so by legitimately bringing Newton's laws (or Lagrange's for that matter) into play.[3] How, then, do airplanes stay afloat?

[3] And the problem is certainly not that a quantum or relativistic or microscopic treatment is needed instead.

Two observations are important. First, we do not need to maintain that no laws obtain where mechanics runs out. Fluid dynamics may have loose overlaps and intertwinings with mechanics. But it is in no way a subdiscipline of basic physics; it is a discipline on its own. Its laws can direct the 1,000-dollar bill as well as can those of Newton or Lagrange. Second, the 1,000-dollar bill comes as it comes, and we have to hunt a model for it. Just the reverse is true of the plane. We build it to fit the models we know work. Indeed, that is how we manage to get so much into the domain of the laws we know.

Many will continue to feel that the wind and other exogenous factors must produce a force. The wind after all is composed of millions of little particles which must exert all the usual forces on the bill, both at a distance and via collisions. That view begs the question. When we have a good-fitting molecular model for the wind, and we have in our theory (either by composition from old principles or by the admission of new principles) systematic rules that assign force functions to the models, and the force functions assigned predict exactly the right motions, then we will have good scientific reason to maintain that the wind operates via a force. Otherwise, the assumption is another expression of fundamentalist faith.

III

Ceteris paribus laws versus ascriptions of natures. If the laws of mechanics are not universal, but nevertheless true, there are at least two options for them. They could be pure *ceteris paribus* laws: laws that hold only in circumscribed conditions or so long as no factors relevant to the effect besides those specified occur. And that's it. Nothing follows about what happens in different settings or in cases where other causes occur. Presumably this option is too weak for our example of Newtonian mechanics. When a force is exerted on an object, the force will be relevant to the motion of the object even if other causes for its motion not renderable as forces are at work as well; and the exact relevance of the force will be given by the formula $F = ma$: the (total) force will contribute a component to the acceleration determined by this formula. For cases like this, the older language of *natures* is appropriate. It is in the nature of a force to produce an acceleration of the requisite size. That means that, *ceteris paribus*, it *will* produce that acceleration. But even when other causes are at work, it will 'try' to do so. The idea is familiar in the case of forces: *trying* to produce an acceleration, F/m, consists in actually producing F/m as a vector component to the total acceleration. In general, what counts as 'trying' will differ from one kind of cause to another. To ascribe a behaviour to the nature of a feature is to claim that that behaviour is exportable beyond the strict confines of the *ceteris paribus* conditions, although usually only as a 'tendency' or a 'trying'. The extent and range of the exportability will vary. Some natures are highly stable; others are very restricted in their range. The point here is that we must not confuse a wide-ranging nature with the universal applicability of the related *ceteris paribus* law. To admit that forces tend to cause the prescribed acceleration (and indeed do so in felicitous conditions) is a long way from admitting that $F = ma$ is universally true.[4] In the next sections I will describe two different metaphysical pictures in which fundamentalism about the experimentally derived laws of basic physics would be a mistake. The first is wholism; the second, pluralism. It seems to me that wholism is far more likely to give rise only to *ceteris paribus* laws, whereas natures are more congenial to pluralism.

IV

Wholism. We look at little bits of nature, and we look under a very limited range of circumstances. This is especially true of the exact sciences. We can get very precise outcomes, but to do so, we need very tight control over our inputs. Most often we do not control them directly, one by one, but rather we use some general but effective form of shielding. I know one experiment that aims for direct control—the Stanford Gravity Probe. Still, in the end, they will roll the spaceship to average out causes they have not been able to command. Sometimes we take physics outside the laboratory. Then shielding becomes even more important. SQUIDS (superconducting quantum interference devices) can make very fine measurements of magnetic fluctuations, which helps in the detection of stroke victims. But for administering the tests, the hospital must have a Hertz box—a small metal room to block out magnetism from the environment. Or, for a more homely example, we all know that batteries are not likely to work if their protective casing has been pierced.

We tend to think that shielding cannot matter to the laws we use. The same laws apply both inside and outside the shields; the difference is that inside the shield we know how to calculate what the laws will produce, but

---

[4] I have written more about the two levels of generalization, laws, and ascriptions of natures, in 1989. See also my 1992.

outside, it is too complicated. Wholists are wary of these claims: if the events we study are locked together, and changes depend on the total structure rather than the arrangement of the pieces, we are likely to be very mistaken by looking at small chunks of special cases.

Consider a scientific example, the revolution in communications technology due to fibre optics. Low-loss optical fibres can carry information at rates of many gigabits per second over spans of tens of kilometres. But the development of fibre bundles which lose only a few decibels per kilometre is not all there is to the story. Pulse-broadening effects intrinsic to the fibres can be truly devastating. If the pulses broaden as they travel down the fibre, they will eventually smear into each other, and destroy the information. That means that the pulses cannot be sent too close together, and the transmission rate may drop to tens or at most hundreds of megabits per second.

We know that is not what happens—the technology has been successful. That's because the right kind of optical fibre in the right circumstance can transmit solitons—solitary waves that keep their shape across vast distances. I'll explain why. The light intensity of the incoming pulse causes a shift in the index of refraction of the optical fibre, producing a slight nonlinearity in the index. The non-linearity leads to what is called a 'chirp' in the pulse. Frequencies in the leading half of the pulse are lowered, while those in the trailing half are raised. The effects of the chirp combine with those of dispersion to produce the soliton. Stable pulse shapes are not at all a general phenomenon of low-loss optical fibres. They are instead a consequence of two different, oppositely directed processes. The pulse widening due to the dispersion is cancelled by the pulse narrowing due to the non-linearity in the index of refraction. We can indeed produce perfectly stable pulses. But to do so, we must use fibres of just the right design, and matched precisely with the power and input frequency of the laser that generates the input pulses. By chance, that was not hard to do. When the ideas were first tested in 1980, the glass fibres and lasers readily available were easily suited to each other. Given that very special match, fibre optics was off to an impressive start.

Solitons are indeed a stable phenomenon. They are a feature of nature, but of nature under very special circumstance. Clearly it would be a mistake to suppose that they were a general characteristic of low-loss optical fibres. The question is, how many of the scientific phenomena we prize are like solitons, local to the environments we encounter, or—more importantly—to the environments we construct? If nature is more wholistic than we are accustomed to think, the fundamentalist's hopes to export the laws of the laboratory to the far reaches of the world will be dashed.

It is clear that I am not very sanguine about the fundamentalist faith. But that is not really out of the kind of wholist intuitions I have been sketching. After all, the story I just told accounts for the powerful successes of the 'false' local theory—the theory that solitons are characteristic of low-loss fibres—by embedding it in a far more general theory about the interaction of light and matter. Metaphysically, the fundamentalist is borne out. It may be the case that the successful theories we have are limited in their authority, but their successes are to be explained by reference to a truly universal authority. I do not see why we need to explain their successes. I am prepared to believe in more general theories when we have direct empirical evidence for them. But not merely because they are the 'best explanation' for something which seems to me to need no explanation to begin with. 'The theory is successful in its domain': the need for explanation is the same whether the domain is small or large or very small or very large. Theories are successful where they are successful, and that's that. If we insist on turning this into a metaphysical doctrine, I suppose it will look like metaphysical pluralism, to which I now turn.

V

The patchwork of laws. Metaphysical nomological pluralism is the doctrine that nature is governed in different domains by different systems of laws not necessarily related to each other in any systematic or uniform way: by a patchwork of laws. Nomological pluralism opposes any kind of fundamentalism. We are here concerned especially with the attempts of physics to gather all phenomena into its own abstract theories. In *How the Laws of Physics Lie* I argued that most situations are brought under a law of physics only by distortion, whereas they can often be described fairly correctly by concepts from more phenomenological laws. The picture suggested was of a lot of different situations in a continuum, from ones that fit not perfectly but not badly to those that fit very badly indeed. I did suggest that at one end fundamental physics might run out entirely ('What is . . . the value of the electric field vector in the region just at the tip of my pencil?'), whereas in transistors it works quite well. But that was not the principal focus. Now I want to draw sharp divides: some features of systems typically studied by physics may get into situations where their behaviour is not governed by the laws of physics at all. But that does not mean that they have no guide for their behaviour or only low-level phenomenological laws. They could fall under a quite different organized set of highly abstract principles.

There are two immediate difficulties that metaphysical pluralism encounters. The first is one we create ourselves, by imagining that it must be joined with views that are vestiges of metaphysical monism. The second is, I believe, a genuine problem that nature must solve.

First. We are inclined to ask: how can there be motions not governed by Newton's laws? The answer: there are causes of motion not included in Newton's theory. Many find this impossible because, although they have forsaken reductionism, they cling to a near-cousin: *supervenience*. Suppose we give a complete 'physics' description of the falling object and its surrounds. Mustn't that fix all the other features of the situation? Why? This is certainly not true at the level of discussion at which we stand now: the wind is cold and gusty; the bill is green and white and crumpled. These properties are independent of the mass of the bill, the mass of the earth, the distance between them.

I suppose, though, I have the supervenience story wrong. It is the microscopic properties of physics that matter; the rest of reality supervenes on them. Why should I believe that? Supervenience is touted as a step forward over reductionism. Crudely, I take it, the advantage is supposed to be that we can substitute a weaker kind of reductionism, 'token–token' reductionism, for the more traditional 'type–type' reductionism which was proving hard to carry out. But the traditional view had arguments in its favour. Science does sketch a variety of fairly systematic connections between micro-structures and macro-properties. Often the sketch is rough; sometimes it is precise; usually its reliability is confined to very special circumstances. Nevertheless, there are striking cases. But these cases support type–type reductionism; they are irrelevant for supervenience. Type–type reductionism has well-known problems: the connections we discover often turn out to look more like causal connections than like reductions; they are limited in their domain; they are rough rather than exact; and often we cannot even find good starting proposals where we had hoped to produce nice reductions. These problems suggest modifying the doctrine in a number of specific ways, or perhaps giving it up altogether. But they certainly do not leave us with token–token reductionism as a fall-back position. After all, on the story I have just told, it was the appearance of some degree of systematic connection that argued in the first place for the claim that micro-structures fixed macro-properties. But it is just this systematicity that is missing in token–token reductionism.

The view that there are macro-properties that do not supervene on micro-features studied by physics is sometimes labelled 'emergentism'. The suggestion is that where there is no supervenience, macro-properties must miraculously come out of nowhere. But why? There is nothing of the newly landed about these properties. They have been here in the world all along, standing beside the properties of physics. Perhaps we are misled by the feeling that the set of properties studied by physics is complete. Indeed, I think that there is a real sense in which this claim is true, but that sense does not support the charge of emergentism. Consider how the domain of properties studied by physics gets set. Here is one caricature: we begin with an interest in motions—deflections, trajectories, orbits. Then we look for the smallest set of properties that is closed (or, closed enough) under prediction. That is, we expand the set until we get all the factors that are causally relevant to our starting factors, and then everything causally relevant to those, and so forth. To succeed does not show that we have gotten all the properties there are. This is a fact we need to keep in mind quite independently of the chief claim of this paper, that the predictive closure itself only obtains in highly restricted circumstances. The immediate point is that predictive closure among a set of properties does not imply descriptive completeness.

Second. The second problem that metaphysical pluralism faces is that of consistency. We do not want colour patches to appear in regions from which the laws of physics have carried away all matter and energy. Here are two stories I have told in teaching the mechanical philosophy of the seventeenth century. Both are about how to write the Book of Nature to ensure that a consistent universe can be created. In one story God is very interested in physics. He carefully writes out all of the law of physics, and lays down the initial distribution of matter and energy in the universe. He then leaves to St Peter the tedious but intellectually trivial job of calculating all future happenings, including what, if any, macroscopic properties and macroscopic laws will emerge. That is the story of reductionism. Metaphysical pluralism supposes that God is instead very concerned about laws, and so he writes down each and every regularity that his universe will display. In this case St Peter is left with the gargantuan task of arranging the initial properties in the universe in some way that will allow all God's laws to be true together. The advantage to reductionism is that it makes St Peter's job easier. God may nevertheless choose to be a metaphysical pluralist.

VI

Conclusion. I have argued that the laws of our contemporary science are, to the extent that they are true at all, at best true *ceteris paribus*. In the nicest cases we may treat them as claims about natures. But we have no grounds in our experience for taking our laws—even our most fundamental laws of physics—as universal. Indeed I should say '*especially* our most fundamental laws of physics', if these are meant to be the laws of fundamental particles. For we have virtually no inductive reason for counting these laws as true of fundamental particles outside the laboratory setting—if they exist there at all. Ian Hacking is famous for his remark 'If you can spray them, they exist'. I have always agreed with that. But I would now be more cautious: '*When* you can spray them, they exist.'

The claim that theoretical entities are created by the peculiar conditions and conventions of the laboratory is familiar from the social constructionists. The stable low-loss pulses I described earlier provide an example of how that can happen. Here I want to add a caution, not just about the existence of the theoretical entities outside the laboratory, but about their behaviour.

Hacking's point is not only that when we can use theoretical entities in just the way we want to produce precise and subtle effects, they must exist; but also that it must be the case that we understand their behaviour very well if we are able to get them to do what we want. That argues, I believe, for the truth of some very concrete, context-constrained claims, the claims we use to describe their behaviour and control them. But in all these cases of precise control, we build our circumstances to fit our models. I repeat: that does not show that it must be possible to tailor our models to fit every circumstance.

Perhaps we feel that there could be no real difference between the one kind of circumstance and the other, and hence no principled reason for stopping our inductions at the walls of our laboratories. But there is a difference: some circumstances resemble the models we have; others do not. And it is just the point of scientific activity to build models that get in, under the cover of the laws in question, all and only those circumstances that the laws govern.[5] Fundamentalists see matters differently. They want laws; they want true laws; but most of all, they want their favourite laws to be in force everywhere. I urge that we resist fundamentalism. Reality may well be just a patchwork of laws.[6]

REFERENCES

Cartwright, N. (1983). *How the Laws of Physics Lie*. Oxford: Clarendon Press.
——(1989). *Nature's Capacities and their Measurement*. Oxford: Oxford University Press.

[5] Or, in a more empiricist formulation that I would prefer, 'that the laws accurately describe'.
[6] Paper first presented at a meeting of the Aristotelian Society, held in the Senior Common Room. Birkbeck College, London, 9 May 1994.

——(1992). 'Aristotelian Natures and the Modern Experimental Method.' In J. Earman (ed.), *Inference, Explanation and Other Philosophical Frustrations*, 44–71. Berkeley: University of California Press.

Giere, R. N. (1988). *Explaining Science: A Cognitive Approach.* Chicago: University of Chicago Press.

Musgrave, A. (1981). 'On Interpreting Friedman.' *Kyklos*, 34/3: 377–87.

Neurath, O. (1933). 'United Science and Psychology.' In B. F. McGuinness (ed.), *Unified Science*, 1987:1–23. Dordrecht: Reidel.

NOTES ON THE CONTRIBUTORS

RICHARD BOYD is Professor of Philosophy at Cornell University, and has written widely on aspects of scientific realism.

NANCY CARTWRIGHT holds the chair in the Department of Philosophy, Logic, and Scientific Method at the London School of Economics. She has written *How the Laws of Physics Lie* (1983) and *Nature's Capacities and their Measurement* (1989).

BRIAN ELLIS is Professor of Philosophy at La Trobe University, and the author of *Rational Belief Systems* (1979).

ARTHUR FINE is Professor of Philosophy at Northwestern University in Evanston, Illinois. He is the author of *The Shaky Game* (1984), and has written widely on the philosophy of physics.

CLARK GLYMOUR is Professor of Philosophy at Carnegie Mellon University in Pittsburgh. He is the author of *Theory and Evidence* (1980) and (with Peter Spirtes and Richard Scheines) of *Causation, Prediction and Search* (1993).

LARRY LAUDAN occupies the chair of Philosophy at the University of Hawaii. His books include *Progress and its Problems* (1977) and *Science and Values* (1984).

PETER LIPTON is a University Lecturer in History and Philosophy of Science at Cambridge University, and the author of *Inference to the Best Explanation* (1991).

ALAN MUSGRAVE is Professor of Philosophy at the University of Otago in New Zealand. He is the author of *Common Sense, Science and Scepticism* (1993).

DAVID PAPINEAU is Professor of the Philosophy of Science at King's College London. He is the author of *For Science in the Social Sciences* (1978), *Theory and Meaning* (1979), *Reality and Representation* (1987), and *Philosophical Naturalism* (1993).

WESLEY SALMON is Professor of Philosophy at the University of Pittsburgh. His books include *The Foundations of Scientific Inference* (1967) and *Scientific Explanation and the Causal Structure of the World* (1984).

LAWRENCE SKLAR is Professor of Philosophy and Nelson Fellow at the University of Michigan, Ann Arbor. He has written *Space, Time and Spacetime* (1973), *Philosophy and Spacetime Physics* (1985), *Philosophy of Physics* (1992), and *Physics and Chance* (1993).

BAS VAN FRAASSEN is Professor of Philosophy at Princeton University. He has written *Introduction to the Philosophy of Time and Space* (1970) and *The Scientific Image* (1980).

JOHN WORRALL is Reader in the Department of Philosophy, Logic, and Scientific Method at the London School of Economics. He has edited *The Ontology of Science* (1994).

ANNOTATED BIBLIOGRAPHY

Compiled by Stathis Psillos and David Papineau

I GENERAL ANTHOLOGIES

GRANDY, R., and BRODY, B. (eds.) (1989). *Readings in Philosophy of Science.* Englewood Cliffs, NJ: Prentice-Hall.
SAVAGE, C. (ed.) (1990). *Scientific Theories.* Minnesota Studies in the Philosophy of Science, vol. 14. Minneapolis: University of Minnesota Press.
BOYD, R., GASPER, P., and TROUT, J. (eds.) (1991). *The Philosophy of Science.* Cambridge, Mass.: MIT Press.
SALMON, M. (ed.) (1992). *Introduction to Philosophy of Science.* Englewood Cliffs, NJ: Prentice-Hall.
LIPTON, P. (ed.) (1994). *Theories, Evidence and Explanation.* Aldershot: Dartmouth.
WORRALL, J. (ed.) (1994). *The Ontology of Science.* Aldershot: Dartmouth.

II VARIETIES OF REALISM

There is a large literature on scientific realism and the various species of non-realism. An important collection of articles is

LEPLIN, J. (ed.) (1984). *Scientific Realism.* Berkeley: University of California Press.

For the demise of the older phenomenalist form of anti-realism see

GRANDY, R. (ed.) (1973). *Theory and Observation in Science.* Atascadero, Calif.: Ridgeview.

Arthur Fine's 'Natural Ontological Attitude' is developed in

FINE, A. (1984). 'And not Anti-Realism Either.' *Noûs*, 18: 51–65.
——(1984). *The Shaky Game.* Chicago: University of Chicago Press.
——(1986). 'Unnatural Attitudes: Realist and Instrumentalist Attachments to Science.' *Mind*, 95: 149–79.
——(1991). 'Piecemeal Realism.' *Philosophical Studies*, 61: 79–96.

Fine's position is discussed in

NEWTON-SMITH, W. (1989). 'Modest Realism.' In A. Fine and J. Leplin (eds.), *PSA 1988*, vol. 2, 179–89. East Lansing, Mich.: Philosophy of Science Association.
MCMULLIN, E. (1991). 'Comment: Selective Anti-Realism.' *Philosophical Studies*, 61: 97–108.

Attempts to clarify scientific realism include

HORWICH, P. (1982). 'Three Forms of Realism.' *Synthese*, 51: 181–202.
PUTNAM, H. (1982). 'Three Kinds of Scientific Realism.' *Philosophical Quarterly*, 32: 195–200.

HELLMAN, G. (1983). 'Realist Principles.' *Philosophy of Science*, 50: 227–49.
BOYD, R. (1983). 'The Current Status of Scientific Realism.' *Erkenntnis*, 19: 45–90. Repr. in Leplin 1984: 41–82.
——(1989). 'What Realism Implies and What it Does Not.' *Dialectica*, 43: 5–29.
——(1990). 'Realism, Conventionality and "Realism About".' In G. Boolos (ed.), *Meaning and Method: Essays in Honour of Hilary Putnam*, 171–95. Cambridge: Cambridge University Press.
LEVIN, M. (1990). 'Realisms.' *Synthese*, 85: 115–38.
HORWICH, P. (1991). 'On the Nature and Norms of Theoretical Commitment.' *Philosophy of Science*, 58: 1–14.
JONES, R. (1991). 'Realism About What?' *Philosophy of Science*, 58: 185–202.
MUSGRAVE, A. (1992). 'Discussion: Realism About What?' *Philosophy of Science*, 59: 691–7.

Recent assessments of the arguments for and against scientific realism include

WYLIE, A. (1986). 'Arguments for Scientific Realism: The Ascending Spiral.' *American Philosophical Quarterly*, 23: 287–97.
KUKLA, A. (1994). 'Scientific Realism, Scientific Practice and the Natural Ontological Attitude.' *British Journal for the Philosophy of Science*, 45: 955–75.

Michael Devitt defends the view that scientific realism is primarily a metaphysical thesis about what kinds of entities exist.

DEVITT, M. (1984). *Realism and Truth*. Oxford: Blackwell. 2nd rev. edn., 1991.
——(1991). 'Aberrations of the Realism Debate.' *Philosophical Studies*, 61: 43–63.

III THE UNDERDETERMINATION OF THEORY BY EVIDENCE AND INFERENCE TO THE BEST EXPLANATION

Articles on the underdetermination thesis include

BOYD, R. (1973). 'Realism, Underdetermination and a Causal Theory of Evidence.' *Noûs*, 7: 1–12.
MALAMENT, D. (1977). 'Observationally Indistinguishable Space-Times.' In J. Earman, C. Glymour and J. Stachel (eds.), *Foundations of Space-Time Theories*, Minnesota Studies in the Philosophy of Science, vol. 8, 61–80. Minneapolis: University of Minnesota Press.
NEWTON-SMITH, W. H. (1978). 'The Underdetermination of Theory by Data.' *Aristotelian Society*, suppl. vol. 52: 71–91.
HORWICH, P. (1982). 'How to Choose between Empirically Indistinguishable Theories.' *Journal of Philosophy*, 79: 61–77.
BERGSTROM, L. (1984). 'Underdetermination and Realism.' *Erkenntnis*, 21: 349–65.
LAUDAN, L. (1990). 'Demystifying Underdetermination.' In C. Savage (ed.), *Scientific Theories*, Minnesota Studies in the Philosophy of Science, vol. 14, 267–97. Minneapolis: University of Minnesota Press.
——, and LEPLIN, J. (1991). 'Empirical Equivalence and Underdetermination.' *Journal of Philosophy*, 88: 449–72.
EARMAN, J. (1993). 'Underdetermination, Realism and Reason.' *Midwest Studies in Philosophy*, 18: 19–38.
NORTON, J. (1993). 'The Determination of Theories by Evidence: The Case for Quantum Discontinuity 1900–1915.' *Synthese*, 97: 1–31.

Bas van Fraassen's 'constructive empiricism' is developed in

VAN FRAASSEN, B. (1980). *The Scientific Image*. Oxford: Clarendon Press.

Constructive empiricism is discussed in

CHURCHLAND, P., and HOOKER, C. (eds.) (1985). *Images of Science*. Chicago: University of Chicago Press. This volume also contains van Fraassen's response to the comments on his views: 'Empiricism in Philosophy of Science', pp. 245–308.

Inference to the best explanation is defended in

HARMAN, G. (1965). 'Inference to the Best Explanation.' *Philosophical Review*, 74: 88–95.

The most comprehensive treatment of inference to the best explanation is

LIPTON, P. (1991). *Inference to the Best Explanation*. London: Routledge and Kegan Paul.

Other pieces which discuss the use of inference to the best explanation to defend realism include

MCMULLIN, E. (1987). 'Explanatory Success and the Truth of Theory.' In N. Rescher (ed.), *Scientific Inquiry in Philosophical Perspective*, 51–73. New York: University Press of America.

NEWTON-SMITH, W. H. (1987). 'Realism and Inference to the Best Explanation.' *Fundamenta Scientiae*, 7: 305–16.

MUSGRAVE, A. (1988). 'The Ultimate Argument for Scientific Realism.' In R. Nola (ed.), *Relativism and Realism in Science*, 229–52. Dordrecht: Kluwer.

PSILLOS, S. (1996). 'On Van Fraassen's Critique of Abductive Reasoning: Some Pitfalls of Selective Scepticism.' *Philosophical Quarterly*, 46.

IV THE PESSIMISTIC META-INDUCTION

Realism is defended against the pessimistic meta-induction in

HARDIN, C., and ROSENBERG, A. (1982). 'In Defence of Convergent Realism.' *Philosophy of Science*, 49: 604–15.

Laudan responds to Hardin and Rosenberg in

LAUDAN, L. (1984). 'Discussion: Realism without the Real.' *Philosophy of Science*, 51: 156–62.

John Worrall develops his 'structural realism' in

WORRALL, J. (1994). 'How to Remain (Reasonably) Optimistic: Scientific Realism and the "Luminiferous Ether".' In D. Hull and M. Forbes (eds.), *PSA 1994*, vol. 1, 334–44. East Lansing, Mich.: Philosophy of Science Association.

Structural realism is criticized in

PSILLOS, S. (1995). 'Is Structural Realism the Best of Both Worlds?' *Dialectica*, 49: 15–46.

Other articles on realism and the pessimistic meta-induction include

CARRIER, M. (1991). 'What is Wrong with the Miracle Argument?' *Studies in History and Philosophy of Science*, 22: 23–36.

ENFIELD, P. (1991). 'Realism, Empiricism and Scientific Revolutions.' *Philosophy of Science*, 58: 468–85.

CUMMINSKEY, D. (1992). 'Reference Failure and Scientific Realism: A Response to the Meta-Induction.' *British Journal for the Philosophy of Science*, 43: 21–40.

PSILLOS, S. (1994). 'A Philosophical Study of the Transition from the Caloric Theory of Heat to Thermodynamics: Resisting the Pessimistic Meta-Induction.' *Studies in History and Philosophy of Science*, 25: 159–90.

V INTERNAL REALISM

Hilary Putnam first develops his 'internal realism' in

PUTNAM, H. (1978). *Meaning and the Moral Sciences*. London: Routledge and Kegan Paul.

See also

PUTNAM, H. (1981). *Reason, Truth and History*. Cambridge: Cambridge University Press.

Related accounts of scientific inquiry include

ELLIS, B. (1979). *Rational Belief Systems*. Oxford: Blackwell.

JARDINE, N. (1986). *The Fortunes of Inquiry*. Oxford: Clarendon Press.

Putnam's position is criticized in

GLYMOUR, C. (1982). 'Conceptual Scheming, or Confessions of a Metaphysical Realist.' *Synthese*, 51: 169–80.

The internal realist view of truth is challenged in

PAPINEAU, D. (1987). *Reality and Representation*. Oxford: Blackwell.

NEWTON-SMITH, W. (1989). 'The Truth in Realism.' *Dialectica*, 43: 31–45.

VI VERISIMILITUDE

The original account of verisimilitude is given in

POPPER, K. (1963). *Conjectures and Refutations*. London: Routledge and Kegan Paul. 3rd rev. edn., 1969.

It was shown to be inadequate by

MILLER, D. (1974). 'Popper's Qualitative Theory of Verisimilitude.' *British Journal for the Philosophy of Science*, 25: 166–77.

TICHY, P. (1974). 'On Popper's Definitions of Verisimilitude.' *British Journal for the Philosophy of Science*, 25: 155–60.

For more recent attempts to characterize verisimilitude see

ODDIE, G. (1986). *Likeness to Truth*. Dordrecht: Reidel.

NIINILUOTO, I. (1987). *Truthlikeness*. Dordrecht: Reidel.

ARONSON, J. (1990). 'Verisimilitude and Type Hierarchies.' *Philosophical Topics*, 18: 5–28.

WESTON, T. (1992). 'Approximate Truth and Scientific Realism.' *Philosophy of Science*, 59: 53–74.
ARONSON, J., HARRÉ, R., and WAY, E. (1994). *Realism Rescued*. London: Duckworth.

VII NATURALIZED PHILOSOPHY OF SCIENCE

Two recent surveys are

KITCHER, P. (1992). 'The Naturalists Return.' *Philosophical Review*, 101: 53–114.
ROSENBERG, A. (1996). 'A Field Guide to Recent Species of Naturalism.' *British Journal for the Philosophy of Science*, 47.

Richard Boyd's naturalized defence of scientific realism has been developed in a series of articles. Apart from those already mentioned, these include

BOYD, R. (1981). 'Scientific Realism and Naturalistic Epistemology.' In P. Asquith and T. Nickles (eds.), *PSA 1980*, vol. 2, 613–62. East Lansing, Mich.: Philosophy of Science Association.
——(1992). 'Constructivism, Realism and Philosophical Method.' In J. Earman (ed.), *Inference, Explanation and Other Frustrations*, 131–98. Berkeley: University of California Press.

Philip Kitcher defends his own variety of naturalism in

KITCHER, P. (1993). *The Advancement of Science*. Oxford: Oxford University Press.

Larry Laudan's means–ends approach to scientific methodology is outlined in

LAUDAN, L. (1984). *Science and Values*. Berkeley: University of California Press.

It is developed and compared with historical evidence in

DONOVAN, A., LAUDAN, L., and LAUDAN, R. (eds.) (1992). *Scrutinizing Science: Empirical Studies of Scientific Change*, 83–105. Baltimore: Johns Hopkins University Press.

Further discussion of Laudan's views can be found in

LAUDAN, L. (1984). 'Explaining the Success of Science.' In J. Cushing *et al.* (eds.), *Science and Reality*. Notre Dame, Ind.: Notre Dame University Press.
WORRALL, J. (1988). 'The Value of a Fixed Methodology.' *British Journal for the Philosophy of Science*, 39: 263–75.
LAUDAN, L. (1989). 'If it Ain't Broke, Don't Fix it.' *British Journal for the Philosophy of Science*, 40: 369–75.
WORRALL, J. (1989). 'Fix it and Be Damned: A Reply to Laudan.' *British Journal for the Philosophy of Science*, 40: 376–88.
DOPPELT, G. (1990). 'The Naturalist Conception of Methodological Standards in Science.' *Philosophy of Science*, 57: 1–19.
LAUDAN, L. (1990). 'Normative Naturalism.' *Philosophy of Science*, 57: 49–59.
——(1990). 'Aim-Less Epistemology?' *Studies in History and Philosophy of Science*, 21: 315–22.
LEPLIN, J. (1990). 'Renormalising Epistemology.' *Philosophy of Science*, 57: 20–33.
ROSENBERG, A. (1990). 'Normative Naturalism and the Role of Philosophy.' *Philosophy of Science*, 57: 34–43.

Resnik, D. B. (1993). 'Do Scientific Aims Justify Methodological Rules?' *Erkenntnis*, 38: 223–32.

The charge that naturalized epistemology of science is circular is addressed in

Friedman, M. (1979). 'Truth and Confirmation.' *Journal of Philosophy*, 76: 361–82.
Papineau, D. (1992). 'Reliabilism, Induction and Scepticism.' *Philosophical Quarterly*, 42: 1–19.
——(1993). *Philosophical Naturalism*. Oxford: Blackwell.

Ronald Giere's version of naturalized philosophy of science is presented in

Giere, R. (1987). 'The Cognitive Study of Science.' In N. Nersessian (ed.), *The Process of Science*, 139–59. Dordrecht: Martinus Nijhoff.
——(1988). *Explaining Science: A Cognitive Approach*. Chicago: University of Chicago Press.

It is discussed in

Giere, R. (1989). 'Scientific Rationality as Instrumental Rationality.' *Studies in History and Philosophy of Science*, 20: 377–84.
Siegel, H. (1989). 'Philosophy of Science Naturalised? Some Problems with Giere's Naturalism.' *Studies in History and Philosophy of Science*, 20: 365–75.

VIII CONFIRMATION THEORY AND BAYESIANISM

Two relatively introductory accounts of Bayesianism are

Horwich, P. (1982). *Probability and Evidence*. Cambridge: Cambridge University Press.
Howson, C., and Urbach, P. (1989). *Scientific Reasoning: The Bayesian Approach*. La Salle, Ill.: Open Court. 2nd rev. edn., 1994.

A more advanced treatment is

Earman, J. (1992). *Bayes or Bust? A Critical Examination of Bayesian Confirmation Theory*. Cambridge, Mass.: MIT Press.

Bayesianism is applied to the issue of scientific realism in

Dorling, J. (1992). 'Bayesian Conditionalization Resolves Positivist/Realist Disputes.' *Journal of Philosophy*, 89: 362–82.

Glymour's 'bootstrap' theory of confirmation is defended in

Glymour, C. (1980). *Theory and Evidence*. Princeton: Princeton University Press.

Glymour's theory of confirmation is discussed in

Earman, J. (ed.) (1983). *Testing Scientific Theories*, Minnesota Studies in the Philosophy of Science, vol. 10. Minneapolis: University of Minnesota Press.

For earlier approaches to confirmation theory see

Achinstein, P. (ed.) (1983). *The Concept of Evidence*. Oxford: Oxford University Press.

IX OTHER APPROACHES TO THE EPISTEMOLOGY OF SCIENCE

Nancy Cartwright and Ian Hacking defend 'entity realism' in

CARTWRIGHT, N. (1983). *How the Laws of Physics Lie.* Oxford: Clarendon Press.
HACKING, I. (1983). *Representing and Intervening.* Cambridge: Cambridge University Press.

Cartwright's views are developed further in

CARTWRIGHT, N. (1989). *Nature's Capacities and their Measurement.* Oxford: Clarendon Press.
—— (1992). 'Aristotelian Natures and the Modern Experimental Method.' In J. Earman (ed.), *Inference, Explanation and Other Frustrations*, 44–71. Berkeley: University of California Press.

For some similar themes see

DUPRÉ, J. (1993). *The Disorder of Things: Metaphysical Foundations of the Disunity of Science.* Cambridge, Mass.: Harvard University Press.

The relevance of artificial intelligence to the philosophy of science is discussed in

THAGARD, P. (1988). *Computational Philosophy of Science.* Cambridge, Mass.: MIT Press.

X SOCIOLOGY OF SCIENCE

Early manifestos for contemporary sociology of science are formulated in

BARNES, B. (1974). *Scientific Knowledge and Sociological Theory.* London: Routledge and Kegan Paul.
BLOOR, D. (1976). *Knowledge and Social Imagery.* London: Routledge and Kegan Paul.
LATOUR, B., and WOOLGAR, S. (1979). *Laboratory Life.* London: Sage.
BLOOR, D. (1981). 'The Strength of the Strong Programme.' *Philosophy of the Social Sciences*, 11: 199–213.
BARNES, B., and BLOOR, D. (1982). 'Relativism, Rationalism and the Sociology of Knowledge.' In M. Hollis and S. Lukes (eds.), *Rationality and Relativism*, 21–47. Oxford: Blackwell.

Philosophical commentaries include

NEWTON-SMITH, W. H. (1982). 'Relativism and the Possibility of Interpretation.' In M. Hollis and S. Lukes (eds.), *Rationality and Relativism*, 106–22. Oxford: Blackwell.
PAPINEAU, D. (1988). 'Does the Sociology of Science Discredit Science?' In R. Nola (ed.), *Relativism and Realism in Science*, 37–57. Dordrecht: Kluwer.
PETTIT, P. (1988). 'The Strong Sociology of Knowledge without Relativism.' In R. Nola (ed.), *Relativism and Realism in Science*, 81–91. Dordrecht: Kluwer.

Some recent sociological and historical works of philosophical interest are

PICKERING, A. (1984). *Constructing Quarks.* Edinburgh: Edinburgh University Press.

SHAPIN, S., and SCHAFFER, S. (1985). *Leviathan and the Air Pump*. Princeton: Princeton University Press.

GALISON, P. (1987). *How Experiments End*. Chicago: University of Chicago Press.

LATOUR, B. (1987). *Science in Action*. Cambridge, Mass.: Harvard University Press.

GOODING, D., PINCH, T., and SCHAFFER, S. (eds.) (1989). *The Uses of Experiment*. Cambridge: Cambridge University Press.

SHAPIN, S. (1994). *A Social History of Truth: Civility and Science in Seventeenth-Century England*. Chicago: University of Chicago Press.

PICKERING, A. (1995). *The Mangle of Practice*. Chicago: University of Chicago Press.

INDEX OF NAMES